Machine Learning in Geosciences

Machine Learning in Geosciences

Edited by Dilan Thomas

Larsen & Keller
www.larsen-keller.com

Machine Learning in Geosciences
Edited by Dilan Thomas
ISBN: 979-8-88836-073-6 (Hardback)

Larsen & Keller
Published by Larsen and Keller Education,
5 Penn Plaza,
19th Floor,
New York, NY 10001, USA

Cataloging-in-Publication Data

Machine learning in geosciences / edited by Dilan Thomas.
p. cm.
Includes bibliographical references and index.
ISBN 979-8-88836-073-6
1. Earth sciences--Data processing. 2. Geophysics--Data processing. 3. Machine learning.
4. Artificial intelligence. I. Thomas, Dilan.
QE48.8 .M33 2023
550.285--dc23

For more information regarding Larsen and Keller Education and its products, please visit the publisher's website www.larsen-keller.com

Table of Contents

Preface

Machine learning is an advanced field of data analytics that teaches computers to learn from their experiences similar to humans and animals. It utilizes two techniques, namely, unsupervised learning and supervised learning. The former makes use of the internal structures or hidden patterns in the input data whereas the latter involves training a model using known input and output data for predicting the future outcomes. Geoscience refers to the study of the Earth and all its natural structures and phenomena including oceans, atmosphere, rivers and lakes, ice sheets and glaciers, soils, complex surface, and rocky interior. Geographic information systems (GISs) are used extensively in studying the Earth. Machine learning is being used in GIS for segmentation, classification and prediction. Machine learning combined with remote sensing can enhance the automation of data analysis, uncover novel insights from large data sets, predict the behavior of environmental systems and lead to better management of resources. This book is a compilation of chapters that discuss the most vital concepts and emerging trends in the use of machine learning in geosciences. It will provide comprehensive knowledge to the readers.

This book unites the global concepts and researches in an organized manner for a comprehensive understanding of the subject. It is a ripe text for all researchers, students, scientists or anyone else who is interested in acquiring a better knowledge of this dynamic field.

I extend my sincere thanks to the contributors for such eloquent research chapters. Finally, I thank my family for being a source of support and help.

Editor

1

A Machine Learning-Based Approach for Wildfire Susceptibility Mapping: The Case Study of the Liguria Region in Italy

Marj Tonini [1,*], Mirko D'Andrea [2], Guido Biondi [2], Silvia Degli Esposti [2], Andrea Trucchia [2] and Paolo Fiorucci [2]

1 Institute of Earth Surface Dynamics, Faculty of Geosciences and Environment, University of Lausanne, CH-1015 Lausanne, Switzerland

2 CIMA Research Foundation, 17100 Savona, Italy; mirko.dandrea@cimafoundation.org (M.D.); guido.biondi@cimafoundation.org (G.B.); silvia.degliesposti@cimafoundation.org (S.D.E.); andrea.trucchia@cimafoundation.org (A.T.); paolo.fiorucci@cimafoundation.org (P.F.)

* Correspondence: marj.tonini@unil.ch

Abstract: Wildfire susceptibility maps display the spatial probability of an area to burn in the future, based solely on the intrinsic local proprieties of a site. Current studies in this field often rely on statistical models, often improved by expert knowledge for data retrieving and processing. In the last few years, machine learning algorithms have proven to be successful in this domain, thanks to their capability of learning from data through the modeling of hidden relationships. In the present study, authors introduce an approach based on random forests, allowing elaborating a wildfire susceptibility map for the Liguria region in Italy. This region is highly affected by wildfires due to the dense and heterogeneous vegetation, with more than 70% of its surface covered by forests, and due to the favorable climatic conditions. Susceptibility was assessed by considering the dataset of the mapped fire perimeters, spanning a 21-year period (1997–2017) and different geo-environmental predisposing factors (i.e., land cover, vegetation type, road network, altitude, and derivatives). One main objective was to compare different models in order to evaluate the effect of: (i) including or excluding the neighboring vegetation type as additional predisposing factors and (ii) using an increasing number of folds in the spatial-cross validation procedure. Susceptibility maps for the two fire seasons were finally elaborated and validated. Results highlighted the capacity of the proposed approach to identify areas that could be affected by wildfires in the near future, as well as its goodness in assessing the efficiency of fire-fighting activities.

Keywords: wildfires; susceptibility mapping; machine learning; random forest; spatial-cross validation

---

1. Introduction

Wildfires represent a hazardous and harmful phenomenon to people and the environment, especially in populated areas. In these areas, the primary cause of ignition is related to human activities. In Europe, for example, 95% of wildfire events are estimated to be induced by humans [1].

Among the weather-induced emergencies, wildfires constitute one of the more complex scenarios. Specifically, in the Mediterranean basin, wildfires are responsible for substantial damaging effects [2]. According to the Advance EFFIS Report on Forest Fires 2017 [3], Italy, Greece, Portugal, and Spain have been affected by a total of 43,733 fires, over a total an area of 89,4244 ha. Despite the financial support of the European Commission and the efforts by local wildfire administrations to prevent and fight wildfires, in the European Union (EU), in 2017, wildfires burnt over 1.2 million ha of natural lands, causing the death of 127 people among fire fighters and civilians. Over 25% of the total burnt area was in the different Natura 2000 sites, undermining the efforts in preserving natural habitats and key biodiversity for future generations. The losses caused by these fires amounted to 10 billion euros, as estimated by the European Forest Fire Information System [3].

Mapping current and past wildfire events by identifying their location, burned area, starting date, and duration, provides a precious tool for addressing prevention-planning programs aiming to reduce human and material losses. These inventories are an important source of information for the elaboration of hazard, risk, and susceptibility maps. Namely, hazard maps represent the zonation of the spatio-temporal probability of events, while the expected damages or losses are assessed and represented on risk maps. Susceptibility maps do not account for the temporal dimension and indicate zones with a potential to experience a particular hazard in the future, based solely on the intrinsic local properties of a site, expressed in terms of relative spatial likelihood. This definition relies on the basic assumption that future events are expected to occur under similar geo-environmental conditions to those observed in past events. Although these concepts are well-consolidated in the research area related with the risk assessment, especially for landslides [4–7], there is a need for elaborating susceptibility and risk maps for other natural hazards and to develop new quantitative and robust methods supporting their production.

In the field of wildfire risk assessment, fire risk has been defined as a quantitative or qualitative indicator of the likelihood that an area would burn in a certain period of time [8–11]. In this context, modeling fire risks represents a modern tool to support forest protection plans and address fuel management strategies in order to reduce fire consequences [12–14]. More generally, risk and susceptibility analyses are of great importance for land use planning, civil protection, and risk reduction programs. A number of techniques were recently developed to monitor and map the spatial distribution of a burned area and to predict the area at risk of wildfires. These often involve the implementation of physically-based models integrated into a Geographic Information System (GIS), relying on expert knowledge or including statistical analyses and modeling to assess the importance of the predisposing factors [15–24]. Lately, the comparison of deterministic physically-/statistically-based and stochastic approaches highlights the benefit of using data driven methods [25–33] that are able to extract knowledge directly from data. Machine learning has proven to be a mature and reliable tool in quantitative wildfire-related studies, as for the estimation of standing crop and fuel moisture content, often outperforming standard statistical approaches [34–36]. In comparison with deterministic methods, which, given a set of initial conditions (i.e., predisposing factors), always give the same results, stochastic models assume that results obtained through the combination of independent predisposing factors can be slightly different, as a consequence of the randomness of the process.

Although the terms "risk" and "susceptibility" are often used as synonymous, hereinafter we refer to "susceptibility mapping", meaning that the probability for an area to burn is assessed with no consideration of the magnitude of the single wildfire event or of its temporal dimension. Indeed, only the spatial probability for an area to burn in the future, assessed by defining a rank from low to high, is evaluated. This quantitative evaluation is performed considering two aspects: where wildfires occurred in the past, in terms of burned area, and which are the geo-environmental and anthropogenic predisposing factors that favor their spread. In this regard, it is worth noting that meteorological factors, such as wind speed and wind direction, temperature, humidity, and rainfall, are considered as triggering and not predisposing factors. The trigger is a local condition that causes a risk to occur if the area is susceptible to that risk, while the susceptibility is assessed based on predisposing factors, stable over time. The proximity to road and pathway networks and to urban and recreation areas are most frequently mentioned as predisposing human factors for wildfire [37–44]. As regards geo-environmental variables, those related to vegetation type and topography are the most significant drivers, especially in Mediterranean-type regions [45–48].

Italy is particularly affected by wildfires because of the high topographic and vegetation heterogeneity of the territory, as well as the favorable climate conditions that characterize the entire

Mediterranean basin. Wildfires are more frequent and larger in the summer (May–October) than in the winter (November–April) season in almost all Mediterranean countries, since it is hotter and drier. The Liguria region in Italy represents an exception, as it is highly affected by wildfires during the entire year and it sees a number of events and burned areas that can be more important in winter than in summer. Nevertheless, the spatial distribution of burned areas differs in the two seasons, probably due to the vegetation phenology at different altitudes in terms of plant senescence. Therefore, it is important to separately assess wildfire susceptibility in Liguria for the summer and winter season.

In this study, we adopted a stochastic approach based on a machine learning (ML) algorithm to elaborate wildfire susceptibility mapping for Liguria. The ML includes a class of algorithms for the analysis, modelling, and visualization of environmental data and performs particularly well to model environmental hazards, which naturally have a complex and non-linear behavior [49,50]. At present, only a few studies exist in literature on the use of this kind of approach for wildfire susceptibility and risk mapping. Moreover, a rigorous procedure to select the predisposing factors and to validate the results needs to be the established. To fill this gap, we developed different models and we evaluated their prediction performance by considering the effect of: (i) including or excluding the neighboring vegetation type as a predisposing factor and (ii) using an increasing number of folds in the spatial cross-validation procedure.

2. Study Area

Liguria has a total area of 5400 km^2, found between the Cote d'Azur (France) and Tuscany (Italy) on the northwest coast of the Tyrrhenian Sea. This Mediterranean region is characterized by a complex topography (with a slope that is higher than 40% over 50% of the total area) and dense and heterogeneous vegetation (with more than 70% of its surface covered by forests, see Figure 1).

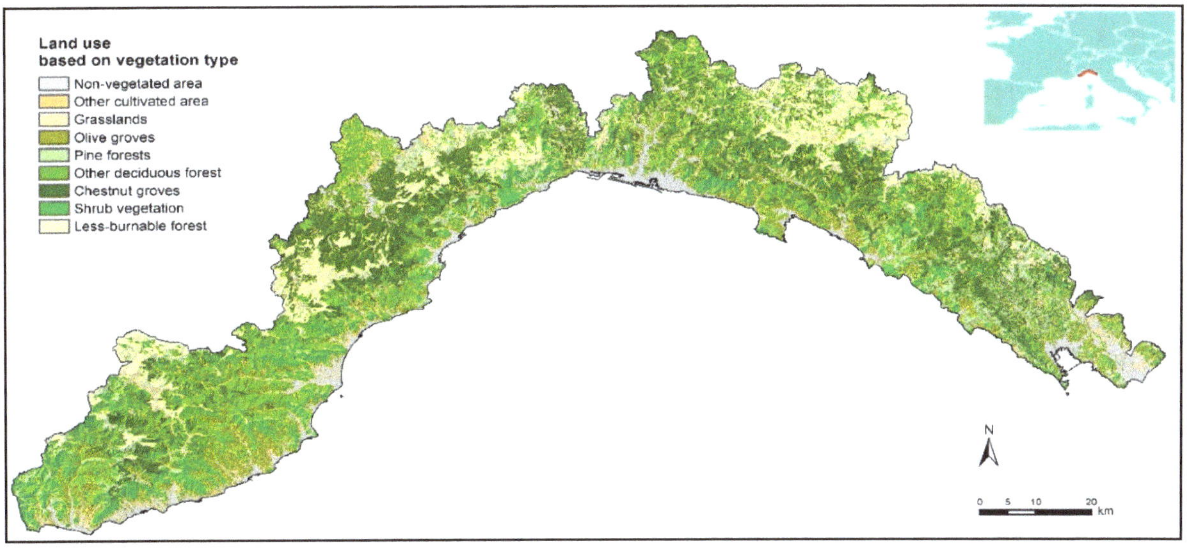

Figure 1. Study areas with the location map. Land use results from the aggregation of the original classes represented on the regional map of forest types, provided by the Liguria region.

During the investigated period, 1997–2017, an average of 365 wildfires burnt an area of 55 km^2 per year and were a recurrent phenomenon both in the summer and winter. The winter fire regime is mainly due to frequent extremely dry winds from the north in conditions of curing, for most of the herbaceous species, resulting in a number of wildfires and burned areas that can be higher than in summer (Figure 2).

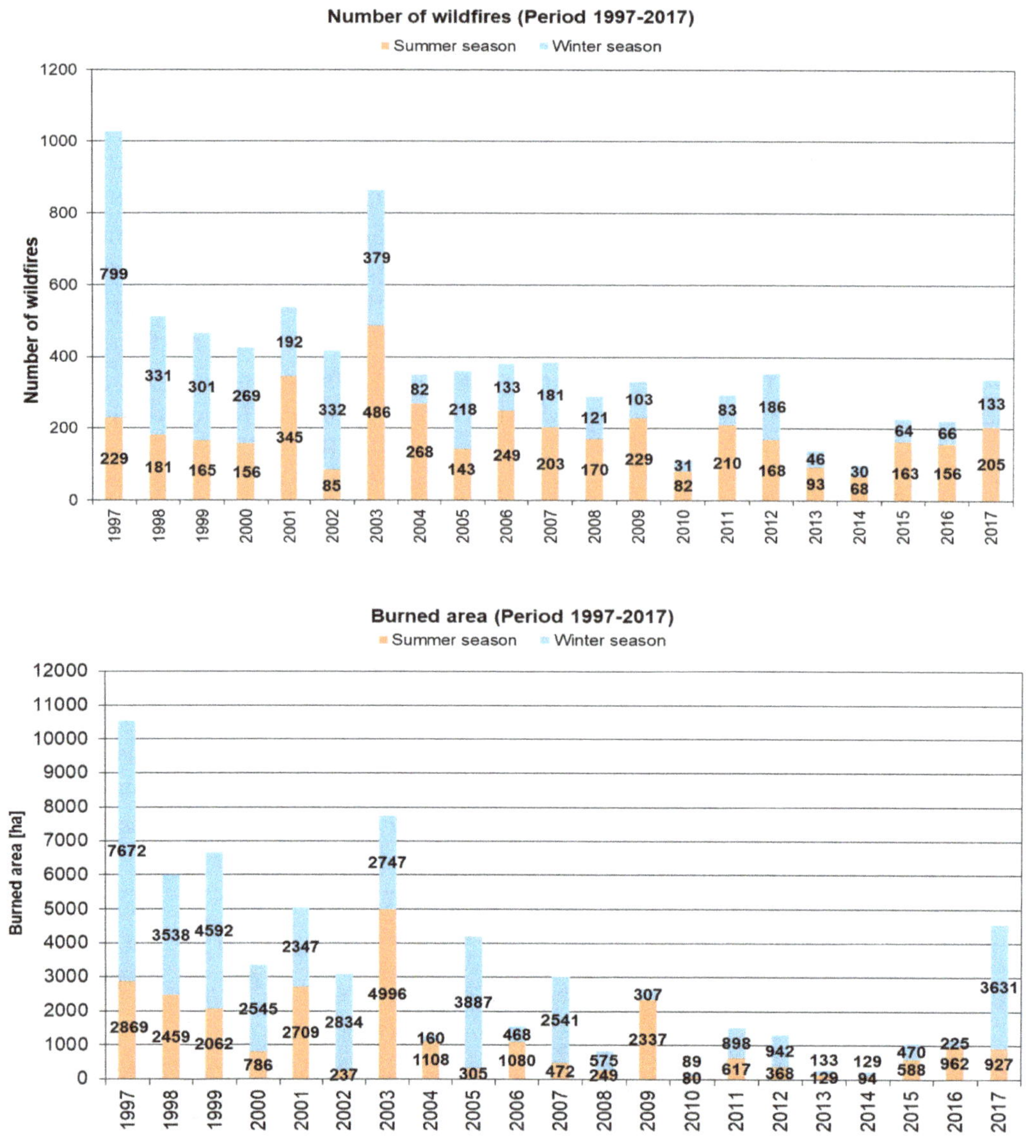

Figure 2. Yearly frequency of wildfires (on the top) and burned area (on the bottom) in the Liguria region (Italy) in the last two decades (1997–2017) during the summer and the winter season.

In 1950, the forest cover was limited to 30% of the regional surface, as most of the areas were subject to grazing activities. After World War II, rural communities were engaged in many reforestation programs, using different Mediterranean pine species. Pines and shrub species that are highly flammable spread quickly as agricultural and grazing activities were abandoned. Furthermore, the rural exodus and urbanization that followed the war greatly extended the Wildland–Urban Interface.

3. Materials and Methods

3.1. The Dataset: Burned Area and Predisposing Factors

The dependent variable of our model, allowing for the assessment of the susceptibility of the region to wildfires, was the burned area available as mapped burned areas and spanning a 21-year period (1997–2017). This dataset was acquired and compiled in shapefile format by the regional forestry service, on the base of GPS-surveying and subsequent digitalization over the cadastral map (scale 1:10,000). It is worth observing that fire perimeters were also affected by the capacity of intervention, which has not been considered in the analysis because of a lack of homogenization of the data.

The seasonality of the fire regime was also considered, partitioning the dataset of burned areas in two macro seasons: winter (November–April) and summer (May–October) (Figure 3)

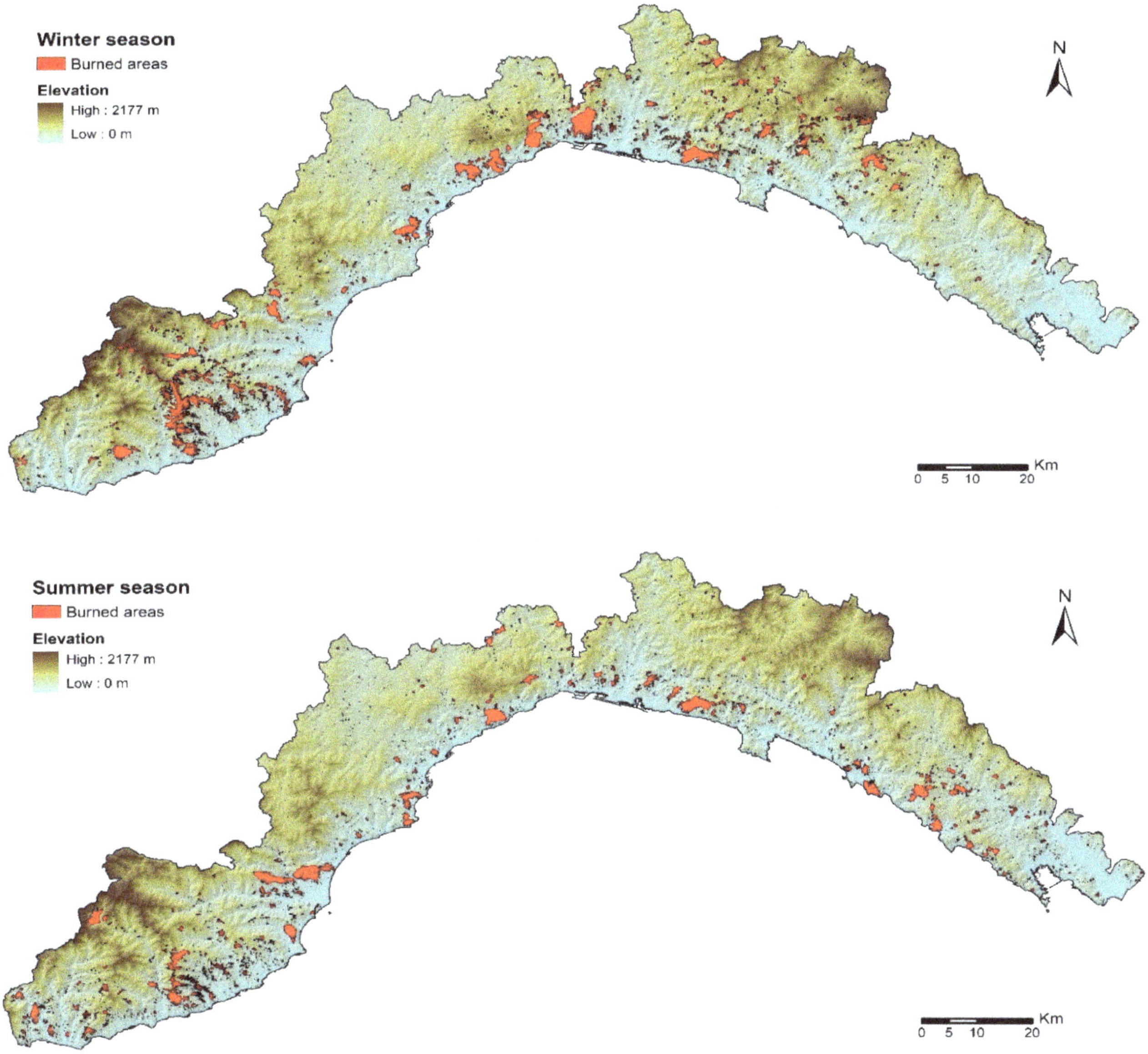

Figure 3. Spatial distribution of burned areas during the winter (on the top) and summer season (on the bottom) in the Liguria region (Italy) during the period 1997–2017.

The following independent variables provided detailed knowledge of topography and land cover, allowing for the understanding of the main features involved in wildfire occurrences and their behavior: DEM (digital elevation model) and derivatives (slope, northness, and eastness), distance to anthropogenic features (urban area, road, pathways, crops), protected areas, vegetation type, and neighboring vegetation.

In more detail, northerness and easterness, corresponding respectively to the cosine and to the minus sign of the aspect angle, were considered instead of the pure aspect angle (i.e., the terrain orientation) to avoid the use of a circular variable. The distance values were evaluated by computing, for each pixel, the Euclidean distance to the closest considered element. Protected areas were introduced as a binary variable, computing for each pixel, if it falls inside or outside of these delimited areas. Information on the type of vegetation came from the land use map (scale 1:10,000); furthermore, the class "forest" was reclassified by splitting the corresponding areal units into the sub-classes identified on the regional map of the forest types (scale 1:25,000). The final map included more than one hundred classes, which, for the purposes of the present study, were aggregated into 37 classes based on the flammability of each vegetation type. The non-vegetated areas, which came from the abovementioned land use map (including urban area, industrial and commercial units, road and rail network, port and airport area, bare rocks, and water bodies), were unified into a unique class under the label "non-flammable area". All these digital layers (Table 1) were provided by the authority of the Liguria region and are available on the official geo-portal (https://geoportal.regione.liguria.it/). The ensemble of the spatial layers, in raster format, were resampled and coregistered (i.e., spatially aligned) to match the same reference image with a spatial resolution of 100 m.

Table 1. Independent variables and their characteristics. (*) The neighboring vegetation was only included in the second model for comparison reasons.

Independent Variables	Acquisition Scale	Variable Type	Range	# of Variables
DEM	1:5000	Numerical (meters)	0–2132	1
Slope	-	Numerical (degree)	0–60	1
Northness and Eastness	-	Numerical	[−1–+1]	2
Distance to anthropogenic features	1:10,000	Numerical (meters)	0–9000	4
Protected area	1:25,000	Binary	0 or 1	1
Vegetation type	1:25,000	Categorical	37 classes	1
Non-flammable area	1:25,000	Categorical	1 class	1
Neighboring vegetation *	-	Numerical (percentage)	[0,100]	38

In addition, a set of new variables, taking into account the vegetation cover in the proximity of each pixel, was estimated as the percentage of each classes (including the 37 vegetation types and the non-flammable area) within a 300 by 300 m neighborhood distance. This resulted in 38 additional variables identifying the neighboring vegetation (Table 1). To prove if the neighboring vegetation type enables a rise in the predictive performance of the model, and consequently if this factor needs to be included in this kind of study, two models were tested: the first did not consider the neighboring vegetation (hereafter defined "standard model") and the second did account for this factor (hereafter defined "neighboring vegetation model").

3.2. Modeling Procedure: Machine Learning Approach

ML is based on algorithms capable of learning from and making predictions on data, through the modeling of the hidden relationships between a set of input and output variables, representing the predisposing factors (independent variables) and the occurrences of the phenomenon (dependent variable).

A well-established procedure to validate a model is to split the dataset into training, validation, and testing. The "training dataset" is needed to calibrate the parameters of the model, which will be used to get predictions on new data, that is the "validation dataset". Thus, the ultimate purpose of the validation dataset was to provide an unbiased evaluation of the model's fitness. Indeed, a good model is the one which gives accurate predictions on new data and avoids overfitting and underfitting.

Lastly, to provide an unbiased evaluation of the final model and to assess its performance (i.e., generalization), results were predicted over unused observations, defined as the "testing dataset". Model generalization refers to the ability of a model to perform good predictions on new/previously unseen data, drawn from the same distribution as the data used to create the model. This concept is closely related with overfitting: this happens when a model performs well on the training data but is unable to make good predictions on new data. Thus, a model that perfectly fits the training data normally displays a large generalization error.

A binary classification problem, such as the prediction of burned and unburned areas in the present case study, can be solved by counting how many times each observation is classified as "one" (burning) or "zero" (non-burning) and normalizing the result over the total number of predictions. This provided "probabilistic outputs", the values of which ranged between 0 and 1, which can be used to elaborate susceptibility maps. In these, the value of each pixel represents the probabilistic predicted value of burning in the future. The proposed approach involves the generation of pseudo-absences to ascertain an unburned area. Indeed, to assure a good generalization of the model, avoiding the overestimation of the low classes (i.e., with a probabilistic value close to 0), pseudo-absences need to be generated in all cases where they are not explicitly expressed (e.g., wildfire location is known, normally as mapped burned areas, but the not-burned areas have to be defined). We solved this problem by randomly generating a number of absences equal to the number of presences (i.e., equal to the number of pixels burning).

Analyses were performed using random forest [51], an ensemble ML algorithm based on decision trees. The most important hyperparameters that need to be specified are the number of decision trees (*ntree*) and the number of variables randomly sampled as candidates at each split (*mtry*). As a general rule, *ntree* should be large enough to ensure that every observation (i.e., input rows) gets predicted at least a few times. The standard value for *mtry* in classification problems is equal to the square root of the number of predictors (i.e., the independent variables), but this value can be optimized. Operationally, the algorithm generated *ntree*-subsets of the training dataset by bootstrapping (i.e., random sampling with replacement), each subset counting about two-thirds of the observations. A decision tree was then generated for each subset, considering a reduced number of variables (*mtry*), randomly selected. At each node, the Gini impurity was computed and the variable that minimized this value was selected as the best one for the split. Specifically, Gini impurity represented the probability of classifying incorrectly an observation in the dataset if it were classified randomly, according to the class distribution in the dataset. This process was iterated up to the maximum level and it could stop when each node contained less than a fixed number of data points. For a classification problem, the prediction of new data was finally computed, taking the maximum voting, which can be converted into a probabilistic output. The hyperparameters of the model were optimized by evaluating the prediction-error on those observations that were not used in the training subsets (called "out-of-bag"). In this study, hyperparameters were set to 750 for *ntree* and to the rounded up square root of the number of predictor factors for *mtry*, both optimized by applying a trial and error process.

3.3. Model Validation

As mentioned above, a well-established procedure in ML to validate a model is to split the dataset into training, validation, and testing.

When dealing with a spatial environmental phenomenon, if the validation dataset is selected randomly, observations can be located close to the training dataset, leading to an over-estimation of the predictive performance of the model. This circumstance is known as "spatial autocorrelation", meaning that observations close to each other hold similar characteristics. To overcome this issue, training and validation data have to be selected far enough apart in the geographic space. To this end, in the present study, we adopted the spatial k-fold cross validation using an increasing number of folds. Methodologically, the k-fold cross validation consisted of splitting the dataset into k folds, holding out a fold at a time, training the model on the remaining k-1 folds and finally validating the model using the hold out fold. The process was repeated k-times and the evaluation scores resulting from each folding were finally averaged. In the same way, the final prediction value was computed as the arithmetical mean among all the predictions estimated from each folding. We compared three models, considering 1-, 5-, and 9-folds. The space was organized into spatial blocks of 15 × 15 km, resulting in 50 blocks covering the entire study area. In the case of 1-fold, the validation dataset was defined by randomly selecting 10 blocks only once, which corresponded to the random selection of the 20% block of observations. In the case of 5- and 9-folds (Figure 4), each single fold included, respectively, 10 blocks (corresponding to the 20% of observations) and 5 or 6 blocks (corresponding to about 10% of the observations).

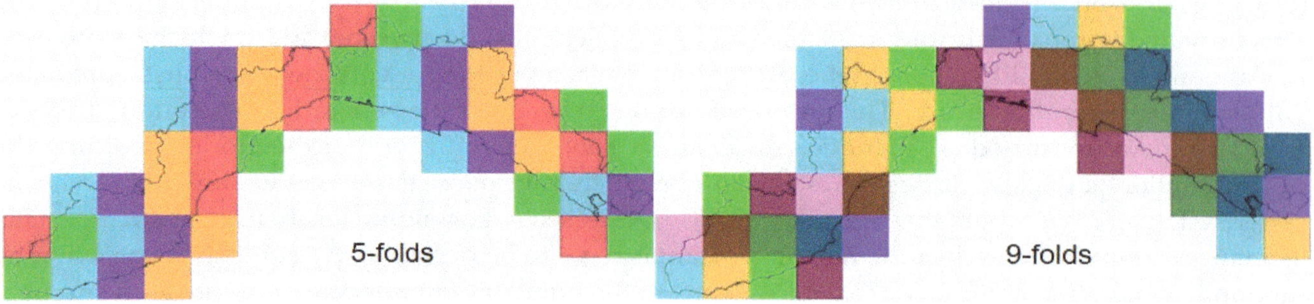

Figure 4. Spatial arrangement of the blocks for the 5- and 9-folds adopted in the work (each color corresponds to a single fold).

The testing dataset was defined by splitting the original dataset by years: Mapped burned areas observed in the period 1997–2011 were used in the training procedure, while the last 6 years (2012–2017), corresponding to about 25% of the entire dataset, were held out for testing the predictive performance of the model. This was assessed by computing the percentage of an area with a probabilistic predicted value of burning within a certain range, falling inside the burned areas in the testing dataset (*Testing BA*, Table 2). Finally, the model that predicted better was the one that resulted in a higher percentage for high range values and a lower percentage for low range values. The probabilistic predicted values were ranked into five classes, defined by the following percentile ranking ranges: 25th, 50th, 75th, 90th, and 95th. It is worth pointing out that the 25th percentile represented 25% of the area with the lowest probabilistic predicted values of burning, while the 95th percentile represented the 5% of the area with the highest probabilistic values, and so on for the intermediate ranges. The rank-limits corresponded to different values of the probabilistic output (*Prob_value*, Table 2) resulting from each model, allowing a flexible interpretation of the results. In addition, keeping in mind the goal of delivering unbiased results by adopting different fitness estimators, the root mean square error (RMSE) was also computed (Table 3). The RMSE is based on the difference between the predicted outputs and the observed true values, where observations can only assume the value "one" (if the area burned) or "zero" (if the area did not burn), while the predicted output is a probabilistic value expressed as a floating number between zero and one.

Table 2. Validation of the models (see Section 3.2 for more details).

Winter Season		1-Fold Cross Validation				5-Fold Cross Validation				9-Fold Cross Validation	
		Standard Model		Neighboring Vegetation		Standard Model		Neighboring Vegetation		Neighboring Vegetation	
Classes	Total Area (%)	Testing BA (%)	Prob_Value	Testing BA (%)	Prob_Value	Testing BA (%)	Prob_Value	Testing BA (%)	Prob_Value	Testing BA (%)	Prob_Value
25%	25	5.07	0.13	3.85	0.09	4.42	0.11	3.5	0.07	3.36	0.07
50%	25	4.95	0.25	3.4	0.22	3.44	0.23	3.27	0.18	3.17	0.17
75%	25	10.52	0.48	8.8	0.47	8.9	0.43	6.22	0.39	6.44	0.39
90%	15	17.64	0.78	14.98	0.74	15.77	0.7	13.05	0.68	11.91	0.69
95%	5	14.38	0.91	15.26	0.87	15.67	0.85	17.7	0.83	16.43	0.85
100%	5	47.26	1	52.86	1	51.78	1	56.26	1	58.69	1
>75%		**79.28**		**83.1**		**83.22**		**87.01**		**87.03**	
Summer Season		**1-Fold Cross Validation**				**5-Fold Cross Validation**				**9-Fold Cross Validation**	
		Standard Model		**Neighboring Vegetation**		**Standard Model**		**Neighboring Vegetation**		**Neighboring Vegetation**	
Classes	**Total Area (%)**	**Testing BA (%)**	**Prob_Value**	**Testing BA (%)**	**Prob_Value**	**Testing BA (%)**	**Prob_Value**	**Testing BA (%)**	**Prob_Value**	**Testing BA (%)**	**Prob_Value**
25%	25	4.71	0.08	1.04	0.04	4.04	0.06	0.8	0.04	0.8	0.04
50%	25	7.52	0.23	4.64	0.17	9.39	0.19	5.08	0.14	5.54	0.14
75%	25	17.94	0.51	18.27	0.41	15.16	0.44	19.77	0.35	18.44	0.35
90%	15	24.45	0.78	26.19	0.7	23.31	0.69	21.51	0.65	22.11	0.66
95%	5	14.06	0.91	14.6	0.87	14.6	0.83	15.06	0.83	14.73	0.85
100%	5	30.66	1	33.43	1	33.5	1	37.71	1	38.31	1
>75%		**69.17**		**74.22**		**71.41**		**74.28**		**75.15**	

Table 3. Root mean square error based on the difference between the predicted and observed value.

Winter Season	1-Fold	5-Folds	9-Folds
Standard model	0.407	0.380	-
Neighboring vegetation	0.377	0.354	0.351
Summer season	1-fold	5-folds	9-folds
Standard model	0.437	0.428	-
Neighboring vegetation	0.411	0.411	0.411

4. Results

The main results of the present study are the following: (i) the models comparison, including the "standard" vs. the "neighboring vegetation" model and the use of 1- vs. 5- vs. 9-folds for the spatial cross validation; (ii) the probabilistic predicted values as the main output of random forest, allowing to elaborate susceptibility maps of wildfires for the winter and summer seasons.

4.1. Models Comparison

The values indicated in Table 2 allow comparing and evaluating the predictive performance of different models: the "standard" vs. the "neighboring vegetation" model and the use of 1- vs. 5- vs. 9-folds spatial cross validation for the winter (values above) and summer seasons (values below). In addition to the percentile ranking ranges (i.e., "Classes"), the corresponding probabilistic predicted value (i.e., Prob_value) is specified. The field "Total area" indicates the surface predicted by the model to fall within each range, normalized over the entire area. The field "Predicted BA" represents the fraction of the *"Total area"* falling inside the burned area detected in the testing dataset (i.e., 2012–2017 mapped burned areas). For example, looking at one-fold standard model, the first class (25th percentile) indicates that only 5.07% of the 25% of the area with the lowest probabilistic predicted values of burning (corresponding to <0.13), fell inside the burned area in the testing dataset. On the opposite extreme, the 95th percentile indicates that about 47% of the 5% of the area with the highest probabilistic predicted values (corresponding to >0.91), fell inside the burned area in the testing dataset. Thus, a way to compare the different models is to evaluate which model predicts the largest area with the highest probability of burning falling inside the burned area in the testing dataset (hereafter defined for brevity as "predicted burned area"), and vice-versa for the lower classes. To facilitate this comparison, the last 75th percentile (i.e., 25% of the area with the highest probabilistic predicted value of burning) was considered and discussed in the following. Results show that the neighboring vegetation model performed better than the standard model in both seasons. Indeed, the predicted burned area was higher for both 1- and 5-folds cross-validation when the model included the information concerning the type of vegetation and non-flammable area in the neighboring of each pixel. The increment in this case is of about three percentage points or more. Instead, when comparing 5- with 5-folds cross-validation, the predicted burned area increased y only 0.02% in the winter season and 0.87% in the summer season. So, using 9- instead of 5-folds allows to slightly increase the predictive performance of the model, facing the fact that the training algorithm is much more computationally intensive. On the other hand, the predictive performance of using 5-folds cross-validation compared with the 1-fold increased by about four-percentage points in the winter season, when considering both the standard and the neighboring vegetation model, and was 2.24% and only 0.07% for the equivalent models in the summer season. Despite this last low value, which will be discussed later, overall, using 5-folds cross-validation results in a better predictive performance of the model than selecting a random 20% of the dataset by blocks for validation (corresponding to 1-fold).

The root mean square errors computed for all the models (Table 3) confirm these results. As expected, all RMSEs were lower than 0.5, as, on average, each model performed good prediction of burned and unburned areas. The neighboring vegetation model performed better than the standard model in both seasons. Increasing the number of folds in the cross validation allowed to increase the model's performance in all the cases, except in the case of the neighboring vegetation model in the summer season, which performed the same, regardless of the number of folds.

All the models performed better in winter than in summer. In the winter, the values of the testing burned area, considering the last 75th percentile, ranged from about 80% (1-fold "standard model") to 87% (5- and 9-folds "neighboring vegetation model"). In the summer, these values were significantly lower, ranging from about 70% (1-fold "standard model") to 75% (5- and 9-folds "neighboring vegetation model"). These performances were confirmed by the RMSE, showing lower values in summer than in winter for the same model.

The neighboring vegetation model using 5-folds cross-validation performed better. This last model was finally evaluated by computing the fraction of predicted burning area above the 75th percentile falling inside the testing burned areas for each single year and by season. Values are expressed both as percentage and number of pixels (Table 4). This allowed for the investigation of the influence of the testing dataset on the predictive performances of the model. In the winter season, the model performed well every year, with values ranging from 83.4% in 2013 to 91.7% in 2016. In the summer season, performances were still quite good in the first five testing periods (years 2012–2016), while, in the last year (2017), the assessed value dropped to 45.7%. At the same time, we noticed that the size of the burned area, which highly differed in each testing period, did not compromise the predictive performance of the model. More specifically, this result seems to imply that something during the 2017 wildfire summer season in Liguria did not work as predicted by the model.

Table 4. Model validation evaluated by computing the percentage of the predicted burning area above the 75th percentile falling inside the testing burned areas for each single year (BA >75%). The number of pixels above and below this value for each season are also shown.

Year	Winter Season			Summer Season			Tot_Year
	BA > 75 % (%)	BA > 75% (*# pixels*)	Tot_Winter (*# pixels*)	BA > 75 % (%)	BA > 75% (*# pixels*)	Tot_Summer (*# pixels*)	(*# pixels*)
2012	84.1	844	1003	77.8	337	433	1436
2013	83.4	121	145	86.9	140	161	306
2014	86.0	117	136	92.8	103	111	247
2015	91.2	465	510	84.9	535	630	1140
2016	91.7	220	240	92.7	936	1010	1250
2017	86.4	3144	3640	**45.7**	449	983	4623
Tot			5674			3328	9002

4.2. *Susceptibility Mapping*

Random forest gives as an output a probabilistic value, expressing the probability for each pixel to burn in the future under the assumption of a set of predisposing variables. These values were used to elaborate wildfire susceptibility maps in the winter and summer seasons (Figure 5). To this end, we retained the results that were obtained by the model that performed better, which was the neighboring vegetation model using 5-folds cross-validation. The two seasons displayed different behavior in relation to the predicted susceptibility to wildfires.

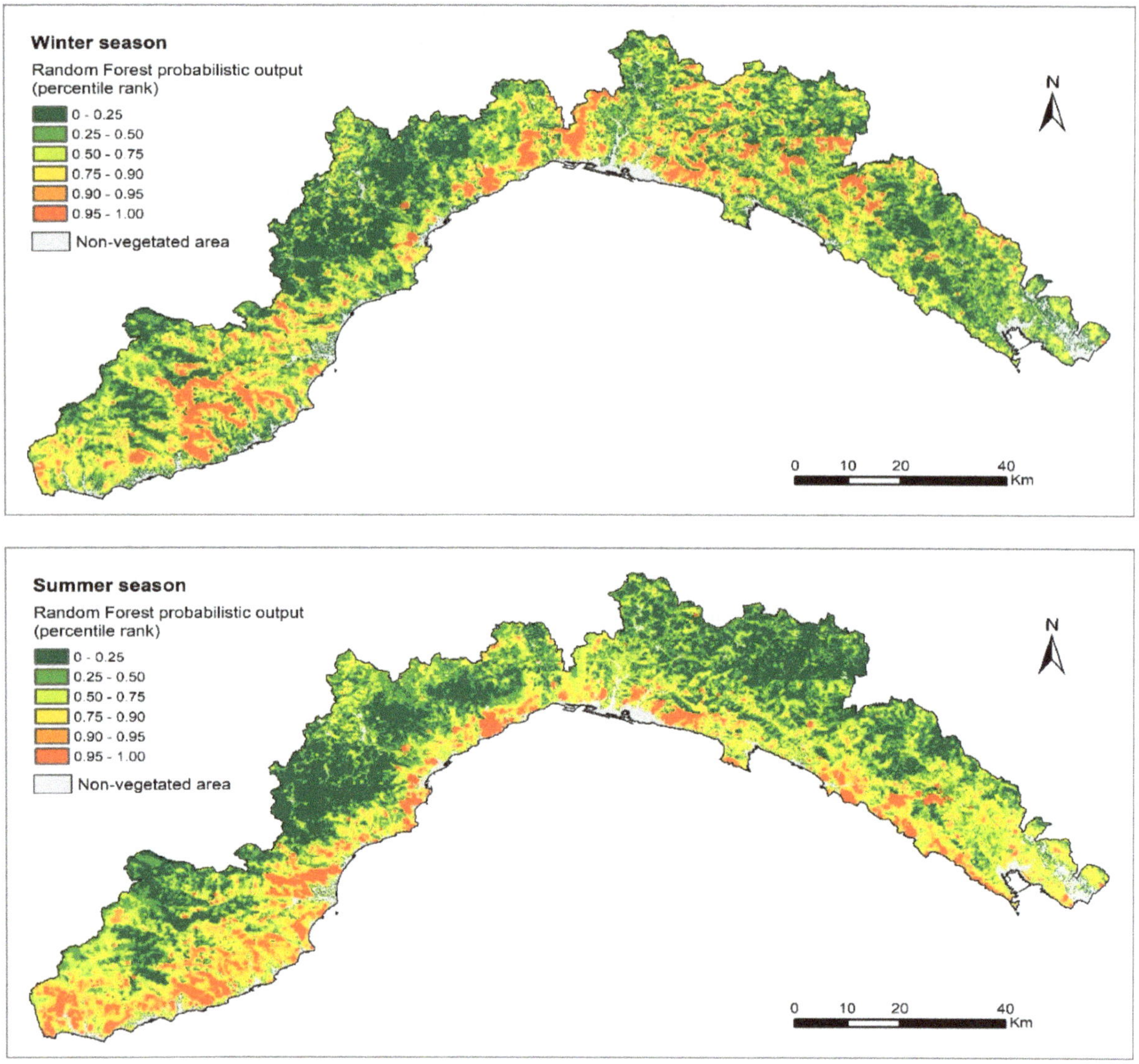

Figure 5. Wildfire susceptibility map in the winter (on the top) and summer seasons (on the bottom) for the Liguria region (Italy).

Higher classes (above the 90th percentile) characterized the coastal area in summer and the more elevated inland areas in winter. This could be due to the state of the vegetation: In winter, vegetation is more stressed and senescent at a higher altitude due to the lower temperature in altitude. On the contrary, in summer, vegetation is dryer and more burnable at lower elevations because of the high temperature and the dry weather. Finally, even though the two models implemented separately for the winter and the summer seasons used the same independent variables as the input, random forest succeeded in discriminating among the two patterns in the distribution of the wildfire-susceptible areas, thanks to the training procedure.

5. Discussion

In a very recent review paper, Jain et al. [52] found a total of 298 publications, from 1996 to 2019, relevant to the topic of machine learning applications in wildfire science and management, with a very strong increase in the last five years of investigation. Among them, a considerable number of references (71) used various ML algorithms to map wildfire susceptibility. Nevertheless, only a few papers considered time series longer than a few years, allowing to assess the predictive performance of the model in the near future [26,31,53] or separately in summer and winter seasons [54].

In the present paper, we introduced an innovative ML approach, based on random forest, which allowed the elaboration of the wildfire susceptibility map for the Liguria region in Italy. Susceptibility was assessed by evaluating the probability for an area to burn in the future, taking into account the spatial extension of past-burned areas and the geo-environmental factors that favor their occurrence (i.e., DEM and its derivatives, distance to an anthropogenic feature, protected area, vegetation type). An alternative model, including the neighboring vegetation type at each location, was developed and compared with the standard model. For validation purposes, we adopted the spatial k-fold cross-validation (with k = 1, 5, and 9) and then we evaluated the predictive performances of the different models over an independent testing dataset. Finally, we compared the standard vs. the neighboring vegetation model and the use of 1- vs. 5- vs. 9-folds spatial cross-validation, both for the winter and summer seasons. The results show that: (1) the neighboring vegetation model performed better than the standard model in both seasons; (2) 5 represents the optimal number of folds; (3) all the models performed better in the winter season than in the summer season.

The predictive performance of the models implemented in the present study was globally good, as proved by their high capacity of discriminating most of the burned area within the 75th percentile for most of the testing periods (2012 to 2017) and in both the wildfire seasons. This attests the good generalization capabilities of our models. The neighboring vegetation model using 5-folds cross-validation gave the best results in terms of predictive performance and was retained to elaborate the winter and the summer wildfire susceptibility maps. The performance of this model in the summer 2017 testing period manifested the only notable exception to the overall good prediction capability and it will be the object of discussion in the following. Instead of considering this fact a downside of the proposed modeling framework, it shifted the attention towards a specific situation: Fire management in summer 2017. As a matter of fact, it is worth considering that the Italian Forestry Corp, in charge of fire management since 1984, was dismissed at the end of 2016 (DLGS 19/08/2016 n. 177). To counter this, the Liguria region named Italian Fire Fighters (CNVVF) in charge of forest fire management. Before 2017, the role of CNVVF was limited to the management of fire in the Wildland–Urban Interface and was mainly restricted to the safeguard of civilians (protection of houses and infrastructures). In January 2017, a week of severe fire danger caused the burning of about 3000 ha, most of them characterized by high or extreme fire risk. However, the summer season of 2017 was atypical, not only in terms of fire management procedures, but also with respect to the meteorological conditions. This season was in fact characterized by a long drought, but with a remarkable average relative humidity higher than 67%. Only a couple of days were characterized by relative humidity lower than 40%. These meteorological conditions favored the mop up phase, which caused each single fire to be reignited several times, extending the fire propagation for many days.

In the period ranging from 1st May to 31st October 2017, 205 wildfires burned a total area of 927 ha. Only six events burned an area greater than 50 ha each, resulting in 554 ha (corresponding to 60% of the total burned area). These six events were highlighted for a better understanding of the implication that the new management can have on the predictive performance of the model. As is evident in Figure 6, most of the burned area within these six largest wildfires was predicted by the model as belonging to a middle-to-low level of susceptibility. This result reflects that random forest could not take into account the management issues of 2017: The underestimation of the mop up phase. However, it successfully predicted the areas to be burned in the future under unchanged geo-environmental and management conditions, as proven by the good performances of the model in the previous testing periods.

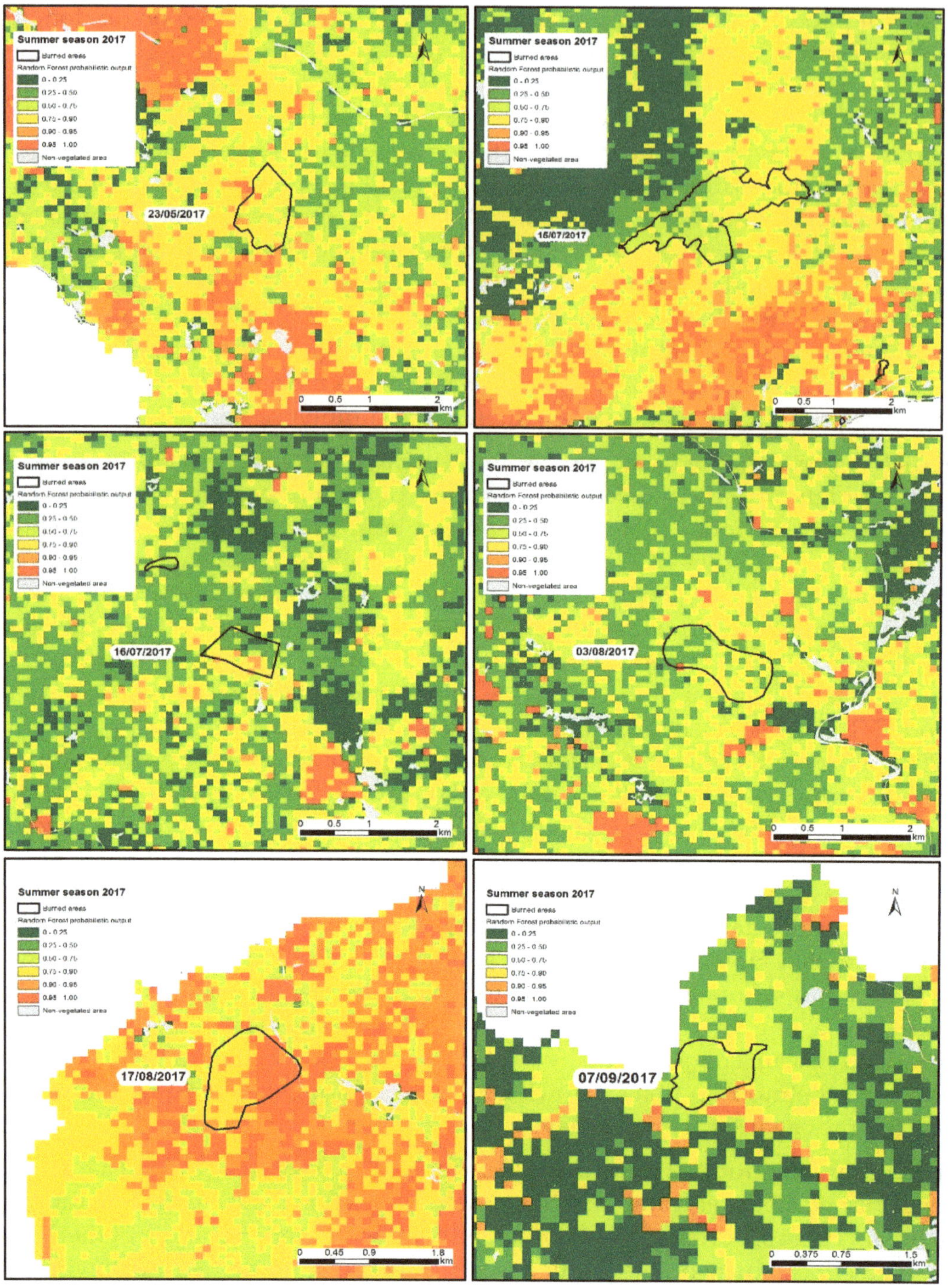

Figure 6. The six largest wildfire occurrences of summer 2017, along with the fire susceptibility map.

6. Conclusions

The proposed approach proved to be globally effective to deal with large, high dimensional spatial data, which reflects the great flexibility of machine learning in general. In particular, one advantage of random forest is its capability of handling directly categorical variables, such as land use classes or vegetation types.

The importance of considering the vegetation cover in the proximity of each pixel, specifically the type of vegetation, as an additional variable is proved by the results of the present research, and we recommend including this factor in similar studies. Moreover, we recommend the use of k-fold cross-validation to improve the predictive performance of machine learning-based models seeking to assess the susceptibility of hazardous events in general and wildfires in particular.

Ultimately, the results of the present study highlight the importance of taking into account possible changes of surrounding conditions, which can affect the predictive performance of the model in space and in time. Indeed, despite the rapid capacity of intervention of the Italian Fire Fighters, the fire brigade did not consider the mop up phase, which, in summer 2017, caused each single fire to be reignited several times. These considerations put in evidence the capacity of the proposed approach to identify the efficiency of fire-fighting activities, specifically if fire extinction procedures are handled with different modalities compared with the procedures used previously and implicitly valued in the model. Before 2017, the different tactics of fire management could have resulted in a lower extension of the burned areas, as identified by the models and results from the susceptibility map.

Author Contributions: Data curation, G.B. and S.D.E.; Formal analysis, M.T. and M.D.; Funding acquisition, M.T. and P.F.; Investigation, M.T. and M.D.; Methodology, M.T.; Project administration, P.F.; Software, M.D.; Supervision, P.F. and M.T.; Validation, M.T. and M.D.; Visualization, G.B.; Writing—original draft, M.T.; Writing—review & editing, M.T., A.T. and P.F. All authors have read and agreed to the published version of the manuscript.

Acknowledgments: The authors acknowledge the Italian Civil Protection Department - Presidency of the Council of Ministers, who founded this research through the convention with the CIMA Research Foundation, for the development of knowledge, methodologies, technologies, and training, useful for the implementation of national wildfire systems of monitoring, prevention, and surveillance. A visiting period of Mirko D'Andrea, from CIMA, at the Institute of Earth Surface Dynamics (UNIL), has been possible thanks to a scientific exchange grant funded by the Swiss National Science Foundation (IZSEZ0_186483). The authors are grateful to Regione Liguria and the Liguria's regional direction of the Italian National Fire Corps for data providing and for the stimulating discussion about the summer fire season 2017. Thanks to the anonymous reviewers for helping to improve the manuscript.

References

1. San-Miguel-Ayanz, J.; Schulte, E.; Schmuck, G.; Camia, A.; Strobl, P.; Liberta, G.; Giovando, C.; Boca, R.; Sedano, F.; Kempeneers, P.; et al. Comprehensive monitoring of wildfires in Europe: The European Forest Fire Information System (EFFIS). In *Approaches to Managing Disaster—Assessing Hazards, Emergencies and Disaster Impacts*; Tiefenbacher, J., Ed.; InTech: London, UK, 2012; ISBN 978-953-51-0294-6.
2. Amraoui, M.; Pereira, M.G.; DaCamara, C.C.; Calado, T.J. Atmospheric conditions associated with extreme fire activity in the Western Mediterranean region. *Sci. Total Environ.* **2015**, *524–525*, 32–39. [CrossRef]
3. San-Miguel-Ayanz, J.; Durrant, T.; Boca, R.; Libertà, G.; Branco, A.; de Rigo, D.; Ferrari, D.; Maianti, P.; Vivancos, T.A.; Schulte, E.; et al. Forest Fires in Europe, Middle East and North Africa 2017. Available online: https://www.driver-project.eu/forest-fires-in-europe-middle-east-and-north-africa-2017-the-new-jrc-report/ (accessed on 17 March 2020).
4. Mihalić Arbanas, S. Landslide hazard, risk assessment and prediction: Landslide inventories and susceptibility, hazard mapping methods, damage potential—Part 2. In *Advancing Culture of Living with Landslides*; Mikos, M., Tiwari, B., Yin, Y., Sassa, K., Eds.; Springer International Publishing: Cham, Switzerland, 2017; pp. 695–698. ISBN 978-3-319-53497-8.
5. Hervás, J.; Bobrowsky, P. Mapping: Inventories, susceptibility, hazard and risk. In *Landslides—Disaster Risk Reduction*; Sassa, K., Canuti, P., Eds.; Springer Berlin Heidelberg: Berlin/Heidelberg, Germany, 2009; pp. 321–349. ISBN 978-3-540-69966-8.
6. Van Den Eeckhaut, M.; Hervás, J. State of the art of national landslide databases in Europe and their potential for assessing landslide susceptibility, hazard and risk. *Geomorphology* **2012**, *139–140*, 545–558. [CrossRef]
7. Fell, R.; Corominas, J.; Bonnard, C.; Cascini, L.; Leroi, E.; Savage, W.Z. Guidelines for landslide susceptibility, hazard and risk zoning for land-use planning. *Eng. Geol.* **2008**, *102*, 99–111. [CrossRef]
8. Finney, M.A. The challenge of quantitative risk analysis for wildland fire. *For. Ecol. Manag.* **2005**, *211*, 97–108. [CrossRef]
9. Hardy, C.C. Wildland fire hazard and risk: Problems, definitions, and context. *For. Ecol. Manag.* **2005**, *211'* 73–82. [CrossRef]

10. Watts, J.M.; Hall, J.R. Introduction to fire risk analysis. In *SFPE Handbook of Fire Protection Engineering*; Hurley, M.J., Gottuk, D., Hall, J.R., Harada, K., Kuligowski, E., Puchovsky, M., Torero, J., Watts, J.M., Wieczorek, C., Eds.; Springer: New York, NY, USA, 2016; pp. 2817–2826. ISBN 978-1-4939-2564-3.
11. Meacham, B.J.; Charters, D.; Johnson, P.; Salisbury, M. Building fire risk analysis. In *SFPE Handbook of Fire Protection Engineering*; Hurley, M.J., Gottuk, D., Hall, J.R., Harada, K., Kuligowski, E., Puchovsky, M., Torero, J., Watts, J.M., Wieczorek, C., Eds.; Springer: New York, NY, USA, 2016; pp. 2941–2991. ISBN 978-1-4939-2564-3.
12. Brillinger, D.R.; Preisler, H.K.; John, W. Benoit risk assessment: A forest fire example. *Lect. Notes-Monogr. Ser.* **2003**, *40*, 177–196.
13. Catry, F.X.; Rego, F.C.; Bação, F.L.; Moreira, F. Modeling and mapping wildfire ignition risk in Portugal. *Int. J. Wildland Fire* **2010**, *18*, 921–931. [CrossRef]
14. Ager, A.A.; Finney, M.A.; Kerns, B.K.; Maffei, H. Modeling wildfire risk to northern spotted owl (*Strix occidentalis* caurina) habitat in Central Oregon, USA. *For. Ecol. Manag.* **2007**, *246*, 45–56. [CrossRef]
15. Eugenio, F.C.; dos Santos, A.R.; Fiedler, N.C.; Ribeiro, G.A.; da Silva, A.G.; dos Santos, Á.B.; Paneto, G.G.; Schettino, V.R. Applying GIS to develop a model for forest fire risk: A case study in Espírito Santo, Brazil. *J. Environ. Manag.* **2016**, *173*, 65–71. [CrossRef] [PubMed]
16. Teodoro, A.C.; Duarte, L. Forest fire risk maps: A GIS open source application—A case study in Norwest of Portugal. *Int. J. Geogr. Inf. Sci.* **2013**, *27*, 699–720. [CrossRef]
17. Gai, C.; Weng, W.; Yuan, H. GIS-based forest fire risk assessment and mapping. In Proceedings of the 2011 Fourth International Joint Conference on Computational Sciences and Optimization, Kunming and Lijiang City, China, 15–19 April 2011; pp. 1240–1244.
18. Pourghasemi, H.R. GIS-based forest fire susceptibility mapping in Iran: A comparison between evidential belief function and binary logistic regression models. *Scand. J. For. Res.* **2016**, *31*, 80–98. [CrossRef]
19. Mohammadi, F.; Bavaghar, M.P.; Shabanian, N. Forest fire risk zone modeling using logistic regression and GIS: An Iranian case study. *Small-Scale For.* **2014**, *13*, 117–125. [CrossRef]
20. Vadrevu, K.P.; Eaturu, A.; Badarinath, K.V.S. Fire risk evaluation using multicriteria analysis—A case study. *Environ. Monit. Assess.* **2010**, *166*, 223–239. [CrossRef]
21. Kant Sharma, L.; Kanga, S.; Singh Nathawat, M.; Sinha, S.; Chandra Pandey, P. Fuzzy AHP for forest fire risk modeling. *Disaster Prev. Manag. Int. J.* **2012**, *21*, 160–171. [CrossRef]
22. Carmel, Y.; Paz, S.; Jahashan, F.; Shoshany, M. Assessing fire risk using Monte Carlo simulations of fire spread. *For. Ecol. Manag.* **2009**, *257*, 370–377. [CrossRef]
23. Chuvieco, E.; Aguado, I.; Yebra, M.; Nieto, H.; Salas, J.; Martín, M.P.; Vilar, L.; Martínez, J.; Martín, S.; Ibarra, P.; et al. Development of a framework for fire risk assessment using remote sensing and geographic information system technologies. *Ecol. Model.* **2010**, *221*, 46–58. [CrossRef]
24. Chuvieco, E.; Salas, J. Mapping the spatial distribution of forest fire danger using GIS. *Int. J. Geogr. Inf. Syst.* **1996**, *10*, 333–345. [CrossRef]
25. Pourtaghi, Z.S.; Pourghasemi, H.R.; Aretano, R.; Semeraro, T. Investigation of general indicators influencing on forest fire and its susceptibility modeling using different data mining techniques. *Ecol. Indic.* **2016**, *64*, 72–84. [CrossRef]
26. Leuenberger, M.; Parente, J.; Tonini, M.; Pereira, M.G.; Kanevski, M. Wildfire susceptibility mapping: Deterministic vs. stochastic approaches. *Environ. Model. Softw.* **2018**, *101*, 194–203. [CrossRef]
27. Arpaci, A.; Malowerschnig, B.; Sass, O.; Vacik, H. Using multi variate data mining techniques for estimating fire susceptibility of Tyrolean forests. *Appl. Geogr.* **2014**, *53*, 258–270. [CrossRef]
28. Oliveira, S.; Oehler, F.; San-Miguel-Ayanz, J.; Camia, A.; Pereira, J.M.C. Modeling spatial patterns of fire occurrence in Mediterranean Europe using Multiple Regression and Random Forest. *For. Ecol. Manag.* **2012**, *275*, 117–129. [CrossRef]
29. Ngoc Thach, N.; Bao-Toan Ngo, D.; Xuan-Canh, P.; Hong-Thi, N.; Hang Thi, B.; Nhat-Duc, H.; Dieu, T.B. Spatial pattern assessment of tropical forest fire danger at Thuan Chau area (Vietnam) using GIS-based advanced machine learning algorithms: A comparative study. *Ecol. Inform.* **2018**, *46*, 74–85. [CrossRef]
30. Satir, O.; Berberoglu, S.; Donmez, C. Mapping regional forest fire probability using artificial neural network model in a Mediterranean forest ecosystem. *Geomat. Nat. Hazards Risk* **2016**, *7*, 1645–1658. [CrossRef]
31. Tehrany, M.S.; Jones, S.; Shabani, F.; Martínez-Álvarez, F.; Tien Bui, D. A novel ensemble modeling approach for the spatial prediction of tropical forest fire susceptibility using LogitBoost machine learning classifier and multi-source geospatial data. *Theor. Appl. Climatol.* **2019**, *137*, 637–653. [CrossRef]

32. Rodrigues, M.; de la Riva, J. An insight into machine-learning algorithms to model human-caused wildfire occurrence. *Environ. Model. Softw.* **2014**, *57*, 192–201. [CrossRef]
33. Ghorbanzadeh, O.; Blaschke, T.; Gholamnia, K.; Aryal, J. Forest fire susceptibility and risk mapping using social/infrastructural vulnerability and environmental variables. *Fire* **2019**, *2*, 50. [CrossRef]
34. Sharma, S.; Ochsner, T.E.; Twidwell, D.; Carlson, J.D.; Krueger, E.S.; Engle, D.M.; Fuhlendorf, S.D. Nondestructive estimation of standing crop and fuel moisture content in Tallgrass Prairie. *Rangel. Ecol. Manag.* **2018**, *71*, 356–362. [CrossRef]
35. Sayad, Y.O.; Mousannif, H.; Al Moatassime, H. Predictive modeling of wildfires: A new dataset and machine learning approach. *Fire Saf. J.* **2019**, *104*, 130–146. [CrossRef]
36. Féret, J.-B.; le Maire, G.; Jay, S.; Berveiller, D.; Bendoula, R.; Hmimina, G.; Cheraiet, A.; Oliveira, J.C.; Ponzoni, F.J.; Solanki, T.; et al. Estimating leaf mass per area and equivalent water thickness based on leaf optical properties: Potential and limitations of physical modeling and machine learning. *Remote Sens. Environ.* **2019**, *231*, 110959. [CrossRef]
37. Ganteaume, A.; Camia, A.; Jappiot, M.; San-Miguel-Ayanz, J.; Long-Fournel, M.; Lampin, C. A review of the main driving factors of forest fire ignition over Europe. *Environ. Manag.* **2013**, *51*, 651–662. [CrossRef]
38. Tonini, M.; Parente, J.; Pereira, M.G. Global assessment of rural–urban interface in Portugal related to land cover changes. *Nat. Hazards Earth Syst. Sci.* **2018**, *18*, 1647–1664. [CrossRef]
39. Conedera, M.; Tonini, M.; Oleggini, L.; Vega Orozco, C.; Leuenberger, M.; Pezzatti, G.B. Geospatial approach for defining the wildland-urban interface in the alpine environment. *Comput. Environ. Urban Syst.* **2015**, *52*, 10–20. [CrossRef]
40. Zumbrunnen, T.; Menéndez, P.; Bugmann, H.; Conedera, M.; Gimmi, U.; Bürgi, M. Human impacts on fire occurrence: A case study of hundred years of forest fires in a dry alpine valley in Switzerland. *Reg. Environ. Chang.* **2012**, *12*, 935–949. [CrossRef]
41. Badia, A.; Serra, P.; Modugno, S. Identifying dynamics of fire ignition probabilities in two representative Mediterranean wildland-urban interface areas. *Appl. Geogr.* **2011**, *31*, 930–940. [CrossRef]
42. Romero-Calcerrada, R.; Barrio-Parra, F.; Millington, J.D.A.; Novillo, C.J. Spatial modelling of socioeconomic data to understand patterns of human-caused wildfire ignition risk in the SW of Madrid (central Spain). *Ecol. Model.* **2010**, *221*, 34–45. [CrossRef]
43. Castillo Soto, M.E. The identification and assessment of areas at risk of forest fire using fuzzy methodology. *Appl. Geogr.* **2012**, *35*, 199–207. [CrossRef]
44. Vilar, L.; Woolford, D.G.; Martell, D.L.; Martín, M.P. A model for predicting human-caused wildfire occurrence in the region of Madrid, Spain. *Int. J. Wildland Fire* **2010**, *19*, 325–337. [CrossRef]
45. Mermoz, M.; Kitzberger, T.; Veblen, T.T. Landscape influences on occurrence and spread of wildfires in Patagonian forests and shrublands. *Ecology* **2005**, *86*, 2705–2715. [CrossRef]
46. Carmo, M.; Moreira, F.; Casimiro, P.; Vaz, P. Land use and topography influences on wildfire occurrence in northern Portugal. *Landsc. Urban Plan.* **2011**, *100*, 169–176. [CrossRef]
47. Harris, L.; Taylor, A.H. Previous burns and topography limit and reinforce fire severity in a large wildfire. *Ecosphere* **2017**, *8*, e02019. [CrossRef]
48. Moreira, F.; Vaz, P.; Catry, F.; Silva, J.S. Regional variations in wildfire susceptibility of land-cover types in Portugal: Implications for landscape management to minimize fire hazard. *Int. J. Wildland Fire* **2009**, *18*, 563–574. [CrossRef]
49. Kanevski, M.; Pozdnoukhov, A.; Timonin, V. *Machine Learning for Spatial Environmental Data: Theory, Applications and Software*; Collection: environmental engineering; Epfel: Lausanne, Switzerland, 2009; ISBN 978-0-8493-8237-6.
50. Lee, S.; Jung, H.-S. *Machine Learning Techniques Applied to Geoscience Information System and Remote Sensing*; MDPI: Basel, Switzerland, 2019; ISBN 978-3-03921-216-3.
51. Breiman, L. Random forest. *Mach. Learn.* **2001**, *45*, 5–32. [CrossRef]
52. Jain, P.; Coogan, S.C.P.; Subramanian, S.G.; Crowley, M.; Taylor, S.; Flannigan, M.D. A review of machine learning applications in wildfire science and management. *arXiv* **2020**, arXiv:2003.00646.
53. Martín, Y.; Zúñiga-Antón, M.; Rodrigues Mimbrero, M. Modelling temporal variation of fire-occurrence towards the dynamic prediction of human wildfire ignition danger in northeast Spain. *Geomat. Nat. Hazards Risk* **2019**, *10*, 385–411. [CrossRef]
54. Vacchiano, G.; Foderi, C.; Berretti, R.; Marchi, E.; Motta, R. Modeling anthropogenic and natural fire ignitions in an inner-alpine valley. *Nat. Hazards Earth Syst. Sci.* **2018**, *18*, 935–948. [CrossRef]

2

An Overview of Opportunities for Machine Learning Methods in Underground Rock Engineering Design

Josephine Morgenroth, Usman T. Khan * and Matthew A. Perras

Department of Civil Engineering, Lassonde School of Engineering, York University, Toronto, ON M3J 1P3, Canada; jmorgen@yorku.ca (J.M.); mperras@yorku.ca (M.A.P.)

* Correspondence: usman.khan@lassonde.yorku.ca

Abstract: Machine learning methods for data processing are gaining momentum in many geoscience industries. This includes the mining industry, where machine learning is primarily being applied to autonomously driven vehicles such as haul trucks, and ore body and resource delineation. However, the development of machine learning applications in rock engineering literature is relatively recent, despite being widely used and generally accepted for decades in other risk assessment-type design areas, such as flood forecasting. Operating mines and underground infrastructure projects collect more instrumentation data than ever before, however, only a small fraction of the useful information is typically extracted for rock engineering design, and there is often insufficient time to investigate complex rock mass phenomena in detail. This paper presents a summary of current practice in rock engineering design, as well as a review of literature and methods at the intersection of machine learning and rock engineering. It identifies gaps, such as standards for architecture, input selection and performance metrics, and areas for future work. These gaps present an opportunity to define a framework for integrating machine learning into conventional rock engineering design methodologies to make them more rigorous and reliable in predicting probable underlying physical mechanics and phenomenon.

Keywords: rock mechanics; geomechanics; rock engineering; machine learning algorithms; Artificial Neural Networks; numerical modelling; rock mass modelling

---

1. Introduction

The study of rock mechanics encompasses the theoretical and applied science of the mechanical behaviour of rock in response to its physical environment and was formalized as a field of study in the 1960s [1]. Due to the high degree of variability in natural materials, a precise rock engineering design within a narrow tolerance is difficult to produce. Experience and expert knowledge are heavily relied upon in rock engineering practice and empirical design charts have become prolific for preliminary stage design. Numerical modelling methods such as continuum and discrete methods, are also conventional tools in rock engineering design. These methods represent important tools for understanding rock mass behaviour and predicting its response to its environment and changes in in situ stress conditions. In practice, it is often difficult to integrate all the data collected into empirical and numerical models effectively due to time and budget constraints, as well as limitations in how constitutive behaviour is defined. Furthermore, it is sometimes not feasible to collect the quality and quantity of data needed, so extrapolation and interpolation techniques are often used.

Prior to the advent of "big data" and advances in machine learning, it was appropriate to base rock engineering design primarily on empirical data and expert knowledge because the geology, discontinuity, and in situ stress data available for projects were generally sparse. Now, increasing amounts of some forms of data such as displacements and pore water pressures, are being collected relatively inexpensively, while others are still infrequently collected because of the complexity and cost of the measurement methods, such as those used for stress measurement. Herein lies an opportunity to integrate machine learning into the existing rock engineering best practices to evaluate datasets more efficiently and to maximize the value (or information) extracted from the data. Machine learning algorithms are especially powerful because simultaneous hypotheses can be tested much quicker than by a human being [2]. This frees up the expert user or engineer to devote their judgment to selecting inputs and validating outputs in a more rigorous way, rather than manipulating data, which can be more time consuming and poses the risk of introducing bias.

Although machine learning has been used in other fields that rely heavily on numerical modelling, such as flood forecasting, for over 20 years [3], comparatively it is in its infancy in rock engineering applications. Some professionals in the mining industry believe the community is reluctant to accept data-driven methods because they are not as interpretable as applying conventional design criteria determined by previous experience and empirical evidence [4]. Work is being done to produce scientifically interpretable data-driven models [5], however, these frameworks have not migrated into engineering practice yet. Expert judgment is given much credence in the rock engineering industry, and empirical relationships form the foundation of fundamental rock engineering methods [6].

The progression of the design process is constrained by the analysis of the data that is collected. Machine learning algorithms have the potential to ease the burden of data analysis. If geotechnical professionals are able to predict geomechanical events or behaviours earlier and more accurately, this could translate to increased safety for underground personnel as well as cost savings in terms of reducing equipment loss, increasing efficiency of resource extraction or excavation rates, and minimizing delays and shutdowns.

This paper aims to summarize the most common practices in rock engineering design and presents the opportunities for integrating machine learning into existing geomechanical design frameworks. It also presents a review of common machine learning algorithms and how they are currently being applied to rock engineering problems in the literature, as well as future opportunities for data-driven approaches.

2. Current Practices in Rock Engineering Design

The inherent variability associated with natural materials makes standardization across geomechanical design processes difficult. Recent work to develop a geomechanical design framework has resulted in the publication of Eurocode 7 [7]. This code provides guidelines for tunnelling and underground excavations using a limit state approach, however, in general, customized designs are developed for most projects. Two of the most common categories of design approaches used in rock engineering design are: (i) empirical methodologies, and (ii) numerical modelling. These usually fall into a workflow together depending on the stage of the design, from prefeasibility to detailed design. Empirical design recommendations are often taken as a first estimate and are then verified and updated using the appropriate numerical tools, which are based on the geometry and complexity of the problem. For both these approaches, rock engineers use their judgement to combine numerical data (e.g., rock mass properties, stress conditions, pore water pressure) and categorical data (e.g., geological mapping, discontinuity conditions) into an empirical or numerical framework. As context for the remainder of this paper, both categories of conventional methodologies are briefly described below.

2.1. Empirical Design

In the early stages of conventional rock engineering design, it is common to turn to an established empirical rock support recommendation method to inform the design. The designed rock support may consist of a combination of rock or cable bolts, shotcrete, steel mesh, steel straps, concrete liner segments, etc. These methods are generally dependent on the size of the excavation and the quality of the rock mass and are used only as a first estimate followed by analytical or numerical methods to verify the design. It is typical to use multiple empirical methods and a range of input parameters representing the best- and worst-case to obtain a range of rock support recommendations. In this section, two common empirical rock support methodologies are described: the Rock Mass Rating (RMR) [8] and the Q Tunnelling Index [9]. These systems have been in wide use in the tunnelling and mining industries for 45 years, and have become engrained in standard rock engineering practice [10]. However, previous researchers [10] advocate for exercising caution when using these empirical systems, and for practicing engineers to use them in the spirit for which they were developed—to assist with the design of excavations, but not as the sole tool for designing underground support. Both these classification systems result in a recommended category for rock support, which is meant to encompass the range of rock mass conditions found within that category.

2.1.1. Empirical Support Recommendation—Rock Mass Rating

Bieniawski was among the first to assert that no single index, such as Deere's Rock Quality Designation [11], was enough to capture that aggregate behaviour of the rock, and so developed the Rock Mass Rating (RMR) system [8] to combine several measureable parameters. This system is the sum of five basic parameters to arrive at a score ranging from 0 to 100:

1. Strength of intact rock
2. Rock Quality Designation (RQD) [11]
3. Spacing of joints
4. Condition of joints
5. Groundwater conditions

This information is obtained through field mapping and site investigations. The site engineer or geologist collects a score for each of the parameters using the scoring scheme set out by Bieniawski [8]. Statistical analysis is conducted to determine the appropriate value, or more likely values, for each parameter and then sums them to obtain an RMR. The RMR corresponds to a rock quality categorisation that ranges from very good to very poor rock with five classes overall based on a linear relationship with the RMR value. Bieniawski created support guidelines based on the quality of the rock mass i.e., the RMR value and the stress condition the excavation is exposed to [8]. For example, low-stress environments only require heavy support if the rock is of poor quality.

2.1.2. Empirical Support Recommendation—Q Tunnelling Index

The Q Tunnelling Index was developed primarily to predict the appropriate support for tunnels [9] and is defined by multiplication of three quotients: the rock block size, the roughness and frictional resistance, and a stress quotient. As with RMR, the parameters are collected during the site investigation stage of a project and subsequently, a statistical analysis is performed to determine the most representative value or values. The Q Tunnelling Index is a value that ranges from 0.001 to 1000, and the formula is given in Equation (1).

$$Q = \left(\frac{RQD}{J_n}\right) * \left(\frac{J_r}{Ja}\right) * \left(\frac{J_w}{SRF}\right) \quad (1)$$

where:

RQD = Rock Quality Designation [11]
J_n = Number of joint/fracture sets
J_r = Roughness of most unfavourable joints
J_a = Alteration or infilling of joints
J_w = Water inflow
SRF = Stress reduction factor, quantifies stress conditions

The major difference between the Q Tunnelling Index support recommendation and others (such as RMR) is the inclusion of a parameter called the equivalent dimension, which uses an excavation support ratio to modify the span or width of the underground opening to capture a factor of safety correlating to the end use of the tunnel.

2.2. Numerical Methods

Numerical modelling is crucial for understanding fundamental intact rock and rock mass behaviour, assessing rock-structure interactions and completing rock engineering designs [12]. These methods offer the tools to capture the mechanisms that are causing observed phenomena in a rock mass or intact rock sample, and subsequently incorporate them into the design. Choosing inputs, geometry, boundary and stress conditions, among other defining aspects of a numerical model, requires significant engineering judgement and the combination of qualitative and quantitative site observations.

A typical approach is to use the back analysis of numerical models to calibrate their behaviour against observed site conditions. These calibrated models are then used to forward predict the excavation behaviour as the excavation is advanced, or as a new excavation is being designed in the same rock mass but at a different location [13]. Figure 1 illustrates this process, summarized as follows:

1. Site observations
2. Measurements and model calibration
3. Conceptual translation of the calibrated behaviour to the new site location
4. Comparison with empirical approaches [14] to validate the new design.

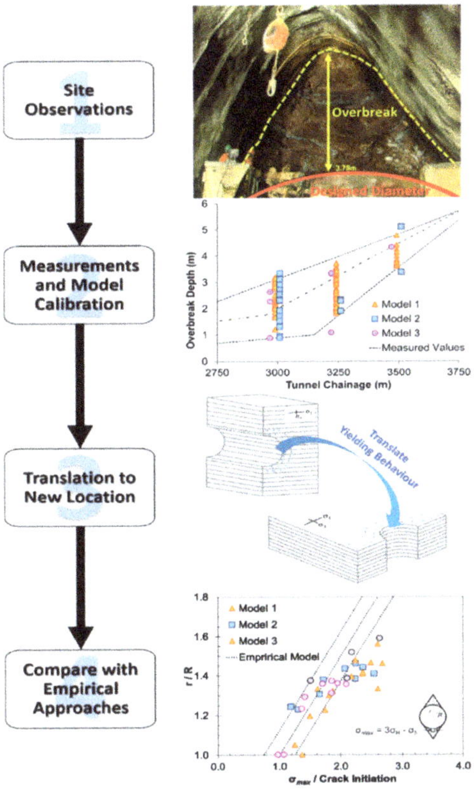

Figure 1. An illustration of a numerical modelling approach used to predict the overbreak depth around a vertical shaft (modified from [13]). Step 1. Observed overbreak (yellow dashed line) beyond the designed perimeter (solid red line) of the tunnel [13]. Step 2. Measured overbreak depths versus various numerical model results ([13,15–17]). Step 3. Translation of the yielding behavior from the tunnel in the horizontal beds to a vertical shaft in the same rock mass. Step 4. Comparison of calibrated models and conceptual behaviour to an empirical approach [14] for predicting normalized overbreak depth (r/R) using the maximum tangential stress (σ_{max}) calculated from the maximum and minimum horizontal stresses, σ_H *and* σ_h, respectively ([13,18]).

For example, the observed overbreak or damage beyond the designed diameter of a tunnel can be back analyzed using a numerical modelling approach. Individual parameters such as tunnel depth, are manipulated independently of the other input variables in order to calibrate them, by taking measurements along the tunnel axis at multiple locations, for example. This calibration process ensures that a robust set of input parameters are determined based on their individual impacts on the measured site conditions at a variety of locations along the tunnel alignment. The empirical knowledge from the tunnel site is translated conceptually to the new site in order to understand what similarities and differences may exist. With this conceptual understanding, a variety of models of the new site can be developed and the results compared to other empirical approaches, such as the damage depth prediction of Diederichs [14].

The most common numerical methods for rock engineering problems are [12]:

1. Continuum methods—finite element modelling (FEM), finite difference modelling (FDM), boundary element modelling (BEM).
2. Discrete methods—discrete element modelling (DEM), discrete fracture network (DFN) modelling.
3. Hybrid continuum/discrete methods.

The choice between methods is made based primarily on the scale of the problem and the geometry of the discontinuity or fracture system in the rock mass [19]. Continuum methods are most appropriate when a detachment of discrete blocks is not a significant factor. Discrete methods are generally chosen when the number of fractures is too large to treat the rock mass as a continuum with fracture elements, or if discrete block detachment is anticipated. Hybrid models are selected to avoid the pitfalls of each of the former two approaches. Each of these methods, their applications and limitations are briefly introduced.

2.2.1. Continuum Methods

The basic concept for continuum methods is to discretize the material being modelled into a grid governed by partial differential equations. The partial differential equations at the grid points are in close enough spatial proximity that the errors introduced between them are insignificant and thus, acceptable.

The FDM is the most direct way to discretize a continuum, where points in space are replaced with discrete equations called finite difference equations. These equations are used to calculate displacement, strain and stress in the material in response to conditional changes in the rock mass [20]. Solutions are formulated at grid points at the local scale, so no global matrix inversion is required, thus saving computational time and intensity. Complex constitutive behaviour can be captured without iterative solutions. However, FDM is inflexible with respect to fractures, complex boundary conditions and material heterogeneity. FDM was historically unsuitable for rock mechanics problems due to these limitations, however, advancements in irregular grid shapes gave rise to related Finite Volume Modelling (FVM) techniques [12]. FVM is more flexible in handling heterogeneity and boundary conditions and has been regarded as the bridge between FDM and FEM [17]. Commonly used commercial FDM software includes FLAC in two-dimensions and FLAC3D in three-dimensions [17].

FEM is the most widely applied numerical method in science and engineering [12]. FEM also involves discretizing a continuum into a grid, however, here, the material is subdivided into parts called finite elements. The partial differential equation at each element is informed by the elements adjacent to it, allowing FEM to handle heterogeneity, plasticity and deformation, complex boundary conditions, in situ stresses and gravity [21]. However, detachment of the elements is not permitted since these models are based on continuum assumptions. The treatment of fractures has been the largest limitation of FEM in the past, and modern software packages include special algorithms to overcome this. Commercial FEM software available are RS2 [22], SIGMA/W [23], Plaxis 2D [24] and ABAQUS [25], while open source options include Adonis [26], OpenSees [27] and Code-Aster [28].

BEM differs from FDM and FEM in that it first seeks an approximate global solution. Initially, BEM was developed for underground stress and deformation analysis, soil-structure interactions, groundwater flow and fracturing processes [12]. BEM approximates the solution of a partial differential equation inside an element by looking to the solution on the boundary and using that to inform the solution inside the element [29]. The main advantage of BEM over FDM or FEM is the simpler mesh generation and decreased computational expense. However, BEM is less efficient than FEM in handling material heterogeneity and plasticity. BEM has been used for: stress analysis of underground excavations, dynamic problems, back analysis of in situ and elastic properties, and borehole permeability tests [12]. Commercially available BEM software includes Examine2D [30] and Map3D [31], while open-source BEM libraries are also available [32].

2.2.2. Discrete Methods

Rock mechanics is one of the fields that originated Discrete Element Method (DEM) modelling, because highly fractured rock masses are not easily described mechanistically by a continuum approach. In discrete element approaches, the material is treated as an assemblage of rigid or deformable blocks or particles, the contacts between which are updated during the modelling process [12]. DEM solutions combine implicit and explicit formulations, based on FEM and FDM discretization, respectively. The main difference between DEM and continuum approaches is that the contacts between the elements are continuously changing, while they remain static for the latter. DEM methods are computationally demanding, however, they offer an advantage when the rock mass experiences loss of continuity (from progressive failure, for example), as continuum constitutive models are inappropriate in that case [33]. DEM methods have been popular for modelling a variety of rock engineering problems including: underground works, rock dynamics, rock slopes, laboratory tests, hard rock reinforcement, borehole stability, acoustic emissions in rock, among others. Commercially available DEM codes include UDEC [34] and PFC [35] for two-dimensional problems, and 3DEC [36] and PFC3D [37] for three-dimensional problems. Open source alternatives include Yade [38] and LAMMPS [39].

The Discrete Fracture Network (DFN) model is a discrete method focused on fracture pattern simulation. It takes in statistical information about the fracture sets and can be used to generate a network for input into DEM codes for use in its behaviour, such as considering fluid flow through a series of interconnected fractures. It is a powerful method for studying fractured materials where an equivalent continuum cannot be established. DFN codes have been applied to the following problems: developments for multiphase fluid flow, hot dry rock reservoir simulations, permeability of fractured rock, and water effects on underground excavations and rock slopes [12]. Commercial DFN softwares include FracMan [40] and MoFrac [41], while open source options include ADFNE [42].

Discrete approaches are limited by the modeller's knowledge of the geometry of the fracture network, which can only be estimated based on geological mapping and interpretation of in situ information.

2.2.3. Hybrid Continuum/Discrete Methods

Hybrid numerical models have gained popularity to overcome the limitations of each of the numerical methods previously described. Hybrid methods are commonly used to address limitations in how fracture growth is addressed by other methods. Some hybrid methods add fractures discretely and the mesh is adjusted dynamically as the crack propagates, while others have a very fine mesh and the fractures propagate along the boundaries. Care must be taken where two methods interact to ensure compatibility of the underlying assumptions.

The most common hybrid models are BEM/FEM [43], DEM/BEM [44] and DEM/FEM [45]. There are many advantages to hybrid numerical models when applied correctly, for example, the coupled DEM/FEM method explicitly satisfies equilibrium conditions of displacement at the interface between two domains. The FEM/DEM approach has been shown to realistically model the dynamic response of rocks, and simulations show good agreement with laboratory observations [46]. New developments in DEM/FEM approaches allow the modelling of rock masses with anisotropic strength and deformation characteristics, as well as the explicit modelling of fracture growth [33,47]. Irazu is a commercially available DEM/FEM hybrid software package that explicitly models fracture processes in brittle materials, capturing complex non-linear behaviour [48]. In addition to Irazu, other commercially available software includes the Hybrid Optimization Software Suite [49] and Elfen [50], while open-source alternatives include Y-GEO [51].

The choice of numerical method depends on the data available and the complexity of the problem being solved, and sometimes multiple methods may be explored before one is selected to use in subsequent design activities. Similar to empirical design approaches, significant expert judgement is required to determine the model inputs and interpret the outputs.

2.3. Discussion of Current Practices

Both empirical and numerical methods are strongly rooted in conventional rock mechanics design and lend insight into anticipated rock mass response to the construction of an excavation. The issues often arise not from the methods themselves, but rather how the data collected is used to make support decisions or how it is input into numerical models. The inherent variability of geomechanical datasets, as well as the variety of types (numerical measurements, categorical descriptions, photos), introduce error to the design process.

When making use of empirical support recommendations, it is common practice to determine the rock mass classification value based on statistical analysis or using the best- and worst-case values. While this is appropriate for a prefeasibility estimate, these values are sometimes carried forward into detailed design. A data-driven method would allow the design engineer to make use of all the data to inform the most appropriate parameters for further use. Some research has been conducted using Artificial Neural Networks (ANNs) to classify rock masses [52], determine strength properties [53] and model stress-strain behaviour [54].

Numerical modelling of rock mass behaviour is difficult because rock is a discontinuous natural material with inherent variability and inhomogeneous properties. These need to be captured by fundamental equations and constitutive models which have geometrical and physical constraints. This can be overcome using data-driven methods such as ANNs because the data trends are not conformed to these constraints, which sometimes understate the complexity of the problem [12]. Research has been performed using ANNs for predicting tunnel convergence [55] (inward radial displacement of the tunnel), rock bursts [56] (an accumulation and sudden release of strain energy), open-pit stability [57] (stability of the mine slope geometry), among other applications.

3. Review of Machine Learning Algorithms

Machine learning is a branch of artificial intelligence that aims to program machines to perform their jobs more skillfully [2]. This is done by using intelligent software that takes inputs to train a model to produce the desired result, thus replicating learning. Machines are better at performing repetitive tasks than humans, so harnessing this potential has been at the forefront of almost every industry since the beginning of the technological age. Sometimes the model is intuitively understandable, and other times the workings of the model cannot be easily explained. The choice of machine learning techniques is informed by what the output is and what data is available [58]:

- Supervised learning: data is labelled, i.e., the training samples contain inputs with a corresponding output
- Unsupervised learning: data is unlabeled, i.e., the training samples do not have an associated output
- Semi-supervised learning: a mixture of labelled and unlabeled data
- Reinforcement learning: no data; the algorithm maps situations to actions to maximize a reward [59]

Before machine learning can be implemented in a project, the data must be acquired from various sources and cleaned for use. Data cleaning involves identifying incomplete, incorrect, inaccurate or irrelevant parts of the dataset and then replacing, modifying, or deleting raw data. Oftentimes, data acquisition and preparation are the most time consuming and onerous part of the process. Unlike other industries where machine learning already has a solid foothold or is widely used in practice, in rock engineering this is not yet the case [6].

Machine learning algorithms present an opportunity to remove the error associated with data manipulation and offer predictive capabilities to increase the efficiency of the design process. As discussed previously, early work has shown their suitability for determining rock mass parameters and constitutive behaviours, as well as predicting geomechanical phenomena and instabilities. This section presents a brief overview of the most common types of machine learning algorithms, with an emphasis on ANNs. These are broadly divided into categorical prediction models and numerical prediction models.

3.1. Categorical Prediction Models

Categorical machine learning algorithms are useful for classifying data or making a categorical prediction. For example, given a dataset comprised of site-specific inputs for RMR or Q classifications, the algorithm can predict what the RMR or Q value is for a point ahead of the excavation face (i.e., the current extent of the excavation).

3.1.1. Decision Trees

The Decision Tree approach is a supervised machine learning algorithm that places data into classes and presents the results in a flowchart. The data flows through a query structure from the "root" or selected attributes through subsequent partitioning until it reaches a "leaf" where no sample remains, no attribute remains, or the remaining samples have the same attribute [60]. The goal of creating a Decision Tree is to have a generalisable model that can classify unlabelled samples. Attribute selection is a crucial step when applying decision tree algorithms, as they must be meaningful to split the dataset at hand in to "purer" subsets [61]. A simple schematic is shown in Figure 2.

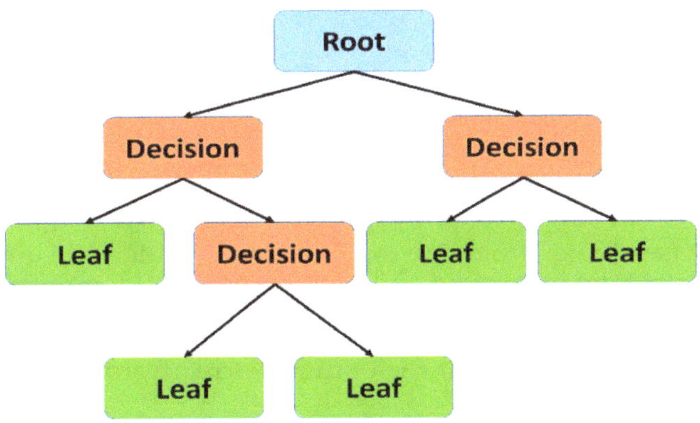

Figure 2. Decision Tree schematic showing root node, decision nodes and leaf nodes.

In one case study, a Decision Tree method was employed to predict rockburst potential in a kimberlite pipe diamond mine [62]. Here, the root is the linear elastic energy, the decision node is the ratio between maximum tangential stress and unconfined compressive strength (UCS), and the decision node is the ratio between the UCS and the uniaxial tensile stress. This led to classification, or leafs of "no rockburst", "moderate rockburst", "strong rockburst" and "violent rockburst". The authors trained the algorithm with 132 training samples from real rockburst cases around the world, and subsequently the accuracy of the validation samples was shown to be 93%. Validation samples are those that have not been used to train the algorithm, and are therefore a metric for generalisability. The results of the study found that the mine under study was susceptible to moderate bursts, which matched the observed conditions.

3.1.2. Naïve Bayesian Classification

Bayesian networks are nodal networks that graphically represent probabilistic relationships between input variables using particular simplifying assumptions. This technique is founded in Bayes' theorem, where the "posterior probability" of an event occurring is updated based on new "evidence" [63]:

$$\text{Posterior Probability} = \frac{\text{Prior Probabiliy} \times \text{Likelihood}}{\text{Evidence}} \tag{2}$$

Naïve Bayesian classifiers are a type of Bayesian network that uses simple probabilistic classifiers that assumes independence between predictor variables (hence "naïve") [64], as shown in Figure 3. To complete the classification, the numerator of Equation (2) is compared to each sample since the evidence remains constant [63]. Though this type of algorithm can be advantageous because it is a simple representation of a problem and therefore easy to implement [6], a common criticism is that they assume independence between the input attributes [65].

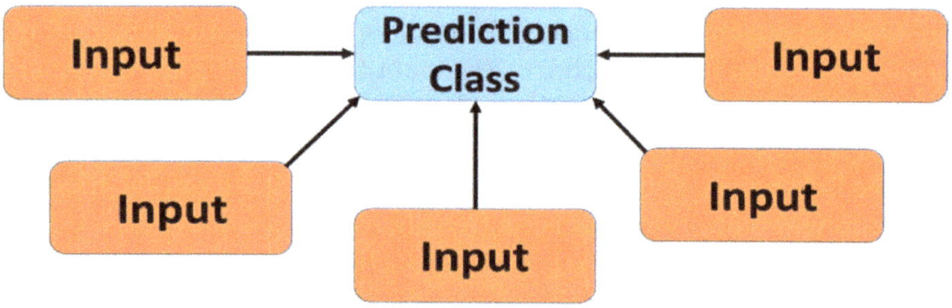

Figure 3. Naïve Bayesian classifier schematic showing predicted class and independent input variables.

A case study comparing different types of Bayesian network classifiers found that all the developed models showed a high accuracy rate when applied to predict the magnitude of rock bursts for the dataset consisting of 60 cases [65]. The naïve inputs included the type and rock strength, geometry, stress state and construction method, which were used to predict the magnitude of the rockburst. The naïve Bayesian model in particular classified 100% of overbreak cases, 83% of strong rock bursts, 25% of moderate rock bursts and 87.5% of slight rock bursts correctly, respectively.

3.1.3. *k*-Nearest Neighbours Classification

The *k*-nearest neighbours (*k*-NN) clustering algorithm is used for classification and is among the simplest of the machine learning algorithms. The *k*-NN approach can also be used to determine a numerical output using regression, as discussed in Section 3.2.1. The input consists of the *k* closest training examples and the output is a class membership. An object is classified by a majority vote of its closest neighbours, where the most common classifier is assigned [66], as shown in Figure 4. The function is approximated locally and computation is deferred until classification is complete. The user can assign a weight to the contributions of the neighbours as a function of distance, for example. This algorithm is sensitive to the structure of the data, which may pose a limitation.

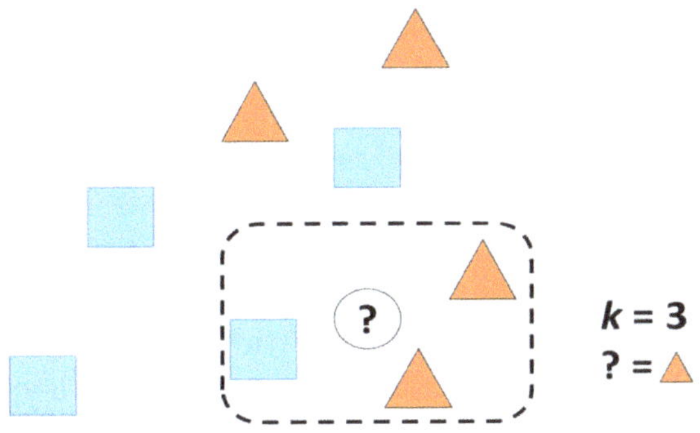

Figure 4. *k*NN classification schematic showing $k = 3$ and, therefore, the three nearest neighbours are used to make the classification.

Work has been done comparing *k*-NN to four other supervised machine learning algorithms to classify geology using remotely sensed geophysical data [67]. Inputs included a Digital Elevation Model, Total Magnetic Intensity and four Gama-Ray Spectrometry channels, and the parameter *k* (number of nearest neighbours used for classification) was varied from 1 to 19. The authors conclude that as the spatial distribution of training data increases, the accuracy of the classifications also increases. They also conclude that explicit spatial information (coordinates) should be combined with geophysical data so that predictions are geologically plausible.

3.1.4. Support Vector Machine

Support Vector Machines (SVMs) are supervised clustering machine learning algorithms used for classification analysis [68]. The SVM maps the labelled training dataset as points in space divided into categories separated by a clear gap. New examples are mapped into the same space and are categorized depending on which side of the gap they fall into. The gap is defined by a hyperplane in two or three dimensions, depending on how many features are being used to classify the data. The main goal of the SVM is to find the hyperplane that maximizes the margin between the classes, as shown in Figure 5.

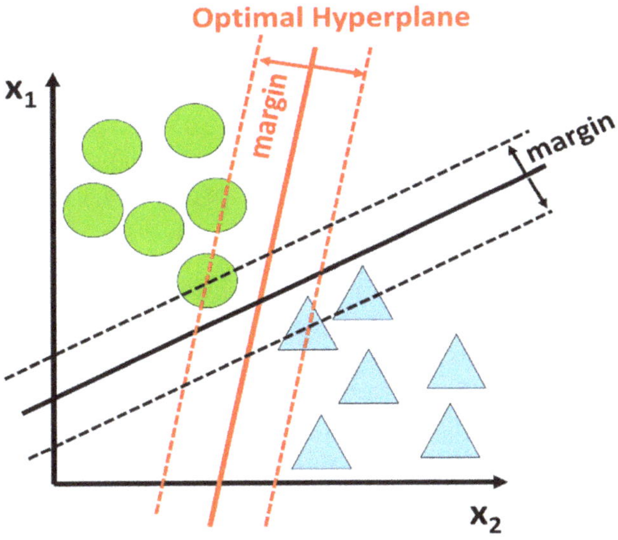

Figure 5. Support Vector Machines (SVM) schematic showing a two-dimensional classification problem and the optimal hyperplane that should be used to classify future unlabeled data based on the smallest margin between the two closest data points.

A recent study applied SVM to predict tunnel squeezing based on four parameters: diameter, buried depth, support stiffness and the Q Tunnelling Index [69]. An 8-fold cross-validation was used to create multiple models (called an ensemble) to get a measure of the performance of the model. Cross-validation is a common technique used in data-driven methods to determine how well generalised the model is to an independent dataset, and how accurately the model will perform in practice. This process highlights whether the model is overfitting or if there is input selection bias. The resulting average performance of the algorithm was 88% and importantly, this performance decreased to 74% when the support stiffness was not included in the SVM. The authors concluded that this method produced better performance in prediction accuracy as compared to existing empirical approaches, similar to other uses of the SVM cited therein. They were also able to estimate the severity of the potential squeezing based on the predicted squeezing class by introducing a multiclass SVM classifier trained using a database of 117 case histories. The multiclassifier was used to classify the database into the severity of the squeezing problem, similar to empirical classification schemes based on the strain or convergence.

3.1.5. Random Forests

Random Forest (RF) is a supervised ensemble classifying method that consists of many decision trees. The output is the majority vote of the classes output by the individual trees, as shown in Figure 6. Each decision tree is an individual learner and the aggregate of each individual yields the prediction of the algorithm [4]. RF is considered to be one of the most accurate classifying algorithms, running efficiently on large databases and effectively estimating missing data [4]. However, RF has been noted to overfit for noisy datasets and tends to be biased towards categories that are over-represented in the training dataset.

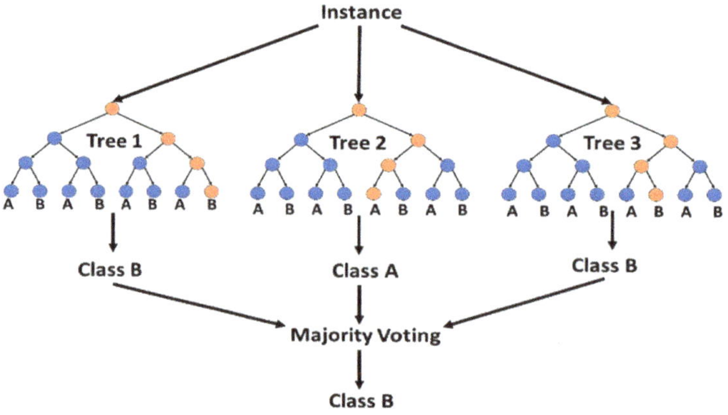

Figure 6. Random Forest schematic, showing an ensemble of tree decision tree classifications and the majority voting that determines the final class.

One case study applied an RF algorithm for predicting hanging wall stability using a training dataset consisting of 115 cases [70]. The inputs were subdivided into hanging wall geometry (stope dip, strike and height), geological properties (RQD, joint set number, joint set roughness, joint set alteration, dilution graph factors A, B, and C) and construction parameters (stope design method, undercut area and stress category). Each of these inputs represents a "branch" of the RF structure. A 5-fold cross-validation method was applied and the grid search method was used to tune the hyperparameters. A common performance metric, the area under the receiver operating characteristic curve (ROC-AUC), was used to evaluate the accuracy of the classifications made by the RF algorithm and was shown to be 0.873 (out of a maximum of 1.0) for the testing dataset (the data subset withheld to ensure the generalized performance of the model). The authors state that this indicates their optimum RF model is excellent at predicting hanging wall stability.

3.2. Numerical Prediction Models

Numerical prediction models are useful for obtaining an estimate where not enough information is available to do a conventional calculation (i.e., a discrete analytical solution), or where the relationships between the inputs and output are too complex for an analytical solution. For example, an algorithm trained on site-specific rock mass parameters and the in situ stress field could predict the radial convergence of a tunnel before it is constructed. Or, a trained algorithm using an existing dataset comprised of typical rock mass parameters could determine the rock mass strength of an unmapped location. While other numerical predictors are briefly discussed, this paper focuses on ANNs as a prediction tool, as they have been the focus of the research performed in machine learning methods for rock engineering design.

3.2.1. Support Vector Clustering

Support Vector Clustering (SVC) is an expansion of Support Vector Machines that is used when data is unlabelled or only some data is preprocessed [68]. SVC maps data points into a multi-dimensional feature space (where each feature is an input variable) using a kernel function [71]. The algorithm then searches for the smallest sphere that encloses the data in the feature space and maps it back to the data space, where the sphere is transformed into contours that enclose the data that form part of the same group. Now unlabelled data has been classified.

Little work has been done applying SVC to geomechanical problems, however, a case study from water quality literature is discussed here. An SVC algorithm was employed to model the electric conductivity and total dissolved solids in a river system and was compared against a more conventional/genetic programming algorithm [72]. The authors concluded that the SVC method has better accuracy for modelling water quality parameters than the genetic programming algorithm [73].

3.2.2. k-Nearest Neighbours Regression

The k-nearest neighbours (k-NN) algorithm is used for regression, where the input consists of the k closest training examples and the output is the property value for the object [66]. Similar to *k*-NN classification, the value is the average of its nearest neighbours, except the output is a numerical value rather than a classification.

Little work has been done applying *k*-NN regression to geomechanical problems, however, an analogous case study completed to forecast the municipal solid waste (MSW) generated by a city [74]. Four machine learning algorithms were compared: SVM, ANN, adaptive neuro-fuzzy inference systems and kNN regression. The authors found that the prediction ability of the kNN regression algorithm was in the middle in terms of matching the observed data and peaks in the trends, however, it was the best at predicting the monthly average values of MSW generated.

3.2.3. Artificial Neural Networks

Artificial Neural Networks (ANNs) are inspired by biological neural networks, where a series of highly interconnected nodes and a series of parallel nonlinear equations are used simultaneously to process data and perform functions quickly [75]. ANNs are known as universal predictors and can approximate any continuous function under certain conditions (e.g., availability of appropriate input parameters). ANNs are powerful for real-time or near real-time scenarios as they can function with high volumes of data as it is collected [76].

In general terms, ANNs are comprised of an input layer, hidden layers and an output layer. Each node is linked to all the nodes in the layer preceding and following it, where each link has a function defined by a weight and bias (Figure 7). Each node also has an activation function, which compares the weighted sum of all the inputs to that node and compares it to a predetermined threshold. If the threshold is exceeded the node fires and the input is transmitted further in the network through the activation function. The most common activation functions are the step, sign, linear and sigmoid functions [75]. Activation functions should be chosen with care as they inform the response of the network to the inputs and, therefore, the resulting output.

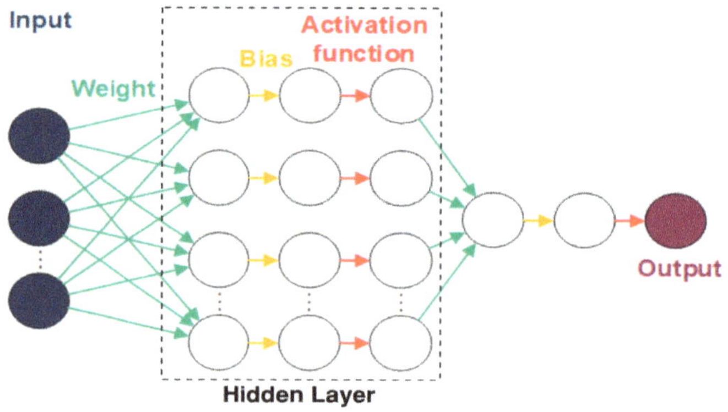

Figure 7. Simple schematic of an Artificial Neural Networks (ANN).

Multi-Layer Perceptron (MLP) is the simplest and most common form of ANN and employs a learning technique called backpropagation [55]. Backpropagation, short for backward propagation of errors, is used to adjust the weights and biases of the ANN by minimizing the error at the output. Bacpropagation is widely used in engineering and science because it is the most versatile and robust technique to find the global minima on an error surface [77]. Backpropagation consists of two steps: the propagation phase and the updating of the weight [75]. In the propagation phase, the inputs are fed into the ANN and the values at the hidden and output nodes are calculated. In the second phase, the error is calculated at the output and then propagated backward to update the weights at the nodes using deterministic optimisation to minimise the error sum [76].

The ANN is trained using a dataset to create a model that is able to make predictions using new information. In the case of an MLP, the network is inputs fed forward and the error is backpropagated to update the ANN iteratively until a solution is converged upon that matches the observed data within the error tolerance [75]. A concern with ANN development is "overfitting", where the ANN predictions match all the observed data too closely and cannot handle new information well [59]. This can be avoided by using data partitioning techniques to avoid creating a biased model, which is a common approach for machine learning methods in general. The partitioned subsets are called the "training," "validation" and "testing" subsets. The testing set is withheld entirely during ANN development. The training and validation sets can be used to create multiple models (called an ensemble) to get a measure of the performance of the model, called *k*-fold cross-validation (Figure 8). This is good practice in ANN model development but currently rarely used in ANN applications in various engineering sub-disciplines [3].

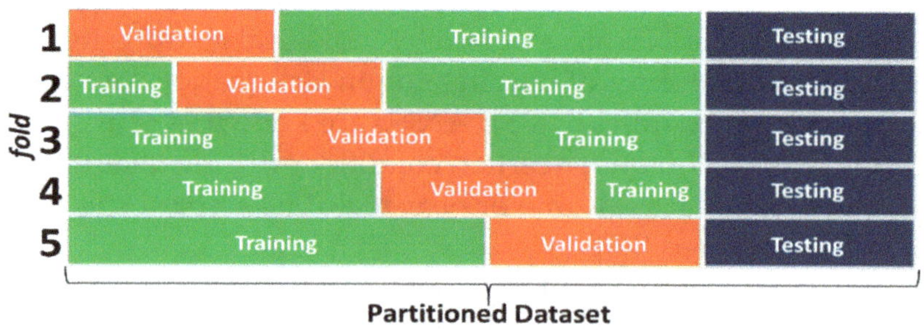

Figure 8. Example of data partitioning showing 5-fold cross-validation.

ANN model uncertainties come from the choice of ANN architecture (i.e., number of hidden layers, number of neurons, choice of activation function, type of training algorithm and data partitioning), as well as the performance metric chosen. Due to the data-driven nature of these models, propagating these uncertainties is easier [78,79]. One method of quantifying this uncertainty is by using fuzzy numbers to quantify the total uncertainty in the weights, biases and output of the ANN [78,80,81]. This technique is useful for dealing with limited or imprecise datasets and can be used to conduct risk analysis [82].

ANNs offer predictive and descriptive capabilities and have been applied to a range of rock property definition and rock engineering problems including: intact rock strength, fracture aperture, rock mass properties, displacements of rock slopes, tunnel support, earthquake analysis, tunnel boring machine performance, among others [12]. Despite its wide applications, ANNs have not yet been proven to be a viable alternative to conventional methods to the rock engineering community.

4. Discussion of Machine Learning for Rock Engineering Design

As with all machine learning problems, the size and quality of the geomechanical datasets should inform the algorithm chosen for a given problem. The desired output is also a determining factor as to which algorithm is best suited to making the prediction. A summary of MLAs that have been researched for application to rock engineering problems are presented in Table 1, column two summarizes the MLAs discussed in Section 3, while columns one and three ties these to various rock engineering problems discussed in Section 2. Note that the majority of the applications of MLAs are for categorical problems. To date, a variety of MLAs have been applied to classify rock mass properties, rock bursts and tunnel performance. The research that has been performed using MLAs to scale lab data and field observations to rock mass properties have been primarily categorical, as the rock mass classification schemes presented in Section 2.1 represent an industry-accepted classification scheme. It is common for a project to have an incomplete dataset due to the cost associated with lab testing or field data collection. In practice, a basic statistical analysis may be performed to obtain conservative estimates of these properties, which are then scaled up to the rock mass scale properties using empirical or analytical methods. MLAs present a more precise method for determining the relationship between the lab-scale and the rock mass scale properties and potentially avoids the bias that can be injected by doing this manually. The algorithms for rockbursts and tunnel performance have been developed using inputs that represent physical properties that have been binned into categories that correspond to the severity of the rockbursts or tunnel deformation being predicted. Based on the classification schemes discussed in Section 2.1, the bins for these and other rock mass properties are already defined in engineering practice, and therefore practical to apply to a classification problem such as these. In practice, these properties are combined to calculate an overall quality score and then the support is designed based on this. MLAs present the opportunity to perform the rock support determination even if the dataset is incomplete, and allows the rock engineer to quantify the uncertainty associated with those predictions.

Table 1. MLAs applied to various rock engineering problems.

Rock Engineering Problem	MLAs	Opportunities
Rock mass properties	Categorical ANNs [52,83,84]	• Backwards predict rock mass properties based on observed site conditions • Predict rock mass scale properties based on lab-scale properties and rock mass behaviour
Laboratory testing and constitutive behaviour	Numerical ANNs [54,85] SVM [85]	• Use geology and peak Unconfined Compressive Strength to predict crack initiation and crack damage thresholds • Use laboratory tests and field observations to predict constitutive behaviour • Use rock mass scale classification (e.g., Q, RMR) to predict lab-scale properties
Slope stability	Categorical ANNs [57,86] SVM [87] RF [88,89] Clustering [90]	• Predict slope movements based on geometry, piezometers, inclinometer data, etc. • Predict volume of structurally controlled failure based on mapped discontinuities

Table 1. *Cont.*

Rock Engineering Problem	MLAs	Opportunities
Point cloud analysis	RF [91] *k*NN [92]	• Use successive tunnel scans to get volume differences and predict time-dependent deformations • Predict stress/strain fields for input into numerical models
Tunnel performance	Numerical ANNs [55,93–95] Categorical ANNS [96,97] SVM [69] RF [61,70]	• Use preliminary/incomplete field mapping to prediction rock mass classification (e.g., Q, RMR) • Predict tunnel support class based on rock mass classification (e.g., Q, RMR) • Predict rock support performance-based on geology, excavation method, environmental conditions, etc. • Use microseismic monitoring arrays to predict rock mass deformation as excavation is developed
Rock bursts	Categorical ANNs [65] Naïve Bayesian classifiers [65] *k*NN [65] RF [98] SVM [65,99] Decision Trees [62]	• Predict magnitude and location of events using 3D excavation geometry, time-series seismic events, mapped geology, etc. • Use previous rockburst events, geology, etc. to predict magnitude of failed material and performance of rock support • Use mapped rock classification (e.g., Q, RMR) to predict probability of rockburst
Blasting	Categorical ANNs [100,101] Numerical ANNS [100] SVM [99,102] RF [102]	• Use blast parameters and damage extent to predict optimum parameters for future blasts • Predict blast parameters using mapped rock classification (e.g., Q, RMR)

Little work in the literature has been done applying MLAs to the numerical modelling methods described in Section 2.2. Although sensitivity analyses are performed in practice, in general, the numerical calibration process consists of manually adjusting the model parameters in a systematic manner until the numerical model outputs match the field observations. This requires careful adjustment of the input parameters followed by computation of the model to check the output against the observation or measured behaviour. For complex problems, running the model repeatedly is time-consuming and therefore not done regularly in practice. MLAs present an opportunity to define the complex relationships between the input parameters, the numerical model behaviour and observed rock mass phenomena, and subsequently to conduct a more precise sensitivity analysis of the model inputs to ensure the rock mechanics relationships are being captured. An avenue for future research is surrogate modelling, where an MLA is used in conjunction with the numerical model outputs to systematically iterate through distribution of inputs until the numerical model result matches the field observations. The MLA model may be used to calibrate the numerical model or simply for predicting an unknown state—hence the term "surrogate model".

As shown in Table 1, several authors have found success applying ANNs to model complex rock mass behaviour over other algorithms, which is why they are emphasized herein. To date, the most common ANNs researchers are using are MLPs with one or two hidden layers [78]. The main advantage of ANNs is that geometrical and physical constraints that govern rock mass constitutive behaviour and cause problems in numerical modelling approaches are not as problematic. This is because the data-driven method does not rely on a function to capture all the anticipated rock mass behaviour, but rather learns from the specific cases it is given to inform future predictions. An additional advantage is that ANNs can incorporate judgments based on empirical methods and can mimic the "perception" the human brain is capable of [59]. However, ANNs are limited to making predictions within the training parameters, meaning that they cannot make predictions outside the dataset they are given to train with. In other words, if there are not catastrophic events, such as a rock burst or falls of ground included in the training dataset, then the ANN may not be able to predict these events given new data. Since ANNs are fundamentally a complex curve fitting algorithm, overfitting or underfitting may pose a concern to the general applicability of the final model. It is up to the developer of the ANN to ensure that the network has been validated and tested appropriately, and that suitable performance metrics (for example, the coefficient of determination (R^2), root mean square error (RMSE), precision/recall, receiver operating characteristic (ROC) curve) have been used to ensure the general applicability of the final network. Since ANNs are relatively new in the field of rock engineering design, there is a lack of verification and validation of ANN outcomes in the literature, and, therefore, cross-disciplinary literature review and independent validation using conventional design tools are necessary to prove their validity.

A unique challenge for rock engineering problems is the inherent spatial dependency of the variables, particularly when dealing with anisotropic and non-homogeneous rock mass properties. This may be considered explicitly as coordinate inputs, or implicitly by including inputs that are spatially variable. Some work has been done to determine the effects of these two input methods for ANNs, and it was noted that the model performed better when the spatial coordinates were included implicitly [67]. An analogous problem has been addressed and is far more advanced in the field of image recognition, where the image is treated as a two-dimensional raster and inputs, or features, are mapped to a point in two-dimensional space on the image that is being processed. The techniques being used in that field may be applicable for encoding 2D and 3D spatial information in the geomechanical context. There is ample opportunity for further research in this area.].

Time-dependent rock mass behaviour can be challenging to capture using numerical models [103 However, there may be an opportunity to couple an ANN with numerical models, particularly because ANNs can take time-lagged variables as an input. Furthermore, convolutional layers may be added between the input node and the first hidden layer to allow a variety of input types (numerical, categorical, time-dependent) to be preprocessed and then combined to make a rock mass behaviour prediction. by one network [104]. A recurrent neural network (RNN) may also be developed, which allows the network to "remember" previous predictions made and use them to make further predictions [105 Further work in this avenue will have applications to solve both the time-dependent and spatial issues inherent to rock engineering modelling problems. 9,

A continuum of data quantity and redundancy for geomechanical datasets is presented in Figure based on this literature review and the authors' collective experiences in both machine learning and rock mechanics. The data methods discussed in this paper and those used in practice are plotted on this continuum to show how much data and how many redundant data points are required to effectively use a particular data manipulation technique. Data redundancy indicates how many samples of a particular phenomenon or rock mass behaviour are required in order for the data method to be able to reliably calculate or predict it. Deterministic methods are located in the top left corner of Figure 9, with the lowest amount and redundancy of data required. This means that, if the correct data is available (through field or lab measurements), a deterministic method (e.g., a physics-based equation) can be easily implemented. However, common methods like regression have a higher data redundancy: several series of observations are required in order to "train" or "calibrate" a regression-based model. The higher the redundancy in the dataset, the more confidence in the predictions of the regression model. The commonly used deterministic or probabilistic methods are usually supplemented by expert judgment or typical values when completing analytical or numerical rock engineering designs. Probabilistic, kNN, SVM, Decision Trees all have a relatively higher amount of data requirements than standard deterministic and are not generally suitable for smaller datasets, however, reinforcement learning techniques (not covered herein) may become applicable as the algorithms become more advanced. Larger, more comprehensive datasets such as those collected in mining environments, are well suited to the machine learning algorithms discussed in this paper, especially ANNs.

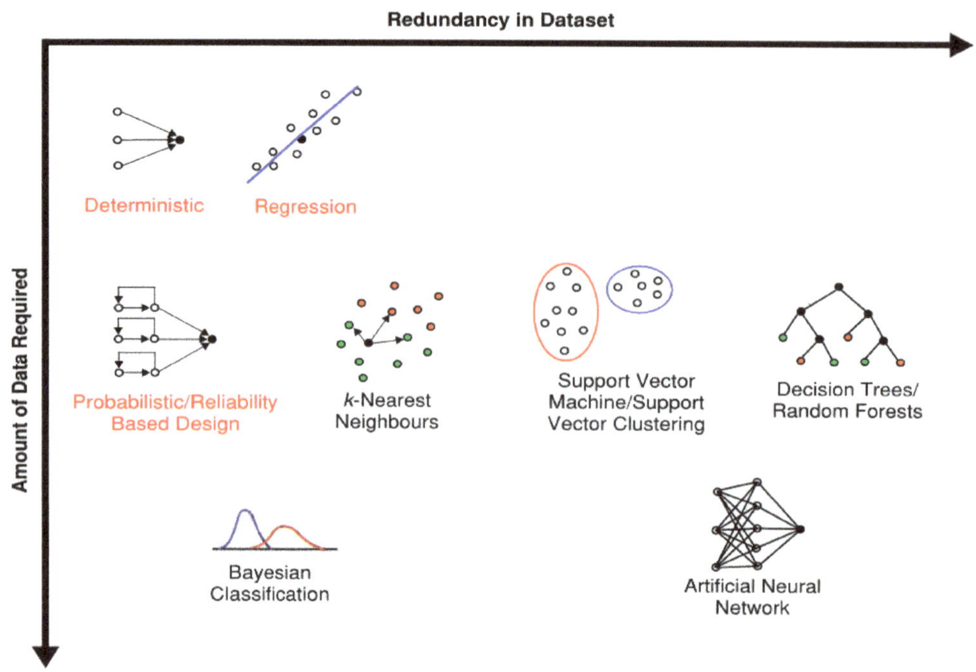

Figure 9. Comparison of data needs for machine learning algorithms included in this review (black) and some conventional rock engineering methods (red) in terms of the amount of data and data redundancy required. Data redundancy indicates the representation of samples of the behaviour the data method should capture.

Operating mines often collect data continuously over time, for example: geological conditions, microseismic events, extensometer or crack meter data, in situ stress measurements, groundwater and pore water pressure monitoring, among many, many others. This kind of multivariate dataset, where each parameter is intertwined and related to the overall stability in a complex way, is ideal for implementing an ANN. This is especially true if there are geomechanical events of interest that are ongoing, such as rockbursts or spalling, that have occurred over the life of the mine and were captured by some or all the instrumentation.

No matter the size or redundancy of the input dataset, it is crucial that the input selection and data partitioning scheme are appropriate for the problem to prevent overfitting and to ensure generalisability of the resulting model [106]. Input variable selection is often done on an ad hoc basis [3], or using expert judgement and simple linear models [107]. Research in other geoscience fields suggests that input selection be done using a systematic approach based on a rigorous input ranking [108]. Formal input selection methods are used to determine which inputs from the larger input dataset are most useful in terms of relevance to the desired output prediction while minimizing redundancies between input variables [108]. The framework for input selection has not been formalized and rarely receives the requisite attention [109–111].

While the importance of appropriate data partitioning is widely accepted [112], this process has not been formalised and is sometimes done arbitrarily [110,113]. Poor data partitioning may result in a model with poor performance, and some work has been done to quantify the variability in the quality of training, validation and testing subsets and their impacts on model performance [111,113,114]. In particular, four main data division methods are prevalent in the literature:

1. Random data division
2. Data division ensuring statistical consistency within subsets
3. Data division using self-organising maps
4. Data division using fuzzy logic methods

A thorough comparison on the impacts of these methods on model performance specifically for rock engineering datasets is needed to develop a framework for further application of machine learning methods in this field.

Model architecture and hyperparameter (i.e., number of hidden nodes, activation function, weights and biases, etc.) optimization are often done informally, despite the impact of these decisions on model behaviour [109]. Given that this approach is the accepted norm, Abrahart et al. (2012) asked if obtaining an optimal model structure is feasible or if the efforts required to acquire it are warranted [3]. The authors go on to state that since many permutations of hyperparameters and architecture yield similar performances, an ad hoc approach may be appropriate for practical applications. However, if a more detailed and complex solution is required, numerous hyperparameter optimisation methods exist, and have been applied in other geoscience fields, such as: *k*-fold cross-validation to produce and ensemble of models [115], parameter regularization [87] and metaheuristic optimisation algorithms [116].

5. Conclusions

This paper briefly summarises the current state of rock engineering design and presents the opportunities to integrate machine learning algorithms into the existing geomechanical design frameworks. A literature review has been conducted on the work that has been done specifically using ANNs on geomechanical datasets, and areas for future work have been identified. In practice, stability analyses are often performed relying on past experience and observational methods rather than on new data being collected on geology and construction progress [96]. ANNs are suitable for modelling complex rock mass behaviour and have been shown to be more efficient than regression functions [96]. ANNs are especially powerful for repetitive construction processes as they can use real-time data to update predictions ahead of the current stage of excavation. Future work on convolutional neural networks and recurrent neural networks will be invaluable in addressing spatial and temporal rock mass behaviour explicitly using ANNs.

Based on the body of literature available at this time, this review has found that there is a lack of standardisation of the input selection process, data partitioning methods, model architecture and hyperparameter optimisation and performance measures in ANNs for rock engineering. These gaps present an opportunity to define a framework for integrating machine learning into rock engineering design to make the process more rigorous and reliable.

Author Contributions: Conceptualization, J.M., U.K. and M.P.; methodology, J.M.; investigation, J.M.; writing—original draft preparation, J.M.; writing—review and editing, J.M., U.K. and M.P.; visualization, J.M.; supervision, U.K. and M.P.; funding acquisition, U.K. and M.P.

Acknowledgments: The authors would like to thank our industry partners Yield Point Inc., RockEng, and Cameco for insightful conversations on the topic. We would also like to thank two anonymous reviewers for their helpful comments and suggestions, which has helped improved the quality and focus of this manuscript.

References

1. Hoek, E. Rock Mechanics-an Introduction for the Practical Engineer. *Min. Mag.* **1966**, 1–67.
2. Mitchell, T.M. Chapter 1: What Is Machine Learning? In *Machine Learning: Hands-on for Developers and Technical Professionals*; John Wiley & Sons, Inc.: Indianapolis, IN, USA, 2015.
3. Abrahart, R.J.; Anctil, F.; Coulibaly, P.; Dawson, C.W.; Mount, N.J.; See, L.M.; Shamseldin, A.Y.; Solomatine, D.P.; Toth, E.; Wilby, R.L. Two Decades of Anarchy? Emerging Themes and Outstanding Challenges for Neural Network River Forecasting. *Prog. Phys. Geogr.* **2012**. [CrossRef]
4. PDAC. *Concepts and Application of Machine Learning to Mining Geoscience: A Practical Course*; PDAC: Toronto, ON, Canada, 2019.

5. Karpatne, A.; Atluri, G.; Faghmous, J.; Steinback, M.; Banerjee, A.; Ganguly, A.; Shekhar, S.; Samatova, N.; Kumar, V. Theroy-Guided Data Science: A New Paradigm for Scientific Discovery from Data. *IEEE Trans. Knowl. Data Eng.* **2017**, *29*, 2318–2331. [CrossRef]
6. Mcgaughey, W.J.; Geoscience, M. Data-Driven Geotechnical Hazard Assessment: Practice and Pitfalls. In Proceedings of the First International Conference on Mining Geomechanical Risk, Perth, Western Australia, 9–11 April 2019; pp. 219–232.
7. LimiteStateInc. Eucocode 7 Design Approach 1: Basic Procedure. Available online: http://www.limitstate.com/geo/eurocode7 (accessed on 7 October 2019).
8. Bieniawski, Z.T. Classification of Rock Masses for Engineering: The RMR System and Future Trends. In *Chapter 22 in Comprehensive Rock Engineering*; Elsevier: Amsterdam, The Netherlands, 1993; pp. 553–573.
9. Barton, N.R.; Lien, R.; Lunde, J. Engineering Classification of Rock Masses for the Design of Tunnel Support. *Rock Mech.* **1974**, *6*, 189–236. [CrossRef]
10. Barton, N.R.; Bieniawski, Z.T. *RMR and Q-Setting Records Straight*; Tunnels & Tunnelling International; Progressive Media Markets Ltd.: London, UK, 2008; pp. 26–29.
11. Deere, D.U. Technical Description of Rock Cores for Engineering Purposes. *Rock Mech. Eng. Geol.* **1963**, *1*, 16–22.
12. Jing, L.; Hudson, J.A. Numerical Methods in Rock Mechanics. *Int. J. Rock Mech. Min. Sci.* **2002**, *39*, 409–427. [CrossRef]
13. Perras, M.A.; Wannenmacher, H.; Diederichs, M.S. Underground Excavation Behaviour of the Queenston Formation: Tunnel Back Analysis for Application to Shaft Damage Dimension Prediction. *Rock Mech. Rock Eng.* **2015**. [CrossRef]
14. Diederichs, M.S. The 2003 Canadian Geotechnical Colloquium: Mechanistic Interpretation and Practical Application of Damage and Spalling Prediction Criteria for Deep Tunnelling. *Can. Geotech. J.* **2007**, *44*, 1082–1116. [CrossRef]
15. Perras, M.; Diederichs, M.S. Predicting Excavation Damage Zone Depths in Brittle Rocks. *J. Rock Mech. Geotech. Eng.* **2016**, *8*, 60–74. [CrossRef]
16. Perras, M. Tunnelling in Horizontally Laminated Ground: The Influence of Lamination Thickness on Anisotropic Behaviour and Practical Observations from the Niagara Tunnel Project. Master's Thesis, Queen's University, Kingston, ON, Canada, 2009.
17. ITASCA Consulting Group Inc. *FLAC 8 Basics*, 1st ed.; ITASCA Consulting Group Inc.: Minneapolis, MN, USA, 2015.
18. NWMO. *Geosynthesis, Technical Report DGR-TR-2011-11*; NWMO: Toronto, ON, Canada, 2011.
19. Jing, L.; Stephansson, O. Constitutive Models of Rock Fractures and Rock Masses-the Basics. In *Developments in Geotechnical Engineering*; Elsevier B. V.: Amsterdam, The Netherlands, 2007; Volume 85, pp. 47–109. [CrossRef]
20. Garza-Cruz, T.; Pierce, M.; Kaiser, P.K. Use of 3DEC to Study Spalling and Deformation Associated with Tunnneling at Depth. In *Deep Mining 2014, Proceedings of the 7th International Conference on Deep and High Stress Mining, Sudbury, ON, Canada, 14–18 September 2014*; Australian Centre for Geomechanics: Crawley, Australia, 2014; pp. 421–436.
21. Hoek, E.; Grabinsky, M.W.; Diederichs, M.S. Numerical Modelling for Underground Excavation Design. *Trans. Inst. Min. Met. Sect. A Min. Ind.* **1990**, *100*, 22–30.
22. Rocscience. *RS2*; Rocscience: Toronto, ON, Cananda, 2019.
23. GEO-SLOPE. *GeoStudio*; GEO-SLOPE: Calgary, AB, Canada, 2016.
24. Bentley. *Plaxis 2D*; Bentley: Exton, PA, USA, 2019.
25. Dassault Systems. *ABAQUS*; Dassault Systems: Vélizy-Villacoublay, France, 2019.
26. Geraili Mikola, R. Adonis. Available online: https://geotechpedia.com/Software/Show/1015/ADONIS (accessed on 18 October 2019).
27. UCRegents. *OpenSees*; UCRegents: San Francisco, CA, USA, 2006.
28. EDF. *Code_Aster*; EDF: Paris, France, 2019.
29. Laforce, T. *PE281 Boundary Element Method*; Stanford University: Stanford, CA, USA, 2006; pp. 1–12.
30. Roscience Inc. *Examine2D*; Roscience Inc.: Toronto, ON, Canada, 2019.
31. Map3D. Map3D. 2019. Available online: https://www.map3d.com/ (accessed on 20 October 2019).
32. Wieleba, P.; Sikora, J. Open Source BEM Library. *Adv. Eng. Softw.* **2009**, *40*, 564–569. [CrossRef]

33. Lisjak, A.; Tatone, B.S.A.; Grasselli, G.; Vietor, T. Numerical Modelling of the Anisotropic Mechanical Behaviour of Opalinus Clay at the Laboratory-Scale Using FEM/DEM. *Rock Mech. Rock Eng.* **2014**, *47*, 187–206. [CrossRef]
34. ITASCA Consulting Group Inc. *UDEC Manual*; ITASCA Consulting Group Inc.: Minneapolis, MN, USA, 1992.
35. ITASCA Consulting Group Inc. *PFC*; ITASCA Consulting Group Inc.: Minneapolis, MN, USA, 2019.
36. ITASCA Consulting Group Inc. *3DEC Manual*; ITASCA Consulting Group Inc.: Minneapolis, MN, USA, 1994.
37. ITASCA Consulting Group Inc. *PFC3D*; ITASCA Consulting Group Inc.: Minneapolis, MN, USA, 2019.
38. Šmilauer, V. Yade User's Manual. Available online: https://yade-dem.org/doc/user.html (accessed on 18 October 2019).
39. Sandia National Labs; Temple University. LAMMPS. 2019. Available online: https://lammps.sandia.gov/ (accessed on 18 October 2019).
40. Golder Associates Inc. *FracMan Geotechnical Edition*; Golder Associates Inc.: Toronto, ON, Canada, 2019.
41. Miraco Mining Innovation. *MoFrac*; Miraco Mining Innovation: Sudbury, ON, Canada, 2019.
42. Alghalandis Computing. ADFNE. 2019. Available online: https://alghalandis.net/products/adfne/ (accessed on 18 October 2019).
43. Varadarajan, A.; Sharma, K.G.; Singh, R.K. Some Aspect of Coupled FEBEM Analysis of Underground Openings. *Int. J. Numer. Anal. Methods Geomech.* **1985**, *9*, 557–571. [CrossRef]
44. Lorig, L.J.; Brady, B.H.G.; Cundall, P.A. Hybrid Distinct Element-Boundary Element Analysis of Jointed Rock. *Int. J. Rock Mech. Min. Sci.* **1986**, *23*, 303–312. [CrossRef]
45. Pan, X.D.; Reed, M.B. A Coupled Distinct Element-Finite Element Method for Large Deformation Analysis of Rock Masses. *Int. J. Rock Mech. Min. Sci.* **1991**, *28*, 93–99. [CrossRef]
46. Mahabadi, O.K.; Cottrell, B.E.; Grasselli, G. An Example of Realistic Modelling of Rock Dynamics Problems: FEM/DEM Simulation of Dynamic Brazilian Test on Barre Granite. *Rock Mech. Rock Eng.* **2010**, *43*, 707–716. [CrossRef]
47. Li, X.; Kim, E.; Walton, G. A Study of Rock Pillar Behaviors in Laboratory and In-Situ Scales Using Combined Finite-Discrete Element Method Models. *Int. J. Rock Mech. Min. Sci.* **2019**, *118*, 21–32. [CrossRef]
48. Geomechanica. *Irazu*; Geomechanica: Toronto, ON, Canada, 2019.
49. Knight, E.E.; Rougier, E.; Lei, Z.; Munjiza, A. *Hybrid Optimization Software Suite*; Los Alamos National Laboratory: Los Alamos, NM, USA, 2014.
50. Rockfield. Elfen. Available online: https://www.rockfieldglobal.com/ (accessed on 24 November 2019).
51. Grasselli's Geomechanics Group. *Y-GEO*; Grasselli's Geomechanics Group: Toronto, ON, Canada, 2019.
52. Sklavounos, P.; Sakellariou, M. Intelligent Classification of Rock Masses. In *WIT Transactions on Information and Communication Technologies*; WIT Press: Billerica, MA, USA, 1995; Volume 8.
53. Singh, V.K.; Singh, D.; Singh, T.N. Prediction of Strength Properties of Some Schistose Rocks from Petrographic Properties Using Artificial Neural Networks. *Int. J. Rock Mech. Min. Sci.* **2001**. [CrossRef]
54. Millar, D.; Clarici, E. Investigation of Back-Propagation Artificial Neural Networks in Modelling the Stress-Strain Behaviour of Sandstone Rock. *Genet. Sel. Evol.* **2002**, *47*, 3326–3331. [CrossRef]
55. Mahdevari, S.; Torabi, S.R. Prediction of Tunnel Convergence Using Artificial Neural Networks. *Tunn. Undergr. Space Technol.* **2012**, *28*, 218–228. [CrossRef]
56. Afraei, S.; Shahriar, K.; Madani, S.H. Developing Intelligent Classification Models for Rock Burst Prediction after Recognizing Significant Predictor Variables, Section 1: Literature Review and Data Preprocessing Procedure. *Tunn. Undergr. Space Technol.* **2019**. [CrossRef]
57. Ferentinou, M.; Fakir, M. Integrating Rock Engineering Systems Device and Artificial Neural Networks to Predict Stability Conditions in an Open Pit. *Eng. Geol.* **2018**, *246*, 293–309. [CrossRef]
58. Mohammed, M.; Khan, M.; Bashier, E.; Bashier, M. Chapter 1 Introduction to Machine Learning. In *Machine Learning: Algorithms and Applications*; CRC Press: Boca Raton, FL, USA, 2016; pp. 1–34. [CrossRef]
59. Marsland, S. *Machine Learning: An Algorithmic Perspective*; Chapman and Hall: New York, NY, USA, 2009.
60. Mohammed, M.; Khan, M.; Bashier, E.; Bashier, M. Chapter 2 Decision Trees. In *Machine Learning: Algorithms and Applications*; CRC Press: Boca Raton, FL, USA, 2016; pp. 1–204. [CrossRef]
61. Einstein, H.H.; Labreche, D.A.; Markow, M.J.; Baecher, G.B. Decision Analysis Applied to Rock Tunnel Exploration. *Eng. Geol.* **1978**, *12*, 143–161. [CrossRef]

62. Pu, Y.; Apel, D.B.; Lingga, B. Rockburst Prediction in Kimberlite Using Decision Tree with Incomplete Data. *J. Sustain. Min.* **2018**, *17*, 158–165. [CrossRef]
63. Khan, M.; Bashier, E.; Bashier, M. Chapter 4 Naïve Bayesian Classification. In *Machine Learning: Algorithms and Applications*; CRC Press: Boca Raton, FL, USA, 2017.
64. Araghinejad, S. Chapter 5: Artificial Neural Networks. In *Data-Driven Modeling: Using MATLAB ®in Water Resources and Environmental Engineering*; Springer: Berlin, Germany, 2014; Volume 67. [CrossRef]
65. Ribeiro e Sousa, L.; Miranda, T.; Leal e Sousa, R.; Tinoco, J. The Use of Data Mining Techniques in Rockburst Risk Assessment. *Engineering* **2017**, *3*, 552–558. [CrossRef]
66. Khan, M.; Bashier, E.; Bashier, M. Chapter 5 The k -Nearest Neighbors Classifiers. In *Machine Learning: Algorithms and Applications*; CRC Press: Boca Raton, FL, USA, 2017.
67. Cracknell, M.J.; Reading, A.M. Geological Mapping Using Remote Sensing Data: A Comparison of Five Machine Learning Algorithms, Their Response to Variations in the Spatial Distribution of Training Data and the Use of Explicit Spatial Information. *Comput. Geosci.* **2014**, *63*, 22–33. [CrossRef]
68. Khan, M.; Bashier, E.; Bashier, M. Chapter 8 Support Vector Machine. In *Machine Learning: Algorithms and Applications*; CRC Press: Boca Raton, FL, USA, 2017.
69. Sun, Y.; Feng, X.; Yang, L. Predicting Tunnel Squeezing Using Multiclass Support Vector Machines. *Adv. Civ. Eng.* **2018**, *2018*, 17–20. [CrossRef]
70. Qi, C.; Fourie, A.; Du, X.; Tang, X. Prediction of Open Stope Hangingwall Stability Using Random Forests. *Nat. Hazards* **2018**, *92*, 1179–1197. [CrossRef]
71. Ben-Hur, A.; Horn, D.; Siegelmann, H.; Vapnik, V. Support Vector Clustering. *J. Mach. Learn. Res.* **2001**, *2*, 125–137. [CrossRef]
72. Bozorg-Haddad, O.; Soleimani, S.; Loáiciga, H.A. Modeling Water-Quality Parameters Using Genetic Algorithm-Least Squares Support Vector Regression and Genetic Programming. *J. Environ. Eng.* **2017**, *143*, 1–10. [CrossRef]
73. Jing, L.; Stephansson, O. Chapter 4 Fluid Flow and Coupled Hydro-Mechanical Behavior of Rock Fractures. In *Developments in Geotechnical Engineering*; Elsevier B. V.: Amsterdam, The Netherlands, 2007; Volume 85, pp. 111–144. [CrossRef]
74. Abbasi, M.; El Hanandeh, A. Forecasting Municipal Solid Waste Generation Using Artificial Intelligence Modelling Approaches. *Waste Manag.* **2016**, *56*, 13–22. [CrossRef]
75. Khan, M.; Bashier, E.; Bashier, M. Chapter 6 Neural Networks. In *Machine Learning: Algorithms and Applications*; CRC Press: Boca Raton, FL, USA, 2017.
76. Bell, J. Chapter 5: Artificial Neural Networks. In *Machine Learning: Hands-on for Developers and Technical Professionals*; John Wiley & Sons, Inc.: Indianapolis, IN, USA, 2015.
77. Trivedi, R.; Singh, T.N.; Mudgal, K.; Gupta, N. Application of Artificial Neural Network for Blast Performance Evaluation. *Int. J. Res. Eng. Technol.* **2015**, *3*, 564–574. [CrossRef]
78. Khan, U.T.; Valeo, C. Optimising Fuzzy Neural Network Architecture for Dissolved Oxygen Prediction and Risk Analysis. *Water* **2017**, *9*. [CrossRef]
79. Khan, U.T.; Valeo, C. Dissolved Oxygen Prediction Using a Possibility Theory Based Fuzzy Neural Network. *Hydrol. Earth Syst. Sci.* **2016**, *20*, 2267–2293. [CrossRef]
80. Khan, U.T.; He, J.; Valeo, C. River Flood Prediction Using Fuzzy Neural Networks: An Investigation on Automated Network Architecture. *Water Sci. Technol.* **2018**, *2017*, 238–247. [CrossRef]
81. Mosavi, A.; Ozturk, P.; Chau, K.W. Flood Prediction Using Machine Learning Models: Literature Review. *Water* **2018**, *10*, 1536. [CrossRef]
82. Deng, Y.; Sadiq, R.; Jiang, W.; Tesfamariam, S. Risk Analysis in a Linguistic Environment: A Fuzzy Evidential Reasoning-Based Approach. *Expert Syst. Appl.* **2011**, *38*, 15438–15446. [CrossRef]
83. Song, Z.; Jiang, A.; Jiang, Z. Back Analysis of Geomechanical Parameters Using Hybrid Algorithm Based on Difference Evolution and Extreme Learning Machine. *Math. Probl. Eng.* **2015**, *2015*. [CrossRef]
84. Ching, J.; Phoon, K.; Li, K.; Weng, M. Multivariate Probability Distribution for Some Intact Rock Properties. *Can. Geotech. J.* **2019**, *18*, 1–18. [CrossRef]
85. Kumar, M.; Samui, P.; Naithani, A.K. Determination of Uniaxial Compressive Strength and Modulus of Elasticity of Travertine Using Machine Learning Techniques. *Int. J. Adv. Soft Comput. Appl.* **2013**, *5*, 1–15.
86. Fakir, M.; Ferentinou, M. *A Holistic Open Pit Mine Slope Stability Index Using Artificial Neural Networks*; ISRM: Lisbon, Portugal, 2017; Volume 2017.

87. Kumar, M.; Samui, P. Analysis of Epimetamorphic Rock Slopes Using Soft Computing. *J. Shanghai Jiaotong Univ.* **2014**, *19*, 274–278. [CrossRef]
88. Hibert, C.; Provost, F.; Malet, J.P.; Maggi, A.; Stumpf, A.; Ferrazzini, V. Automatic Identification of Rockfalls and Volcano-Tectonic Earthquakes at the Piton de La Fournaise Volcano Using a Random Forest Algorithm. *J. Volcanol. Geotherm. Res.* **2017**, *340*, 130–142. [CrossRef]
89. Mayr, A.; Rutzinger, M.; Geitner, C. Multitemporal Analysis of Objects in 3D Point Clouds for Landslide Monitoring. *Int. Arch. Photogramm. Remote Sens. Spat. Inf. Sci.* **2018**, *42*, 691–697. [CrossRef]
90. Janeras, M.; Jara, J.A.; Royán, M.J.; Vilaplana, J.M.; Aguasca, A.; Fàbregas, X.; Gili, J.A.; Buxó, P. Multi-Technique Approach to Rockfall Monitoring in the Montserrat Massif (Catalonia, NE Spain). *Eng. Geol.* **2017**, *219*, 4–20. [CrossRef]
91. Weidner, L.; Walton, G.; Kromer, R. Classification Methods for Point Clouds in Rock Slope Monitoring: A Novel Machine Learning Approach and Comparative Analysis. *Eng. Geol.* **2019**, 263. [CrossRef]
92. Li, X.; Chen, Z.; Chen, J.; Zhu, H. Automatic Characterization of Rock Mass Discontinuities Using 3D Point Clouds. *Eng. Geol.* **2019**, *259*, 105131. [CrossRef]
93. Bizjak, K.F.; Petkovšek, B. Displacement Analysis of Tunnel Support in Soft Rock around a Shallow Highway Tunnel at Golovec. *Eng. Geol.* **2004**, *75*, 89–106. [CrossRef]
94. Santos, O.J.; Celestino, T.B. Artificial Neural Networks Analysis of São Paulo Subway Tunnel Settlement Data. *Tunn. Undergr. Space Technol.* **2008**. [CrossRef]
95. Koopialipoor, M.; Fahimifar, A.; Ghaleini, E.N.; Momenzadeh, M.; Armaghani, D.J. Development of a New Hybrid ANN for Solving a Geotechnical Problem Related to Tunnel Boring Machine Performance. *Eng. Comput.* **2019**. [CrossRef]
96. Leu, S.S.; Chen, C.N.; Chang, S.L. Data Mining for Tunnel Support Stability: Neural Network Approach. *Autom. Constr.* **2001**, *10*, 429–441. [CrossRef]
97. Xue, Y.; Li, Y. A Fast Detection Method via Region-Based Fully Convolutional Neural Networks for Shield Tunnel Lining Defects. *Comput. Civ. Infrastruct. Eng.* **2018**. [CrossRef]
98. Dong, L.J.; Li, X.B.; Peng, K. Prediction of Rockburst Classification Using Random Forest. *Trans. Nonferrous Met. Soc. China* **2013**, *23*, 472–477. [CrossRef]
99. Zhou, J.; Li, X.; Shi, X. Long-Term Prediction Model of Rockburst in Underground Openings Using Heuristic Algorithms and Support Vector Machines. *Saf. Sci.* **2012**, *50*, 629–644. [CrossRef]
100. Liu, K.; Liu, B. Optimization of Smooth Blasting Parameters for Mountain Tunnel Construction with Specified Control Indices Based on a GA and ISVR Coupling Algorithm. *Tunn. Undergr. Space Technol.* **2017**. [CrossRef]
101. Vallejos, J.A.; McKinnon, S.D. Logistic Regression and Neural Network Classification of Seismic Records. *Int. J. Rock Mech. Min. Sci.* **2013**, *62*, 86–95. [CrossRef]
102. Dong, L.; Li, X.; Xu, M.; Li, Q. Comparisons of Random Forest and Support Vector Machine for Predicting Blasting Vibration Characteristic Parameters. *Procedia Eng.* **2011**, *26*, 1772–1781. [CrossRef]
103. Paraskevopoulou, C.; Diederichs, M. Analysis of Time-Dependent Deformation in Tunnels Using the Convergence-Confinement Method. *Tunn. Undergr. Space Technol.* **2018**. [CrossRef]
104. Spence, E. *CO Summer School Central: Introduction to Neural Networks with Python I*; CO Summer School Central: Toronto, ON, Canada, 2018.
105. Mandic, D.; Chambers, J. *Recurrent Neural Networks for Prediction: Learning Algorithms, Architectures, and Stability*; John Wiley & Sons, Inc.: Indianapolis, IN, USA, 2001.
106. Solomatine, D.P.; Ostfeld, A. Data-Driven Modelling: Some Past Experiences and New Approaches. *J. Hydroinform.* **2007**, *10*, 3–22. [CrossRef]
107. Martins, F.F.; Miranda, T.F.S. Prediction of Hard Rock TBM Penetration Rate Based on Data Mining Techniques. In *Proceedings of the 18th international Conference on Soil Mechanics and Geotechnical Engineering*; Presses des Ponts: Paris, France, 2013; pp. 1751–1754.
108. Snieder, E.; Shakir, R.; Khan, U.T. A Comprehensive Comparison of Four Input Variable Selection Methods for Artificial Neural Network Flow Forecasting Models. *J. Hydrol.* **2019**, 124299. [CrossRef]
109. Maier, H.R.; Jain, A.; Dandy, G.C.; Sudheer, K.P. Methods Used for the Development of Neural Networks for the Prediction of Water Resource Variables in River Systems: Current Status and Future Directions. *Environ. Model. Softw.* **2010**, *25*, 891–909. [CrossRef]
110. Maier, H.R.; Dandy, G.C. Neural Networks for the Prediction and Forecasting of Water Resources Variables: A Review of Modelling Issues and Applications. *Environ. Model. Softw.* **2000**. [CrossRef]

111. May, R.J.; Maier, H.R.; Dandy, G.C. Data Splitting for Artificial Neural Networks Using SOM-Based Stratified Sampling. *Neural Netw.* **2010**, *23*, 283–294. [CrossRef]
112. Shu, C.; Burn, D.H. Artificial Neural Network Ensembles and Their Application in Pooled Flood Frequency Analysis. *Water Resour. Res.* **2004**, *40*, 1–10. [CrossRef]
113. Shahin, M.A.; Maier, H.R.; Jaksa, M.B. Data Division for Developing Neural Networks Applied to Geotechnical Engineering. *J. Comput. Civ. Eng.* **2004**, *18*, 105–114. [CrossRef]
114. Daszykowski, M.; Walczak, B.; Massart, D.L. Representative Subset Selection. *Anal. Chim. Acta* **2002**, *468*, 91–103. [CrossRef]
115. Xie, Q.; Peng, K. Space-Time Distribution Laws of Tunnel Excavation Damaged Zones (EDZs) in Deep Mines and EDZ Prediction Modeling by Random Forest Regression. *Adv. Civ. Eng.* **2019**, *2019*. [CrossRef]
116. Chou, J.S.; Thedja, J.P.P. Metaheuristic Optimization within Machine Learning-Based Classification System for Early Warnings Related to Geotechnical Problems. *Autom. Constr.* **2016**, *68*, 65–80. [CrossRef]

3

Modeling of CO Emissions from Traffic Vehicles using Artificial Neural Networks

Omer Saud Azeez [1], Biswajeet Pradhan [2,*], Helmi Z. M. Shafri [1], Nagesh Shukla [2], Chang-Wook Lee [3,*] and Hossein Mojaddadi Rizeei [2]

1. Department of Civil Engineering, Faculty of Engineering, University Putra Malaysia (UPM), Serdang 43400, Malaysia; baghdad.eagle2016@gmail.com (O.S.A.); helmi@upm.edu.my (H.Z.M.S.)
2. Centre for Advanced Modelling and Geospatial Information Systems (CAMGIS), Faculty of Engineering and Information Technology, University of Technology Sydney, Sydney 2007, NSW, Australia; nagesh.shukla@uts.edu.au (N.S.); h.mojaddadi@gmail.com (H.M.R.)
3. Division of Science Education, Kangwon National University, 1 Kangwondaehak-gil, Chuncheon-si 24341, Gangwon-do, Korea

* Correspondence: Biswajet.Pradhan@uts.edu.au (B.P.); cwlee@kangwon.ac.kr (C.-W.L.)

Abstract: Traffic emissions are considered one of the leading causes of environmental impact in megacities and their dangerous effects on human health. This paper presents a hybrid model based on data mining and GIS models designed to predict vehicular Carbon Monoxide (CO) emitted from traffic on the New Klang Valley Expressway, Malaysia. The hybrid model was developed based on the integration of GIS and the optimized Artificial Neural Network algorithm that combined with the Correlation based Feature Selection (CFS) algorithm to predict the daily vehicular CO emissions and generate prediction maps at a microscale level in a small urban area by using a field survey and open source data, which are the main contributions to this paper. The other contribution is related to the case study, which represents the spatial and quantitative variations in the vehicular CO emissions between toll plaza areas and road networks. The proposed hybrid model consists of three steps: the first step is the implementation of the correlation-based Feature Selection model to select the best model's predictors; the second step is the prediction of vehicular CO by using a multilayer perceptron neural network model; and the third step is the creation of micro scale prediction maps. The model was developed using six traffic CO predictors: number of vehicles, number of heavy vehicles, number of motorbikes, temperature, wind speed and a digital surface model. The network architecture and its hyperparameters were optimized through a grid search approach. The traffic CO concentrations were observed at 15-min intervals on weekends and weekdays, four times per day. The results showed that the developed model had achieved validation accuracy of 80.6 %. Overall, the developed models are found to be promising tools for vehicular CO simulations in highly congested areas.

Keywords: traffic CO; traffic CO prediction; neural networks; GIS; land use/land cover (LULC)

---

1. Introduction

The transport infrastructure like expressways and roads has a significant importance in the development of any country's economy by linking cities. These infrastructures are rapidly developing due to the changing in the traffic modes, leading to congested roads. Hence, road traffic emissions are increasing, creating many negative impacts on air quality on roadways, intersections and toll roads. Traffic emissions, such as carbon monoxide (CO), are the primary contributor to overall air pollution from this infrastructure, and the primary source of traffic emissions is vehicular exhausts.

Spatial prediction models are effectively used as a decision-making support tool for prediction and simulation of traffic emissions on road networks [1–3]. There are various negative impacts that can result from inappropriate traffic levels, including high levels of noise and high concentrations of gaseous pollutants [4,5]. Several diseases e.g., cancers, heart diseases, respiratory problems and preterm births, can occur when human beings are exposed to high concentrations of CO [6–8].

The measurement of vehicular emissions on roadways and toll gates may be costly, risky and requires a lot of time and effort. Moreover, the designers do not have the opportunity to determine the vehicular emissions through the design process. In the most recent planning techniques for design of highways and road networks, traffic emission models are often required to support sustainable transportation planning and the reduction of traffic emissions from sources such as congestion and tollgate areas. Thus, the GIS-based modeling of traffic emission and intra-urban air pollution exposure can be an effective tool in the environmental assessment for sustainable road planning. This tool can distinguish the areas affected by different types of pollutants and the related ecological and social factors. This would be able to determine the best strategy to support the decision makers [9]. On the other hand, GIS can save costs and time in the traffic emission modeling and can therefore be used in sustainable planning.

Different types of traffic CO prediction approaches are mentioned in the literature [10–12]. Early methods of traffic CO modelling were based on traditional techniques using data sampling and global position system (GPS) techniques. Several thematic maps and vehicle emission equations are combined to model the traffic emission distribution in a region and produce informative maps that could help in effective decision making [10]. Recent methods are mostly based on land-use regression analysis using statistical and soft computing algorithms [10,13]. These statistical and computing techniques allow the input of various traffic and road geometry factors. Almost all these models are designed by using experimental samples; consequently, these models are highly influenced by the traffic flow condition and the measurement style and the geographic locations [14]. The main drawback of these models is that they can not be generalized because of the local environment like vehicle model and type and the weather [15,16]. Ref. [17] presented an approach of recognizing the road geometric features from positioning information surveyed by collecting vehicle data.

2. Previous Works

Many models have been developed to predict CO emissions and other traffic emissions, such as NOx, NO_2, CO_2 and SO_2. In a paper, Ref. [18] presented a methodology by integrating the spatial analysis techniques and the neighborhood statistic function algorithm to evaluate the spatial diffusion of the gaseous pollutant in north of Italy by using the air pollutant records obtained from monitoring stations and GIS data (i.e., administrative borders, built-up areas, emission sources and road networks). Their results were illustrated on grids with a cell size of (4×4) km. Although this method showed a significant spatial representation of air pollution, the methodology was constrained by the limited spatial resolution. Therefore, it cannot be used for high-resolution data. Ref. [19] developed a GIS-based tool by combining the operational street pollution model (OSPM) and a multi-agent-based transportation model (MATSIM) to estimate the air pollutant concentrations in Munich, Germany. Their results showed hourly prediction of NOx from traffic. This approach can be used as an effective tool for air quality studies in urban areas. Nevertheless, its disadvantages appear in the complexity of a system that comprises different models where the non-expert users are not able to use it. Ref. [20] developed a model based on land-use regression algorithm and land-use types, meteorological variables and vehicle-kilometers-travelled (VKTs) and linear regression algorithm to estimate the concentrations of Nitrogen Dioxide (NO_2) in Seoul, Korea. The results showed the significant impacts of the residential, commercial land use, wind speed, temperature and humidity on the concentrations of NO_2. The air pollutants recorded by the fixed air quality monitoring stations can be affected by several factors such as terrain and buildings altitude. Moreover, the weather factors are not suitable to model and produce high-resolution products such as roadmaps. Ref. [21] presented a statistical model based on the fuzzy logic system to predict CO concentrations in Tehran, Iran. This model mainly relied on historical data, which were obtained from monitoring stations. Fuzzy logic algorithms were applied to combine the parameters. Their results showed that lowest Room Mean Square Error (RMSE) was recorded at 2.13. Another study related to statistical modeling was conducted by [22] to forecast air pollutants in Hong Kong based on the integration of two statistical models, i.e., the generalized additive models and the Global Forecast System, which linked the air pollution with meteorological data. Results showed a contrast in the air pollutant levels between urban and suburban areas. This model is useful for predicting air quality in complex terrain areas. These models lack the spatial aspect and could not be used to produce prediction maps. Ref. [23] developed a methodology by using two commercial programs to estimate the traffic emissions in small area in Madrid, Spain. The VISSIM program was used for traffic simulation to calculate a velocity-time profile. Then, the related emissions at the vehicle level were completed using the VERSIT + micro program. Results showed the spatial variation in NOx and PM_{10} concentrations are based on microscale maps with high resolution, cell size (5×5) m. This model depends on the estimated emissions data based on prediction simulations without using actual samples based on sampling equipment.

Recently, machine-learning technologies have attracted researchers. The neural network (NN) models are the most popular models in the Artificial Intelligence models. Ref. [24] developed a model by integrating the artificial neural network (ANN) algorithm and evolutionary polynomial regression (EPR) to estimate the CO concentrations in Tabriz City, Iran. The EPR is one of the data mining algorithms developed based on evolutionary computing and the integration of numerical regression and genetic algorithm. The EPR model involves two stages: a genetic algorithm is used in the first stage based on the numerical regression to search for symbolic structures, whereas in the second stage, symbolic structure parameters are determined based on the linear least squares techniques. Their results showed that the ANN model is more reliable than the EPR model. The highest value of the correlation coefficient was measured at 0.85 based on NN and 0.41 using EPR. This study indicated that NN modeling can be efficiently utilized for air quality forecasting. On the other hand, ref. [25] developed a model based on the NN algorithm and data obtained from field survey to estimate the hourly traffic emissions near roads. The authors used different parameters such as traffic data, meteorology, proximity to roads and road direction. This model is considered as an efficient approach for predicting pollutant near a road. Although they used geographic information as a parameter, their results did not contain spatial prediction results such as maps. They only presented a statistical analysis. Results showed that the highest correlation coefficient for the CO prediction was 0.879. Ref. [26] conducted the most relevant studies that combined the NN model and the spatial prediction model. They presented an approach that combined the linear-chain conditional random field algorithm and ANN model to generate real-time air pollution maps. They utilized the data recorded from monitoring stations and the traffic data collected from the field while geographic parameters like land use and road network were derived using GIS data. Their results showed the air pollutants prediction on maps with (1×1) km spatial resolution. However, the developed model did not consider many issues like uncertainty, modeling multifactor and nonlinearity. Although several authors have attempted to overcome these issues, they principally focused on the integration of big data and the large scale modeling. On the other hand, most of these models deal with large quantities of data, expensive equipment, and complex data processing models, which require substantial time, cost and other resources.

In this paper, we presented a hybrid model to produce microscale prediction maps considering toll gate locations, as well as the other parameters listed in the literature. The model is developed by combining the metaheuristic optimization technique and ANN algorithm to predict traffic emissions based on a small number of training data and avoiding transferability issues. The metaheuristic optimization algorithms like correlation-based feature selection models which have the ability to find best model's predictors in a short time were compared to other optimization techniques. Also, ANN algorithms are suitable for prediction based on few training data. The major contributions of this work lie in producing highly accurate predictive maps and providing a description of the high variation of traffic emissions on roads and tollgate areas. Other significant advantages may include easy implementation of the proposed model in open source GIS software where the non-expert users can utilize the model for rapid simulations and assessments of vehicular emissions based on microscale prediction maps. Also, the users can design GIS models based on their needs. We proposed that the combination of metaheuristic optimization and machine learning algorithms could help improving the forecasting of CO emissions on roads, highways and in tollgate areas.

3. Materials and Methods

3.1. Study Area

This study was conducted near of Subang jaya toll plaza which links the New Klang Valley Expressway (NKVE) and the federal expressway in Peninsular Malaysia. Subang toll plaza is located within a highly dense populated area in Petaling Jaya, Selangor, Malaysia (Figure 1). The total length of NKVE is 35 km, which connects urban and industrial areas in the capital city of Kuala Lumpur. It is a major highway for citizens who are living in the main cities like Kuala Lumpur, Subang, Shah Alam, Damansara, Sungai Buloh, Klang and Petaling Jaya. The vehicular speed limits are standardized to 110 and 90 km/h on Bukit Raja to Bukit Lanjan stretch and Bukit Lanjan to Jalan Duta stretch, respectively. The study area contains different types of land use such as tollgate area, commercial, industrial and residential areas, making it well suited for vehicular emissions studies.

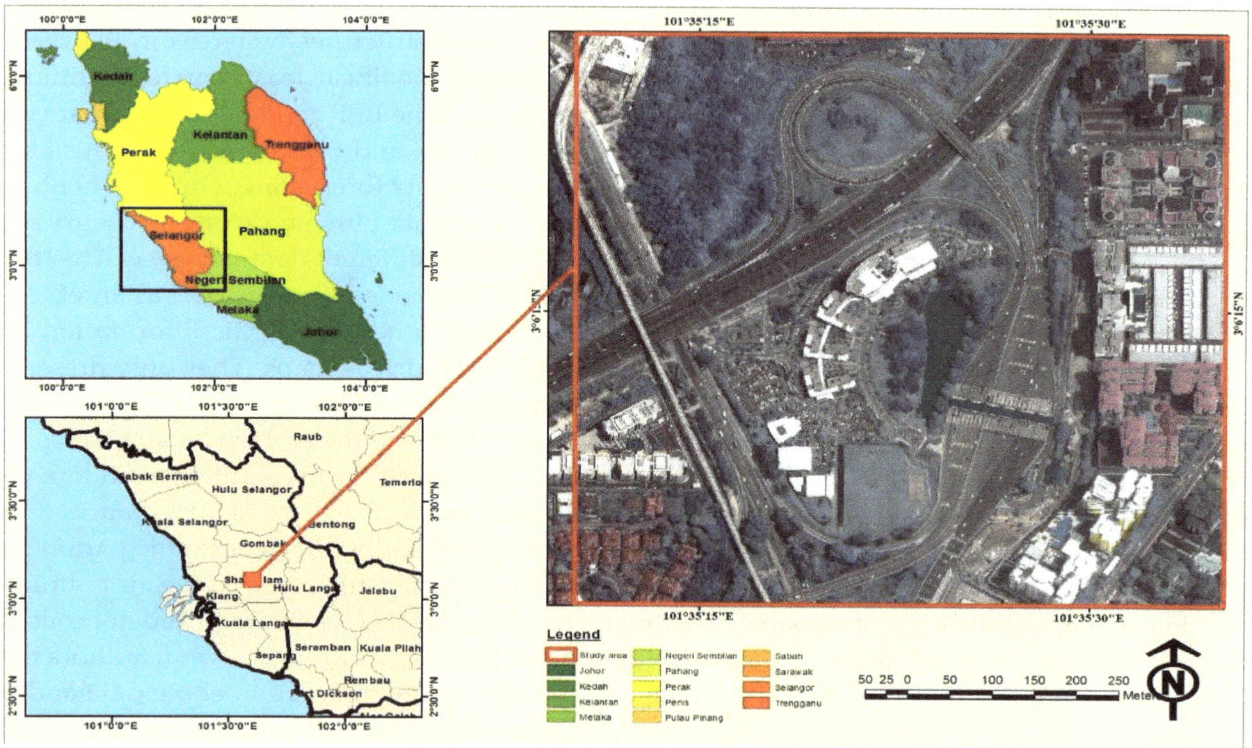

Figure 1. Location map of the study area.

3.2. Data and Method

Several data (i.e., vehicular CO samples, meteorology and traffic flow data) were collected from the field during April 2017. Light Detection and Ranging (LiDAR) data were collected in March 2017, and the Worldview3 satellite image was captured in May 2017. Figure 2 shows the overall methodology, which consists of several steps. The hybrid model is designed to predict CO emissions at a specific time and location, for example, prediction maps based on different times of a day. The first step is data collection, which was achieved based on a gas analyzer and data loggers to simultaneously collect CO, temperature, humidity, and traffic information. The LULC map was extracted from the Worldview-3 image with spatial resolution of 0.3 m. The digital surface model (DSM) was derived from the Airborne

LiDAR point clouds by using Environment for Visualizing Images (ENVI) software. The second step is the statistical modelling, which was applied by combining two models i.e., the correlation based feature selection (CFS) and Multilayer Perceptron (MLP) by using Weka software. The final step is the spatial modelling based on the regression equations derived from regression analysis and GIS techniques to generate microscale prediction maps for the traffic CO emissions during different times of the day.

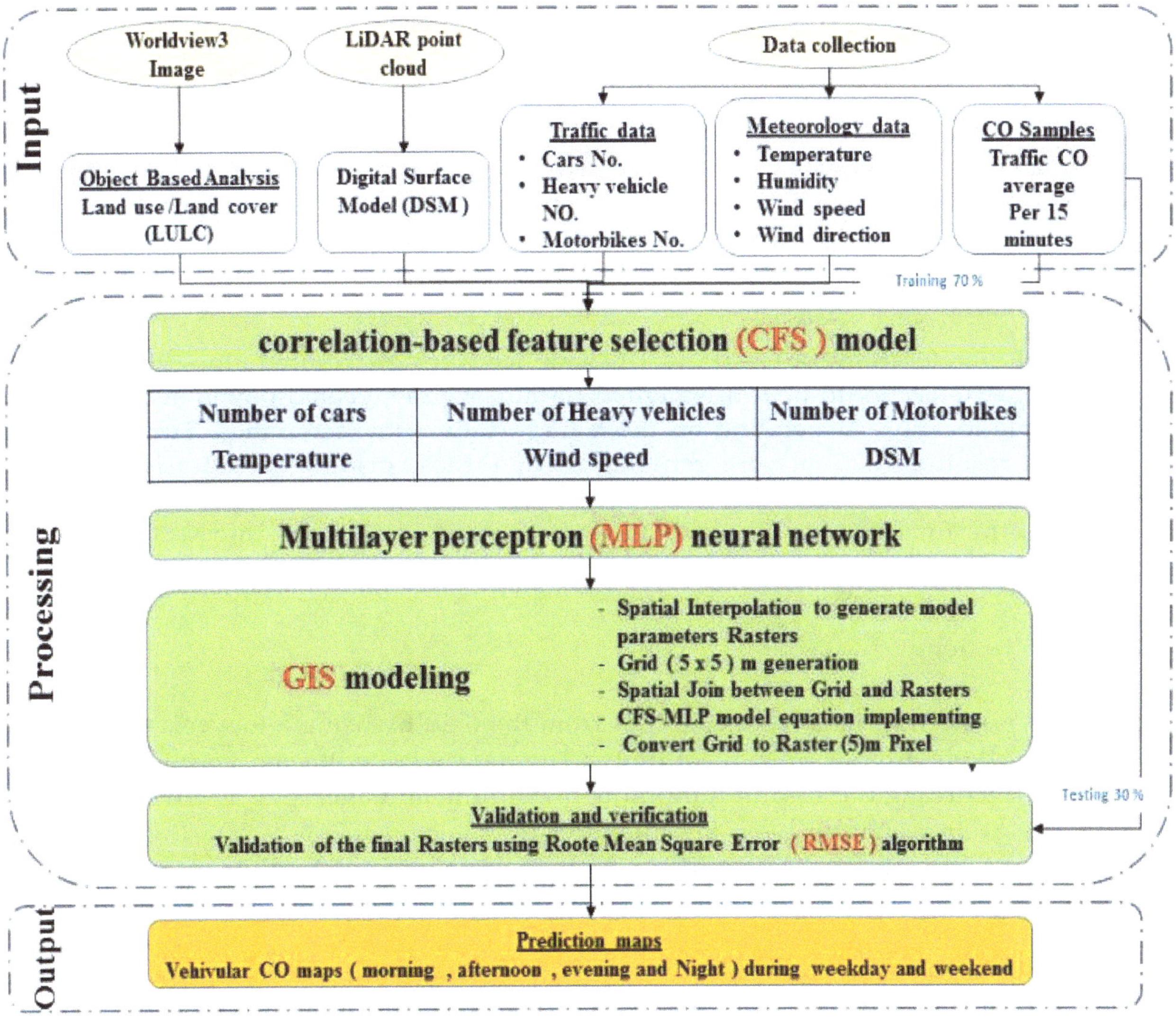

Figure 2. The overall methodological flow chart employed in this study.

3.3. Field Surveying

3.3.1. Sampling Selection

The vehicular CO, traffic condition and meteorology data are important for the development of vehicular CO emissions models [27]. Many studies have described approaches of collecting traffic flow data from the field, and the most important aspect has been determined to be the distribution of air pollutants samples and their suitability [28,29]. The accuracy of the spatial interpolation is highly affected by the sampling design and the variations between traffic CO samples [30]. Moreover, the density of points must be good enough to achieve high accuracy of interpolated data. Conversely, a large number of samples should be avoided to decrease the processing time. Most importantly, the samples density should be adjusted by considering the vehicular CO diffusion characteristics. In this paper, traffic CO data were collected according to a procedure given by [31]. Their method was implemented by creating sampling locations based on a random selection method by using spatial analysis techniques. This approach generates the optimum number of samples compared to the study area [31].

First, three layers (residential, commercial and industrial) from the land-use map were extracted. These layers were converted into points with a spatial constraint to force them inside the land-use polygons. Next, the density of points was estimated using a 150 m search radius and the resolution of the output density raster was set to 25 m. The density rasters were then integrated based on different coefficients (industrial = 1, commercial = 2, and residential = 3). The next step is the rescaling of the final density raster from 0 to 1 depending on linear approach led to the creation of the probability raster which was used to select the samples of the traffic CO. Next, the geospatial balanced points were then created within the study area by using the probability raster. The total number of points were generated based on the length of the road network, the cost of the project and the capability of traffic emissions detection equipment used in the data gathering.

As a result, the locations of the traffic CO samples were distributed along the study area. Therefore, extra steps should be implemented to improve the created points in order to select the final sample locations depending on the transportation characteristics. Tessellated grids with a spatial resolution 25 m were generated. Then, these grids were intersected with the created points and the road network layer and the remaining tessellated grids were removed. Subsequently, the final traffic CO samples were selected within the intersected tessellated grids and road network.

3.3.2. Data Collection

Traffic CO concentrations were collected from the field by using a low cost Gas Analyzer device model Micro-clip5 (1 ppm resolution). The data were collected continuously in 15-min intervals (recording the 15-min minimum, maximum, and average). The traffic flow data and meteorology information were simultaneously collected using a data logger and GPS device (Garmin GPS etrex 10, Olathe, KS, USA; available at University Putra Malaysia). Figure 3 shows the sampling procedure. The traffic CO analyzer was installed in the location of samples by using a Global Navigation System (GPS), at least 2 m from the road edge. The GPS was used to determine the geographic location of samples and to manually verify the locations using land-use maps. The traffic CO was measured four times a day on weekends and weekdays, in the morning (6.30 a.m. to 8.30 a.m.), afternoon (11.30 a.m. to 1.30 p.m.), evening (6.30 p.m. to 8.30 p.m.), and at night (11 p.m. to 12 midnight). In addition, the traffic data were collected by using digital cameras installed in the road's side in the sample's location. The traffic flow data were classified into several types (the number of cars, the number of heavy vehicles and the number of motorbikes), where the cars were private cars and taxi cars.

On the other hand, the heavy vehicles refer to the following: medium truck, heavy truck, super-heavy/special duty truck and buses, while the motorbikes refer to any type of motorbikes. The meteorological data (temperature, humidity, wind speed and wind direction) were collected for two days, including on weekdays and weekends, to examine various scenarios in the study area for accurate and effective study of traffic CO modelling and mapping (e.g., hazard maps, risk maps, and further analysis). Figure 4 shows the data collection procedure adopted in this study.

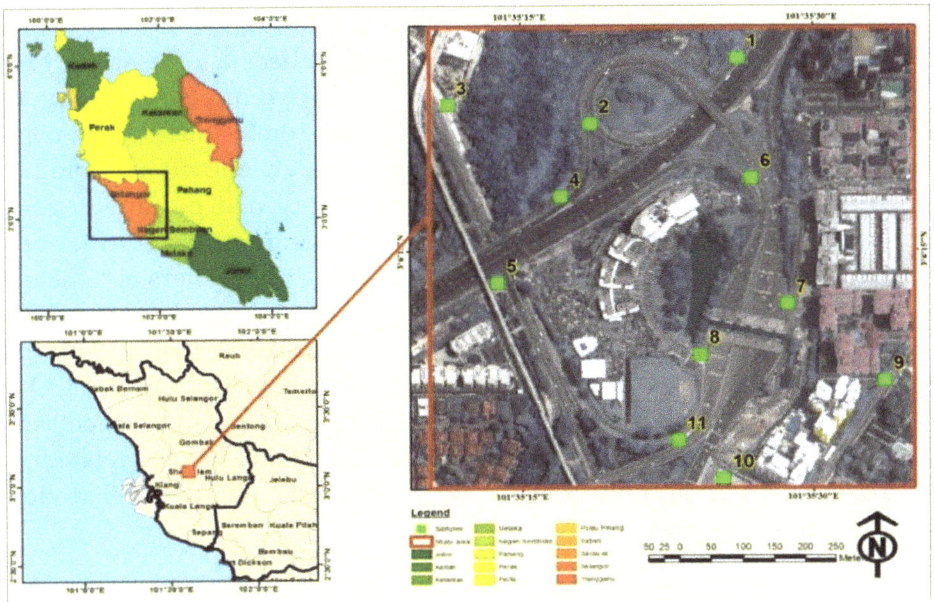

Figure 3. The sampling locations.

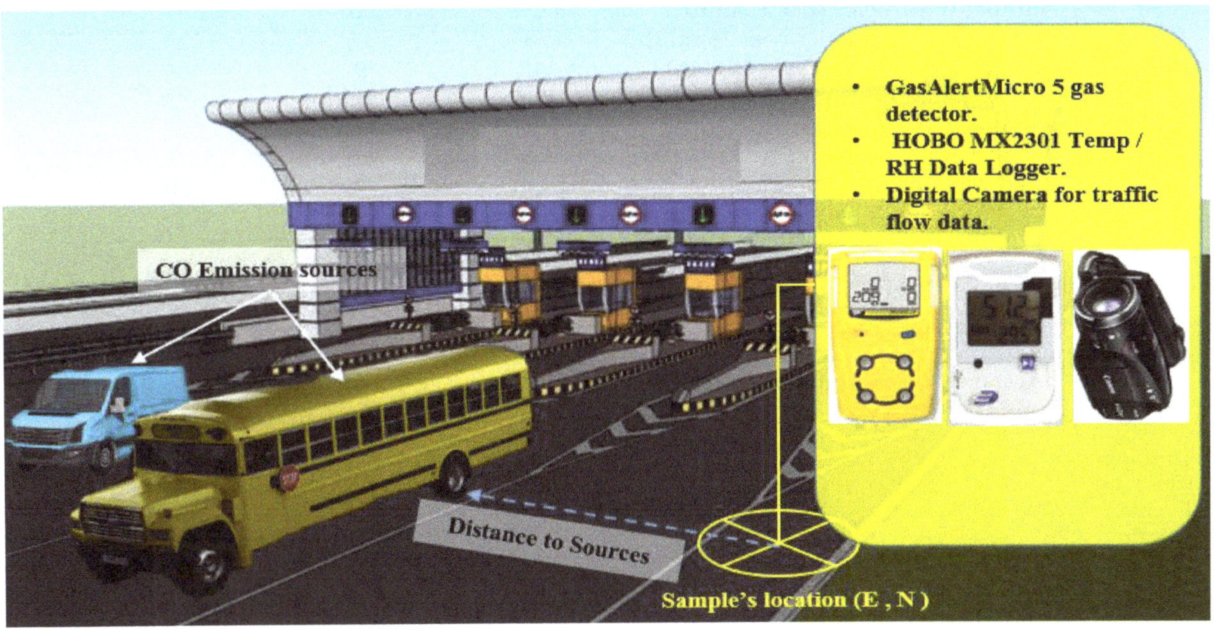

Figure 4. Location of traffic CO and meteorological parameter measurement on a highway section.

3.4. *Vehicular CO Prediction Model*

3.4.1. Vehicular CO Model Parameters

The presented model aims to predict the daily traffic CO emissions and create prediction maps at different times of a day by using GIS techniques. The traffic CO descriptor (i.e., the dependent parameter) in the current study is the vehicular CO emissions measured every 15 min. The contributing parameters to vehicular CO emission were first selected depending on the previous studies and with consideration of traffic condition, weather characteristics, the surrounding (LULC), topography, and the building heights in the study area. These parameters are the number of vehicles, number of heavy vehicles, the number of motorbikes, temperature, humidity, wind speed, wind direction, LULC and digital surface model (DSM). Many studies have been conducted based on these parameters [32–36]. There are many factors that could affect the vehicular emissions such as engine condition and the fuel type (gasoline, diesel); however, these data cannot be detected through field survey. On the other hand, the data collection did not contain vehicle speed information because the study area is very small and the variation between the vehicles speed is difficult to differentiate. Consequently, the vehicle speed profile is stable during the day. There are many vehicular emissions studies conducted without adopting vehicle speed, such as [25]. The parameters' statistics are shown in Table 1. The NN model was used to find out the degree of contribution of each mentioned parameter for estimating the vehicular CO emission. The NN has the ability to model the complex simulations of non-linear problems such as vehicular CO emission. However, contributing parameters to vehicular CO emission may not be directly used as inputs in the NN model, because of high levels of correlation between the factors resulting in multicollinearity, which can reduce the precision of estimating the vehicular CO emission. Moreover, using a large number of contributing parameters to vehicular CO emission (traffic CO predictors) as input layers of NN can generate over-fitting problems and increase the complexity ratio to run NN model. In the proposed NN model, only relevant and low-correlated parameters were used as inputs. The relevant and significant factors were selected using the CFS model. Table 2 shows the traffic CO measurements results.

Table 1. Summary statistics of traffic CO predictors.

Parameter	Average	Minimum	Maximum
Number of vehicles (per 15 min)	1172	126	2762
Number of heavy vehicles (per 15 min)	78	16	325
Number of motorbikes (per 15 min)	112	9	489
Temperature (°C)	29.9	25.6	37.7
Humidity (%)	73.5	54.3	94.5
Wind Speed (mph)	16.87	16	18.20
Wind Direction (angle)	247.1	0	350
DSM (m)	25.7	10.03	129.5

Table 2. Summary statistics of traffic CO measurements.

Time	Weekend			Weekday		
	Average CO Concentration (per 15 min) (ppm)			Average CO Concentration (per 15 min) (ppm)		
	Min	Max	Mean	Min	Max	Mean
Morning	0	8	2.36	0	30.5	8.5
Afternoon	0	14.5	3.5	0	12.8	4.5
Evening	0	9.3	3.92	0	27.3	5.84
Night	0	3.6	1.47	0	5.6	1.9

3.4.2. Correlation-Based Feature Selection (CFS) Model

The CFS algorithm is one of the machine learning algorithms which is considered as a filter algorithm that choose the features based on correlation concepts [37]. A major characteristic of the correlation-based function is the ability to choose sub-groups that include features that are unusually correlated to the targeted class but unassociated with each other. On the other hand, this algorithm neglects the features with low correlation with the targeted class, and this algorithm is used to delete the duplicated features because they will be correlated with one of the rest of the features at least. The acknowledgement of a feature will rely on the degree to which it predicts classes in territories of the instance space not currently anticipated by different features.

The CFS's feature subset assessment function is presented in Equation (1):

$$Ms = k\overline{r_{cf}} / \sqrt{k + k(k-1)\,\overline{r_{ff}}} \tag{1}$$

where Ms is the heuristic "merit" of a feature subset S containing k features, $\overline{r_{cf}}$ is the mean of the feature-class correlation ($f \in$ S), and $\overline{r_{ff}}$ is the average of the feature-feature intercorrelation.

3.4.3. Multilayer Perceptron (MLP) Neural Network

The MLP algorithm is one of the ANN algorithms which results from adding hidden layers to the simple perceptron. In this algorithm, the structure of the NN is generally trained based on backpropagation algorithm and some related variants. Therefore, the models are designed based on the integration of the MLP algorithm and a backpropagation algorithm called backpropagation neural network [38–43]. The multilayer perceptron algorithm was developed due to the computational limitations that resulted from single-layered perceptron models. According to the experiments, the multilayer perceptron algorithm has the ability to represent complex simulations and mapping and process high level non-linear problems. Also, the MLP algorithm has the ability to process the nonlinear features, thus allowing the representation of a continuous function of non-linear activation functions such as sigmoid functions [44], which has a clear analogy with the conventional representation of a periodic function such as a Fourier series (i.e., as the sum of simple sine waves). Therefore, the MLP can be considered as a universal functional approximation. Figure 3 shows the architecture of an MLP with several layers of neurons and nonlinear activation functions.

The MLP is considered as a feed-forward system with single or multiple layer of segments among network output and input layers [38–45]. With the assumption of L-layer MPL, the system can signified by $N^{L}_{n_0,n_1,\ldots,\ n_L}$, where n_l , $l = 0,\ 1,\ldots, L$ indicate the number of segments in the input layer ($l = 0$), the l is number of hidden layers that can range $(l = 0,\ 2,\ldots, L-1)$, L is the number of layers, and the output layer is $(l = L)$. $X^{(0)} = \left[x_0^{(0)} u_1 u_2 \ldots u_{n0}\right]^T$ is the input vector and $X^{(l)} = \left[x_0^{(l)} x_1^{(l)} \ldots x_{ni}^{(l)}\right]^T$ denotes the lth layer output vector in the interval [0, T]. At this time, $\{u_i\}, j = 1,2,\ldots n_0$ represents the input attribute pattern while $x_j^{(l)}$ indicates the output of the jth segments of lth network layer. The threshold input is represented by $x_0^{(l)}$ with a fixed value at one. The neuron weight of the jth segment of lth layer from the ith segment of $(l-1)$ can be represented by $\omega_{ji}^{(l)}$. The activation function, which is connected with all the segments of the system except the input layer, is the tanh function specified by $\rho(S) = \tan\mathrm{h}(S) = \left(\frac{1-e^{-2S}}{1+e^{-2S}}\right)$. The restricted derivative of the $\rho(S)$ based on S is signified by $\rho\prime(S)$ that is known as $\rho\prime(S) = \left(1-\rho^2(S)\right)$. Also, the linear sum of the jth segment of lth layer is symbolized by $S_j^{(l)}$.

In the forward part, at the kth time direct, the input attribute pattern vector of $X^{(0)}$ is implemented in the system, while the corresponding preferred output is $\{y_j\}$, for $j = 1,2,\ldots,n_L$. Since no calculation is applied, the input layer of the MLP is known by $x_j^{(0)} = u_j$ for $j = 1,2,\ldots,n_0$. For other layers, $l = 1,2,\ldots,L$, and $j = 1,2,\ldots,n_1$, the outputs are computed as:

$$x_j^{(0)} = \rho(S_j^{(l)}) \text{ and } S_j^{(l)} = \sum\nolimits_{i=0}^{ni-1} \omega_{ji}^{(1)} x_i^{(l-1)}.$$

The probable output is identified by $\{\hat{y}_J\}$ and is assumed as $\{\hat{y}_J\} = \left\{x_j^{(L)}\right\}$ for all $j = 1,2,\ldots,n_L$. The mean square error for the system can be formulated as $\frac{e^2}{nL}$ where $e_j = y_j - \hat{y}_J$ is the error signal for the jth output. Furthermore, the instantons squared error can be computed by $e^2 = \sum_{j=1}^{n\ L} e_j^2$.

However, in the learning part, the BP procedure reduces the squared error by varying $\{\omega_{ji}^{(l)}\}$ according to the gradient search method, recursively. The squared error derivatives connected with the jth segment in layer l are described as Equation (2):

$$\delta_j^{(1)} = -\frac{1}{2}\frac{\delta e^2}{\delta S_i^{(1)}} \tag{2}$$

Then, these derivatives can be formulated as in Equation (3):

$$\delta_j^{(1)} = \begin{cases} \rho'\left(S_j^{(l)}\right).e_j \; for \; l = L \\ \rho'\left(S_j^{(l)}\right).\sum_{i=1}^{n\; l+1} \omega_i^{(l+1)}.\delta_i^{(l+1)} \; for \; l = L-1, L-2, \ldots, 1. \end{cases} \tag{3}$$

Eventually, the weights of the MLP are reorganized at the kth instant as in Equation (4):

$$\omega_{ji}^{(l)}(k+1) = \omega_{ji}^{(l)}(k) + \Delta_{ji}^{(l)}(k) \; and \; \Delta_{ji}^{(l)}(k) = \mu\delta_i^{(l)}x_i^{(l-1)} + \gamma\Delta_{ji}^{(l)}(k-1) \tag{4}$$

where μ is the learning rate and γ signifies the momentum rate hyper-parameters.

The final proposed network architecture for traffic CO prediction is illustrated in Figure 5. The proposed network is designed based on the results of the best traffic predictors that resulted from the CFS model, optimization and hyper-parameter, which are used to select the best predictors. By using the open source machine learning software (Weka), different MLP structures have been used to select the optimal MLP neural network model for the traffic CO prediction model. The proposed methodology was designed by combining the correlation-based feature selection model and multilayer perceptron.

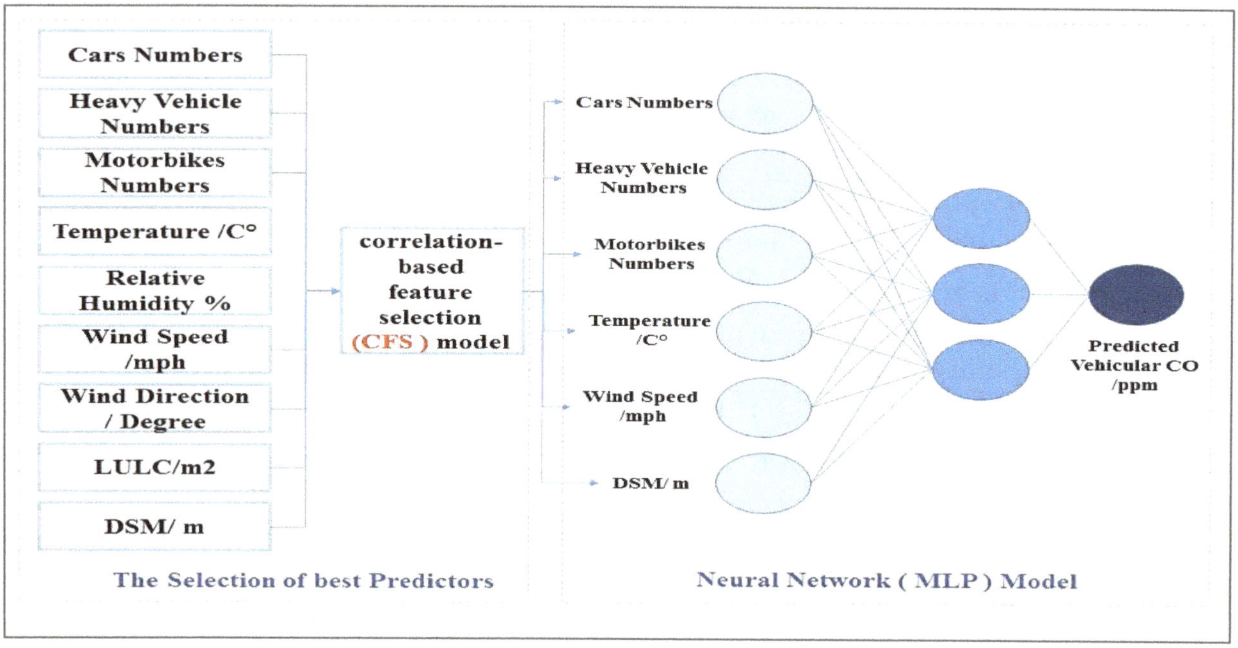

Figure 5. The architecture of the proposed neural network for traffic CO prediction (6-3-1).

3.4.4. Optimization Method

The capacity of the prediction in MLP model relies on its hyperparameter and structure. In this paper, several network structure hyperparameter combinations were tested to determine a sub-optimal network model for modeling vehicular CO. Table 3 shows the structures and hyperparameters evaluated in the current study and their search space domain. In general, there are two main categories of NN, MLP and radial basis function (RBF). The former uses dot products between inputs and weights and monotonic activation functions such as sigmoid. The MLP uses dot products between inputs and weights and monotonic activation functions such as sigmoid while the RBF uses Euclidean distances between inputs and weights and usually Gaussian activation functions. Both of the networks can be trained with the back-propagation algorithm. In the RBF model, it is not necessary to use multiple hidden layers whereas with MLP, multiple hidden layers are used. In addition, the RBF model is less sensitive to noise than the MLP model. Other parameters of the network are the number of hidden units, training algorithm, error function, activation function, learning rate, and momentum. The number of hidden layers controls the complexity of the designed network. A small number of hidden units may result in low prediction capacity due to insufficient learning whereas a large number of hidden layers can reduce the ability of the model to be generalized and can also create overfitting problems. The training algorithm is the optimization method for calculating the weight for each node in the network. There are many training algorithms for NN based on back-propagation; the Broyden–Fletcher–Goldfarb–Shanno (BFGS) and radial basis function training algorithm (RBFT) are the most recommended back-propagation algorithms used for optimization of NN architecture [46]. During training of the network, an objective function or optimization score function was minimized according to the labeled training dataset. The optimizer usually has a learning rate and gradient momentum parameters. In addition, many activation functions such as identity, logistic, and Gaussian could be used in the hidden layers and the output layers of the neural network.

Table 3. Hyper parameters of the proposed model for traffic CO prediction and their search spaces used for fine-tuning.

Parameter	Search Domain
Type of network	MLP, RBF
Number of hidden units	(3–40)
Training Algorithm	BFGS, RBFT
Hidden Activation	Identity, Logistic, Tanh, Exponential, Gaussian
Output Activation	Identity, Logistic, Tanh, Exponential, Gaussian
Learning rate	(0.1, 0.5)
Momentum	(0.1–0.9)

The total number of instances was 352, which was separated into training 70% (246) and testing 30% (106). This sample size is still small compared to other studies. However, it needs a good approach to handle overfitting problems. One approach is to design a cross-validation evaluation procedure. Other methods are data augmentation or collecting new samples. There are also methods such as transfer learning which requires retrained networks. Overall, training of the neural networks with small datasets requires careful analysis and evaluation before using them in practice. There are also some tricks to improve the performance of neural networks for small datasets. Those include batch normalization, rectified linear unit (relu) activations and regularization methods such as l1 and l2. On the other hand, the hyperparameters of the NN for predicting vehicular CO were chosen based on the implementation of systematic grid-based search that can be applied with the Scikit-Learn algorithm using 100 epochs. Although this method demands high quality of computation, more accurate outputs could result by fine tuning the hyperparameter values. Many models based on various integrations of parameters were generated. Cross validation (10-fold) was applied to validate each model. Therefore, the parameters that resulted in higher accuracy are the best parameters.

3.4.5. GIS Modelling

Collecting traffic CO data, especially on the highways, is dangerous and expensive. Therefore, predicting traffic CO concentrations on highways helps to generate traffic CO data that can be used for further studies. In this research, the measured and predicted traffic CO concentrations are used for GIS modelling based on a grid analysis. GIS modelling and mapping are mainly applied to assess affected people and environments due to inappropriate traffic emissions from traffic activities. The observed traffic CO samples are an important factor in the model, for computing the relationship between the predictive factors and carbon concentrations which were applied in the training data, and for the validation process. After training the NN model with the CO measurements from the field, the NN model produced a regression equation based on weighted values for each predictor factor, in order to calculate the predicted values based on the predictor factors that can be easily applied in a GIS platform; this is the main contribution of the proposed model. This equation has been applied using GIS to produce a spatial prediction of traffic CO concentration in the study area. This step was applied by implementing the final prediction equations on the 5 × 5 m GRID, using ArcGIS tools which are very efficient for spatial representation [23].

4. Results and Discussion

4.1. Contribution of Traffic CO Predictors

As shown in Table 4, the traffic CO predictors make different contributions to the traffic CO values that resulted from this study. The statistical analysis based on the Chi-square method shows that the parameters that contributed the most are the number of heavy vehicles (F = 32.784) and the number of vehicles (F = 18.277). Conversely, the findings indicated that the other traffic CO predictors (number of motorbikes, DSM, wind speed and temperature) did not make a significant contribution (Table 4).

Table 4. Results of assessing the contribution of traffic CO predictors using the Chi-square method.

Road Traffic CO Predictors	R-Squared	F-Statistic
Number of heavy vehicles	0.7546	32.784
Number of vehicles	0.5322	18.277
Number of motorbikes	0.0472	1.951
DSM (m)	0.0168	1.231
Wind speed (mph)	0.0016	0.124
Temperature (°C)	0.0014	0.1178

4.2. Traffic CO Prediction Results

Two MLP models were trained and tested. The first model was trained based on MLP algorithm using nine parameters (those listed in Table 1 plus LULC) to predict the traffic CO concentration. The second model was trained based on the combination with the correlation-based feature selection (CFS-MLP) and six parameters that resulted from the CFS model. These parameters were: number of vehicles, number of heavy vehicles, number of motorbikes, temperature, wind speed, and DSM. The CFS algorithm was implemented to select the highly correlated parameters and best parameters to predict the traffic CO that led to increased accuracy of the prediction process.

Table 5 shows the proposed model's results when the input parameters were filtered and reduced from nine parameters to six based on the CFS algorithm which finds features that have higher correlation with the class but are uncorrelated with each other. Therefore, the highest correlated parameters were used for the prediction analysis, which resulted in improving the prediction accuracy. The relative absolute error decreased from 30.94% to 21.99% and the root relative square error also decreased from 23.48% to 19.40%. On the other hand, the correlation coefficient increased from 0.866 ppm to 0.98 ppm. The mean absolute error (MAE) was reduced from 0.99 ppm to 0.89 ppm. The root mean square error (RMSE) also decreased from 1.29 ppm to 1.27 ppm. The prediction results showed that the prediction improvement occurred after the implementation of the CFS algorithm which is able to reduce the high dimensionality, remove the low correlated data and improve the learning accuracy.

Table 5. Results of predictions with MLP model and the proposed model (CFS-MLP) model.

MLP Model		CFS-MLP Model	
Best structure	9-4-1	Best structure	6-3-1
Correlation coefficient	0.8657	Correlation coefficient	0.980
Mean absolute error (ppm)	0.991	Mean absolute error (ppm)	0.8925
Root mean squared error (ppm)	1.2862	Root mean squared error (ppm)	1.2736
Relative absolute error %	30.94%	Relative absolute error %	21.99%
Root relative squared error %	23.48%	Root relative squared error %	19.40%
Total number of instances	247	Total number of instances	247

4.3. Traffic CO Prediction at Different Times of Day

Regression equations were created, based on the results from the CFS-MLP model; the coefficients of vehicular emissions predictors are calculated to formulate regression models to easily predict vehicular emission in the study area using a set of predictors that can be gathered from the field and existing databases in order to facilitate connection with the GIS-based model by applying parameter coefficients in the spatial model. The decision makers can notice the effect of causative parameters on vehicular emissions occurrence, which was assessed by the corresponding coefficient that appears in the regression function [47]. Regression equations were simulated at different times during the day (morning, afternoon, evening and night) to predict traffic CO at these times. The highest RMSE was 2.9817 ppm during evening observations, while the lowest RMSE was 0.387 ppm at night. Table 6 shows the results of the regression analysis at different times.

Table 6. Regression models for traffic CO prediction based on the CFS-MLP model.

Traffic CO Predictors	Estimated Coefficient			
	Morning	Afternoon	Evening	Night
Number of vehicles	−0.0016	0.0142	0.0108	0.0147
Number of heavy vehicles	0.0622	0.01	0.0319	−0.0216
Number of motorbikes	0.0135	−0.0378	−0.0376	−0.0093
Temperature °C	−0.4501	0.5512	0.4888	−0.0333
Wind speed mph	0.0752	−0.194	−0.4084	0.0135
DSM m	−0.2085	0.213	0.0812	0.1116
Intercept	16.8559	−22.2525	−15.8113	−2.1367
RMSE	2.914 ppm	2.0347 ppm	2.9817 ppm	0.387 ppm

4.4. GIS Modelling Results

A GIS model was applied to generate prediction maps at different times a day (Figures 6 and 7). The resultant spatial prediction maps showed that the concentrations of traffic CO increased more on weekdays than on weekends. These maps also showed that there was significant variation between traffic CO concentrations according to the time of day, with the traffic CO ranging from a high of 35.23 ppm per 15 min during the weekday morning to a low of 4.76 ppm per 15 min during the weekend night. The prediction maps showed that the highest values of traffic CO are located near the toll areas compared to other areas such as residential green areas, because of the traffic congestion at toll gates. The lowest values of traffic CO are concentrated near residential areas which reached zero values.

4.5. Comparison with Other Models

The developed model was compared with other popular models such as the support vector machine for regression (SVR) and the linear regression (LR) models. These models generated two statistical equations based on model parameters. Testing data were used for model validation; these equations are shown below:

$$\begin{aligned}\text{Traffic CO} = \; & 0.0022 \times \text{Number of vehicles} + 0.0403 \times \text{Number of heavy vehicles} \\ & - 0.0187 \times \text{Number of Motorbikes} + 0.1957 \times \text{Temperature} \\ & - 0.0984 \times \text{Wind speed} - 0.0102 \times \text{DSM} - 4.4382\end{aligned} \tag{4}$$

For the LR model:

$$\text{Traffic CO} = 0.05 \times \text{Number of heavy vehicles} + 0.21 \tag{5}$$

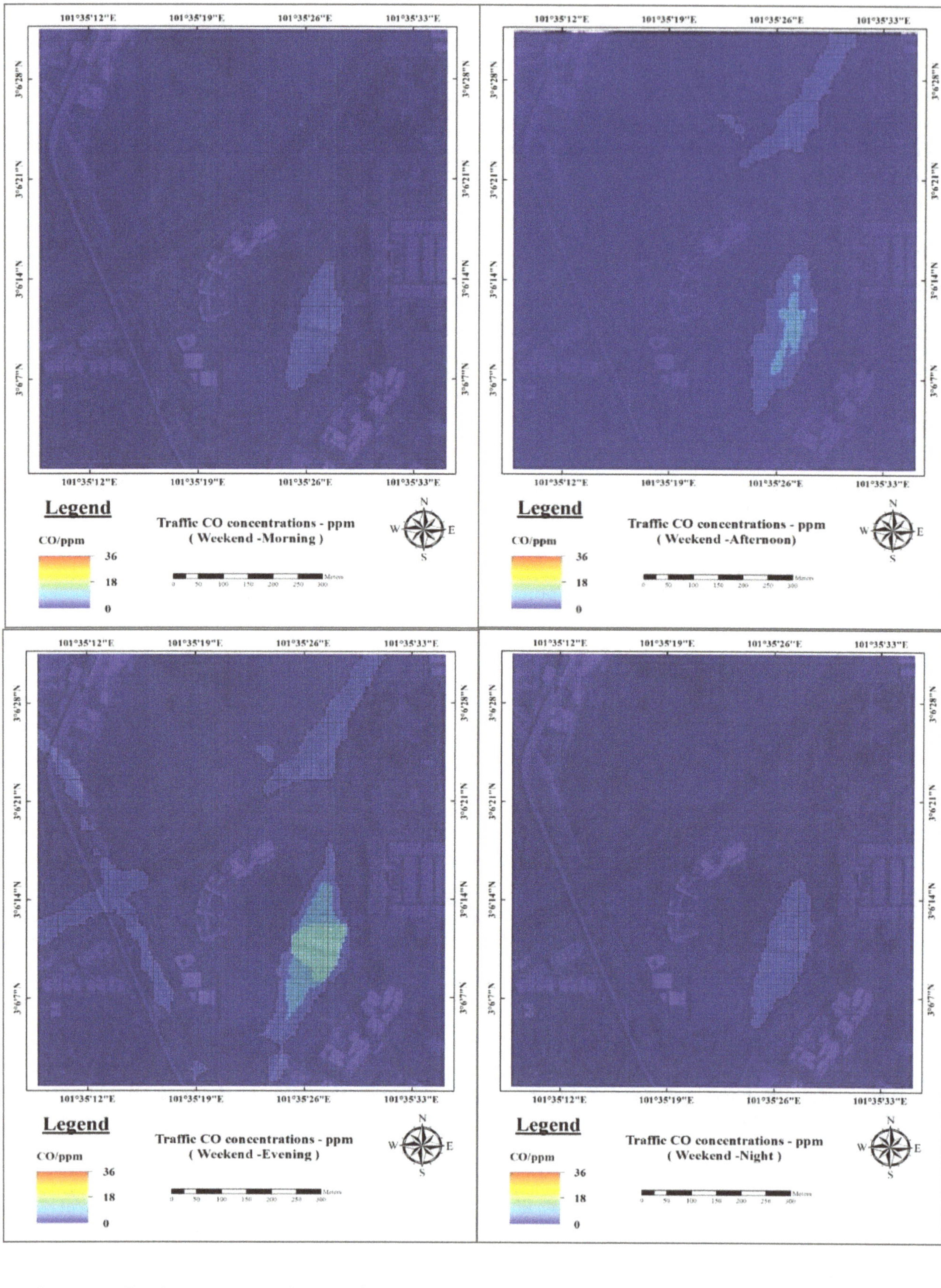

Figure 6. Prediction maps during the weekend at different times (morning, afternoon, evening and night).

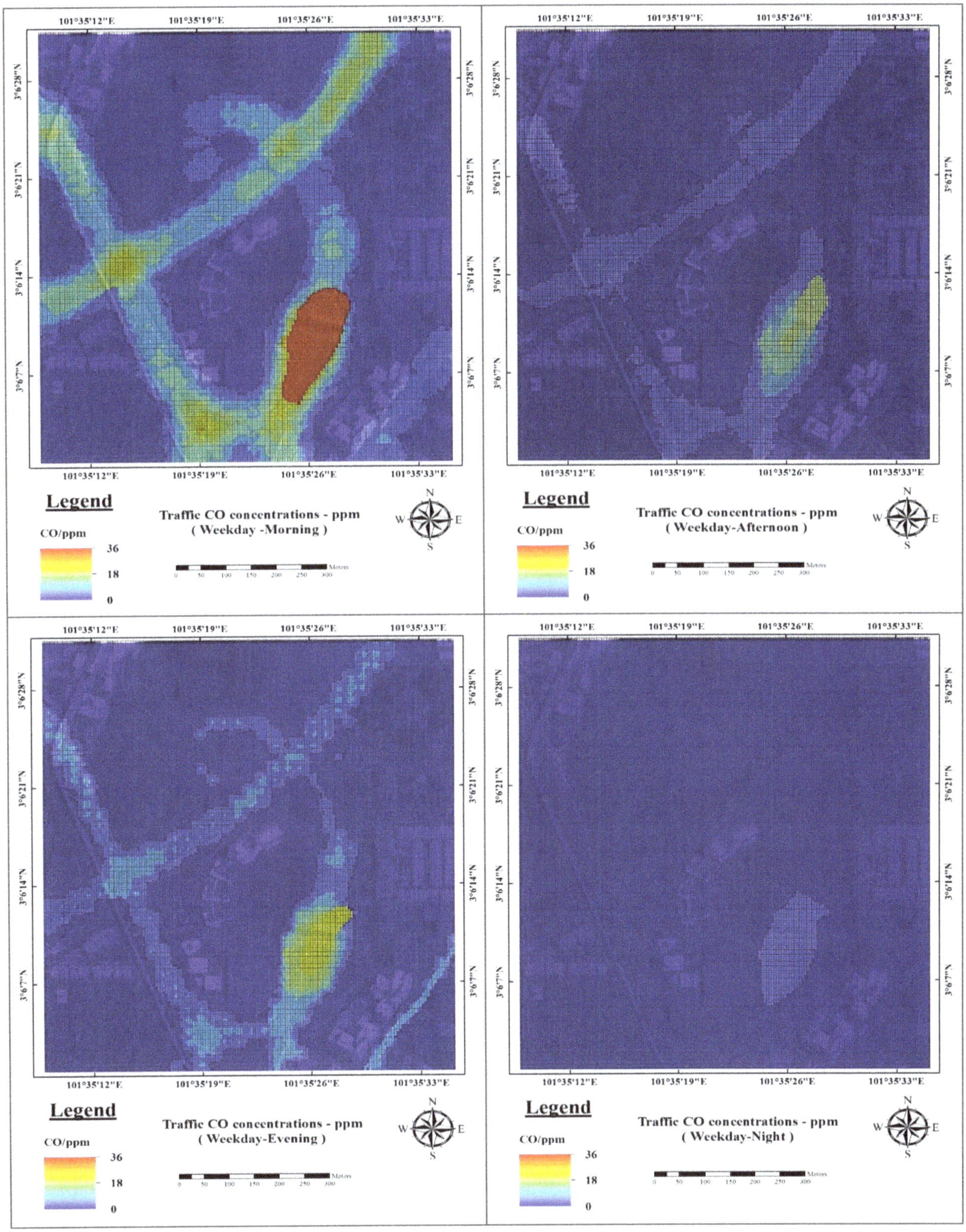

Figure 7. Prediction maps during weekdays at different times (morning, afternoon, evening and night).

4.6. Validation of Traffic CO Prediction Maps

The traffic CO spatial prediction maps were verified using the test sites of traffic CO samples, and the verification method was then performed by comparing the traffic CO test data and the traffic

CO spatial prediction maps. The lowest accuracy of the validation was at night time (72.48%) and the highest accuracy was during the evening (92.75%).

Table 7 shows the comparison between CFS-MLP, SVR and LR model. The comparative analysis was conducted by using the training data and the proposed model, support vector regression model, linear regression, land use regression model and dispersion model i.e., CALINE4 model based on many criteria not only Root mean square RMSE, but we also compared the proposed model with other baseline models based on mean absolute error (MAE), relative absolute error, root relative squared error and correlation coefficient. Results showed that the proposed model is superior to the compared models. The correlation coefficient based on our proposed model was 0.980, which is higher than the SVR (0.8668) and LR (0.851). The MAE of the proposed model was 0.896 ppm which is lower than SVR (1.640 ppm) and LR (1.851 ppm). On the other hand, the RMSE results indicated that the proposed model has the lowest RMSE (1.286 ppm) compared to SVR (2.752 ppm) and LR models (2.849 ppm). The RAE ratio for the proposed model was calculated to be 21.99%, which was lower than the RAE ratio of SVR (51.646%) and LR (55.048%). The root relative squared error indicated that the proposed model has the lowest value (19.40%) among the other models (49.784% and 48.292%).

Table 7. The comparison between CFS-MLP, SVR and LR models.

CFS-MLP Model		SVR Model		LR Model	
Correlation coefficient	0.980	Correlation coefficient	0.8668	Correlation coefficient	0.851
Mean absolute error (ppm)	0.896	Mean absolute error (ppm)	1.640	Mean absolute error (ppm)	1.851
Root mean squared error (ppm)	1.286	Root mean squared error (ppm)	2.752	Root mean squared error (ppm)	2.849
Relative absolute error (%)	21.99	Relative absolute error (%)	51.646	Relative absolute error (%)	55.048
Root relative squared error (%)	19.40	Root relative squared error (%)	49.784	Root relative squared error (%)	48.292
Total number of instances	247	Total number of instances	247	Total number of instances	247

Table 7 shows the comparison between CFS-MLP, SVR and LR model. The comparative analysis was conducted by using the training data and the proposed models, i.e., SVR, LR, land use regression model and dispersion model, i.e., CALINE4 model based on many criteria, not only RMSE. We also compared the proposed model with other baseline models based on mean absolute error (MAE), relative absolute error, root relative squared error and correlation coefficient. Results showed that the proposed model is superior to the other models. The CO emission prediction rate can be justified as highly accurate, where the accuracy is more than 90%; 90% to 80% is a good forecast; 80% to 50% is a reasonable forecast; and more than 50% is an inaccurate forecast [48–50].

The correlation coefficient based on our proposed model was 0.980, which is higher than the SVR (0.8668) and LR (0.851). The MAE of the proposed model was 0.896 ppm which is lower than SVR (1.640 ppm) and LR (1.851 ppm). On the other hand, the RMSE results indicated that the proposed model has the lowest RMSE (1.286 ppm) compared to SVR (2.752 ppm) and LR models (2.849 ppm). The RAE ratio for the proposed model was calculated to be 21.99%, which was lower than the RAE ratio of SVR (51.646%) and LR (55.048%). The root relative squared error indicated that the proposed model has the lowest value (19.40%) among the other models (49.784% and 48.292%).

Table 8 shows the results of the prediction models when combining the correlation-based feature selection; the correlation coefficients of the CFS-SVR and CFS-LR are 0. 0.7578 and 0.82, respectively. These values indicated that there is decrease in the correlation coefficient. The MAE increased for the CFS-SVR and CFS-LR and reached 1.972 ppm and 1.9713 ppm, respectively. The EMSE also increased to 3.7109 ppm and 3.1057 ppm. Figure 8 illustrates the variations of the traffic CO concentrations in the testing data.

Table 8. The comparison between CFS-MLP, CFS-SVR and CFS-LR models.

CFS-MLP Model		CFS-SVR Model		CFS-LR Model	
Correlation coefficient	0.980	Correlation coefficient	0.7578	Correlation coefficient	0.82
Mean absolute error (ppm)	0.896	Mean absolute error (ppm)	1.972	Mean absolute error (ppm)	1.9713
Root mean squared error (ppm)	1.286	Root mean squared error (ppm)	3.7109	Root mean squared error (ppm)	3.1057
Relative absolute error (%)	21.99	Relative absolute error (%)	64.3605	Relative absolute error (%)	64.333
Root relative squared error (%)	19.40	Root relative squared error (%)	67.2464	Root relative squared error (%)	56.2795
Total number of instances	247	Total number of instances	247	Total number of instances	247

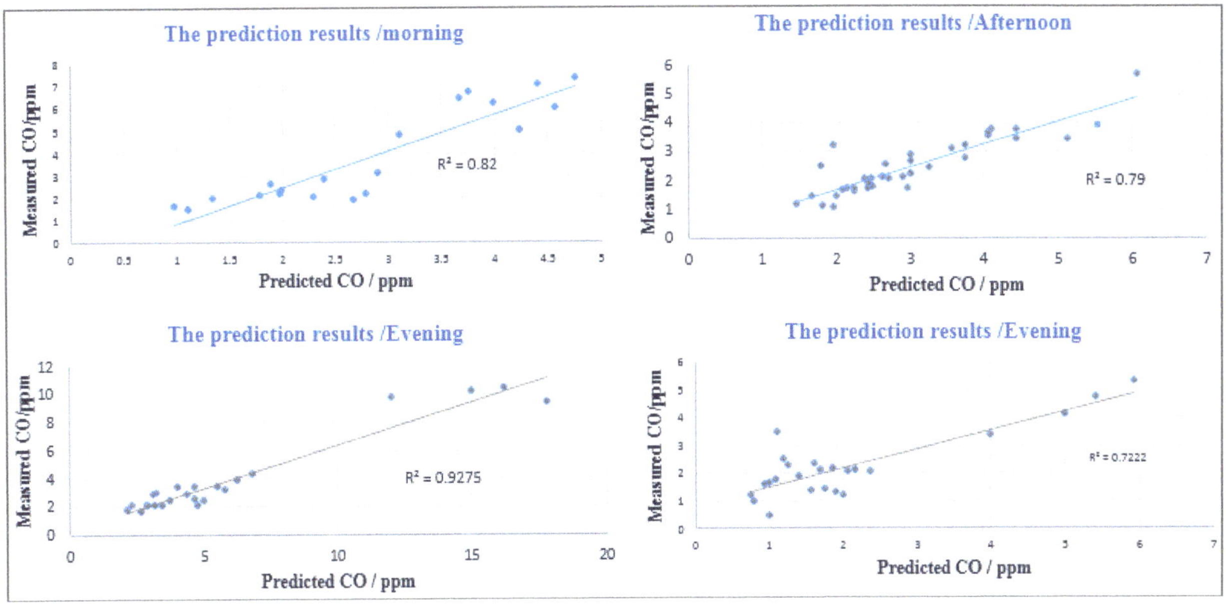

Figure 8. Accuracy assessment of the predicted maps (morning, afternoon, evening and night).

The first model that we compared with one of the baseline models is the CALINE4 model. The ALINE4 model is a dispersion model, which depends on a plume dispersion model used to predict the vehicular CO on roadways [48]. The CALINE4 model simulates the data based on a Gaussian diffusion algorithm and characterizes the pollutants dispersion on roads. We defined the proposed road network, weather data, traffic flow information, and receptor locations, and the prediction of traffic CO emissions was obtained. The MAE and RMSE values were 2.376 ppm and 4.2254 ppm, respectively, whereas and the correlation coefficient value was 0.6504. The prediction results appeared worse than our proposed model. This may due to the fact that the Gaussian diffusion, which was assumed in CL4, is not very realistic.

We also compared our work with the Land Use Regression (LUR) Model. This model is generally applied to predict air pollutants depending on the land use, traffic flow information, meteorology data and combined them based on a linear regression algorithm. The LUR model showed the following values: MAE 2.21 ppm, RMSE 4.50 ppm and a correlation coefficient of 0.5989 (Figure 9).

The final LUR equation used is given below:

$$\text{Predicted CO} = 0.0018 \times \text{Car} + 0.0423 \times \text{HV} - 0.0219 \times \text{Motorbike} + 0.2211 \times \text{Temp/C} + 0.0312 \times \text{Relative Humidity} - 0.1315 \times \text{Wind speed} + 0.0018 \times \text{Wind Angle Degree} - 0.0232 * \text{DSM} + 0.0006 \times \text{Builtup area} + 0.0064 \times \text{Highway} - 8.6627.$$

There are many models developed based on GIS and machine learning, for example [26] designed a model based on the integration of the ANN algorithm and the linear-chain conditional random field algorithm to produce real-time and fine-grained air pollution prediction maps.

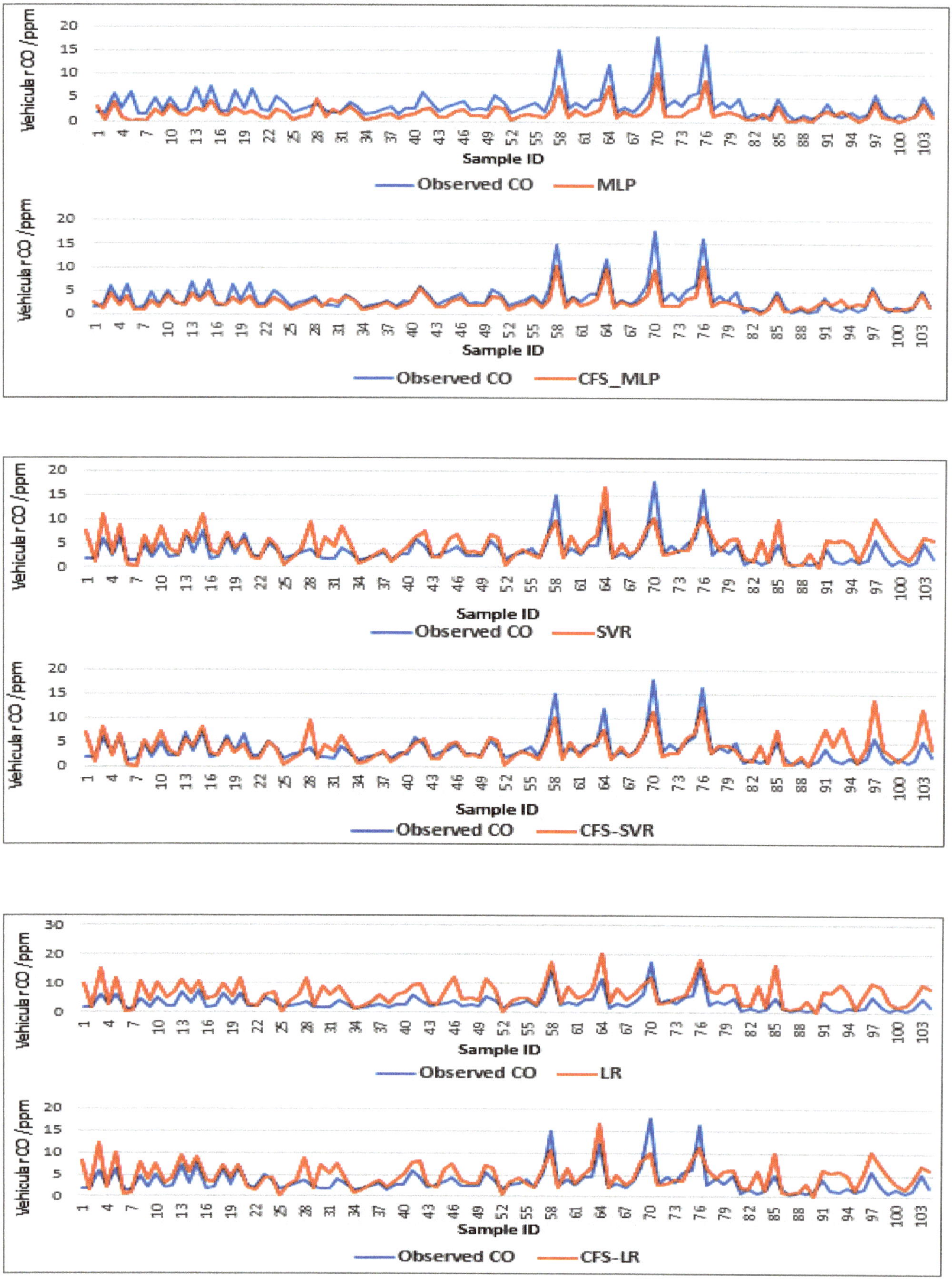

Figure 9. Comparison between the proposed and other models.

The air quality data were obtained from fixed air quality stations, the traffic data were collected from vehicle trajectory, and the meteorology data were collected from monitoring stations. Other data used were land use, road geometry and social information. Their results presented the prediction maps with low resolution, cell size (1×1) km^2 in local scale. The limitation of this model can be summarized in some points. The data collection from the standard fixed air quality monitoring stations may not be able to measure the air quality that people are exposed on the ground level due to the limitation of monitoring location and height. On the other hand, the data obtained from the fixed stations are not suitable for high-resolution prediction maps such as microscale maps. Moreover, this study did not contain information about the terrain and buildings.

The proposed model in this research is different from the aforementioned study based on some points. We developed a GIS-based NN and data were obtained from a field survey. Moreover, the land use and DSM were extracted from a very high resolution LiDAR data clouds. The proposed methodology is designed to predict the vehicular CO and produce high-resolution maps at a microscale level (5×5) m^2 whereas the aforementioned paper estimated air pollutants based on a low-resolution grid (1×1) km^2.

Regarding the meteorology parameters shown in Table 4, it is evident that the correlation between wind speed and vehicular emissions is 0.0016, which is considered weak in the short-term prediction and in a small area compared to other factors. In addition, the temperature has the lowest correlation with vehicular emissions 0.0014; therefore, the variation in the vehicular emissions (i.e., CO) may not be significantly affected by the weather condition in the prediction in small areas. On the other hand, the DSM, which was used to extract the terrain and building's altitude in the study area, has a good correlation with vehicular emissions (more than meteorology parameters). Therefore, the geographic factors are important in prediction studies. This study adopted high-resolution elevation data in order to extract results that are more accurate. In urban areas, building altitude is an important parameter because it can resist vehicular emissions and prevent the distribution of pollutants in urban areas.

5. Conclusions

Traffic emissions (e.g., traffic CO) are considered the major source of air pollution in congested urban areas, including road corridors in toll plaza areas. Traffic emission prediction models are utilized to evaluate the impacts of traffic CO emissions on the population and environment and some models are used to illustrate the spatial prediction of these emissions. In this paper, a hybrid prediction model was proposed by combining three models (CFS, MLP and GIS) to predict the traffic CO emissions and create micro-scale prediction maps in a small area at different times during the day. The final findings have shown that the proposed model scored an accuracy of 80.6% and the correlation coefficient of 0.980, RMSE of 1.2736 ppm and mean absolute error of 0.8925 ppm. We used CFS to identify and remove highly correlated parameters so that redundancy was reduced to choose the optimum parameters used in the prediction model through MLP. The simulation results showed that nine predictors were reduced to six, which contributed to an increase in the prediction accuracy.

The data were collected from the field and remote sensing data (i.e., LiDAR and very high-resolution WorldView-3 satellite image), and modelling was performed in a GIS environment.

In this study, we produced microscale maps for vehicular emissions. The simulated traffic CO emissions ranged from 35 ppm inside the toll plaza area to 0 ppm for areas that were located far away from the toll area. As per the microscale-prediction maps, high spatial variation in the traffic CO emissions was identified. The highest value of CO concentrations is found in traffic jam areas. Conversely, the lowest values of traffic CO emissions are distributed far from traffic activities. The highest concentrations of traffic CO were located inside the toll plaza because of the traffic jam that occurs daily in the toll areas these results give a clear indication about the relationship between the traffic activities and the traffic CO emissions. The traffic CO emissions may have a significant impact on the health of toll plaza workers, drivers and the passengers. Therefore, such a prediction model can aid decision makers to implement plans to mitigate the traffic emissions that can protect people who are working, passing through or living near toll plaza areas; this will be the main advantage of the proposed forecasting system.

Traffic CO emission prediction models and GIS modeling are both efficient tools for transportation planning and traffic emission assessment. The prediction maps produced by the proposed model can be used as an effective tool in the decision-making process to identify optimum solutions which can be used to mitigate traffic jams in toll plaza areas as well as on highways and road networks. As traffic emission pollution assessment by decision makers is very complicated and usually expensive because of the high level of requirements for expert knowledge and the developed support systems, the presented traffic CO assessment model is not expensive and can easily implemented. Moreover, the traffic CO pollution concentrations vary based on the traffic condition and the number of vehicles, which requires a periodic monitoring of traffic emissions by government agencies or relevant departments. The best parameter selection analysis could be used to reduce the data collection requirements, which can lead to reduced time required, resources utilized and processing time needed. GIS modeling is a useful tool for non-expert users to implement the traffic CO impact assessments in various applications. Finally, these models can be improved by using more advanced algorithms such as deep learning algorithms and a large number of samples that can be used to increase the accuracy of the prediction process.

Author Contributions: B.P. conceptualized, supervised, and obtained the grant for the study. B.P. and O.S.A. collected and analyzed the data, performed the analyses and validation, wrote the manuscript and contributed to the re-structuring and editing of the manuscript. B.P., O.S.A., N.S., H.Z.M.S., H.M.R. and C.-W.L. professionally optimized the manuscript.

Acknowledgments: The authors acknowledge and appreciate the provision of airborne laser scanning data (LiDAR), satellite images and logistic support by the PLUS Berhad. In addition, the second author, Biswajeet Pradhan, gratefully acknowledges the financial support from the UPM-PLUS industry project grant.

Abbreviations

GIS	A Geospatial Information System is a system designed to collect, manage, analyze, store and produce different types of spatial data.
CO	Carbon Monoxide is a toxic gas and it has no has no color, taste, or smell, resulting from the incomplete combustion of fuel.
RMSE	The algorithm of the root mean square is used to calculate the differences between values estimated by a model and the observed values.
VISSIM	Software designed for traffic flow simulation at a micro-scale level, which is designed by Planning Transport Verkehr (PTV), Germany.
EPR	The evolutionary polynomial regression, EPR, is one of the data-mining algorithms developed based on evolutionary computing and the integration of numerical regression and genetic algorithm.
CFS	A correlation-based feature selection algorithm, which is a type of filter algorithm that selects features based on a heuristic (correlation-based) function.
LiDAR	Light Detection and Ranging is an advanced surveying technology usually used to create 3D models by measure the distance between targets and the Laser Sensor.
ENVI	Environment for Visualizing Images: professional software used for image analysis and remote sensing applications.
MLP	A multilayer perceptron (MLP) is a class of feedforward artificial neural networks. An MLP consists of, at least, three layers of nodes: an input layer, a hidden layer and an output layer. Except for the input nodes, each node is a neuron that uses a nonlinear activation function.
LULC	Land Use and Land Cover are data files that describe the land surfaces such as water, vegetation and cultural features.
CFS-MLP	The proposed model that is the combination of two models, the correlation based feature selection algorithm and multilayer perceptron Neural Network algorithm.

CALINE4	California Line Source Dispersion is one of the dispersion models used to estimate carbon monoxide emissions near roads based on various parameters related to geographic locations.
MAE	Mean Absolute Error, MAE, measures the average magnitude of the errors in a set of predictions, without considering their direction. It is the average over the test sample of the absolute differences between prediction and actual observation where all individual differences have equal weight.
RAE	Relative Absolute Error is defined as the absolute error relative to the size of the measurement, and it depends on both the absolute error and the measured value. The relative error is large when the measured value is small, or when the absolute error is large.
ANN	An Artificial Neural Network is a computational model based on the structure and functions of biological neural networks. Information that flows through the network affects the structure of the ANN because a neural network changes—or learns, in a sense—based on that input and output.

References

1. Bastien, L.A.; McDonald, B.C.; Brown, N.J.; Harley, R.A. High-resolution mapping of sources contributing to urban air pollution using adjoint sensitivity analysis: Benzene and diesel black carbon. *Environ. Sci. Technol.* **2015**, *49*, 7276–7284. [CrossRef] [PubMed]
2. Fameli, K.M.; Assimakopoulos, V.D. Development of a road transport emission inventory for Greece and the Greater Athens Area: Effects of important parameters. *Sci. Total Environ.* **2015**, *505*, 770–786. [CrossRef]
3. Borge, R.; Narros, A.; Artinano, B.; Yagüe, C.; Gomez-Moreno, F.J.; de la Paz, D.; Quaassdorff, C. Assessment of microscale spatio-temporal variation of air pollution at an urban hotspot in Madrid (Spain) through an extensive field campaign. *Atmos. Environ.* **2016**, *140*, 432–445. [CrossRef]
4. Oftedal, B.; Krog, N.H.; Pyko, A.; Eriksson, C.; Graff-Iversen, S.; Haugen, M.; Aasvang, G.M. Road traffic noise and markers of obesity–a population-based study. *Environ. Res.* **2015**, *138*, 144–153. [CrossRef]
5. Ancona, C.; Badaloni, C.; Mattei, F.; Cesaroni, G.; Stafoggia, M.; Forastiere, F. Health Impact Assessment of Air Pollution, Noise, and Lack of Green in Rome. *J. Transp. Health* **2017**, *5*, S42–S43. [CrossRef]
6. Garshick, E.; Laden, F.; Hart, J.E.; Caron, A. Residence near a major road and respiratory symptoms in US veterans. *Epidemiology* **2003**, *14*, 728. [CrossRef] [PubMed]
7. Delfino, R.J.; Tjoa, T.; Gillen, D.L.; Staimer, N.; Polidori, A.; Arhami, M.; Longhurst, J. Traffic-related air pollution and blood pressure in elderly subjects with coronary artery disease. *Epidemiology* **2010**, *21*, 396–404. [CrossRef]
8. Crouse, D.L.; Goldberg, M.S.; Ross, N.A.; Chen, H.; Labrèche, F. Postmenopausal breast cancer is associated with exposure to traffic-related air pollution in Montreal, Canada: A case–control study. *Environ. Health Perspect.* **2010**, *118*, 1578. [CrossRef]
9. Domene, E.; Lopez, R.; Fauro, B.; Rojas-Rueda, D.; Conill, C.; Alsina, G.; Marull, J. Modelling Impacts of Mobility on Urban Air Quality and Health: Scenario Analysis for the Barcelona Metropolitan Area (Metropolitan Mobility Plan). *J. Transp. Health* **2017**, *5*, S60–S61. [CrossRef]
10. Singh, D.; Kumar, A.; Kumar, K.; Singh, B.; Mina, U.; Singh, B.B.; Jain, V.K. Statistical modeling of O3, NOx, CO, PM2. 5, VOCs and noise levels in commercial complex and associated health risk assessment in an academic institution. *Sci. Total Environ.* **2016**, *572*, 586–594. [CrossRef] [PubMed]
11. Behera, S.N.; Sharma, M.; Mishra, P.K.; Nayak, P.; Damez-Fontaine, B.; Tahon, R. Passive measurement of NO_2 and application of GIS to generate spatially-distributed air monitoring network in urban environment. *Urban Clim.* **2015**, *14*, 396–413. [CrossRef]
12. Johnson, M.; Isakov, V.; Touma, J.S.; Mukerjee, S.; Özkaynak, H. Evaluation of land-use regression models used to predict air quality concentrations in an urban area. *Atmos. Environ.* **2010**, *44*, 3660–3668. [CrossRef]
13. Kanaroglou, P.S.; Adams, M.D.; De Luca, P.F.; Corr, D.; Sohel, N. Estimation of sulfur dioxide air pollution concentrations with a spatial autoregressive model. *Atmos. Environ.* **2013**, *79*, 421–427. [CrossRef]
14. Vandaele, N.; Van Woensel, T.; Verbruggen, A. A queueing based traffic flow model. *Transp. Res. Part D Transp. Environ.* **2000**, *5*, 121–135. [CrossRef]
15. Tomić, J.; Bogojević, N.; Pljakić, M.; Šumarac-Pavlović, D. Assessment of traffic noise levels in urban areas using different soft computing techniques. *J. Acoust. Soc. Am.* **2016**, *140*, EL340–EL345. [CrossRef]

16. Hamad, K.; Khalil, M.A.; Shanableh, A. Modeling roadway traffic noise in a hot climate using artificial neural networks. *Transp. Res. Part D. Transp. Environ.* **2017**, *53*, 161–177. [CrossRef]
17. Di Mascio, P.; Di Vito, M.; Loprencipe, G.; Ragnoli, A. Procedure to determine the geometry of road alignment using GPS data. *Procedia-Soc. Behav. Sci.* **2012**, *53*, 1202–1215. [CrossRef]
18. Righini, G.; Cappelletti, A.; Ciucci, A.; Cremona, G.; Piersanti, A.; Vitali, L.; Ciancarella, L. GIS based assessment of the spatial representativeness of air quality monitoring stations using pollutant emissions data. *Atmos. Environ.* **2014**, *97*, 121–129. [CrossRef]
19. Hülsmann, F.; Gerike, R.; Ketzel, M. Modelling traffic and air pollution in an integrated approach–the case of Munich. *Urban Clim.* **2014**, *10*, 732–744. [CrossRef]
20. Kim, Y.; Guldmann, J.M. Land-use regression panel models of NO2 concentrations in Seoul, Korea. *Atmos. Environ.* **2015**, *107*, 364–373. [CrossRef]
21. Zarandi, M.F.; Faraji, M.R.; Karbasian, M. Interval type-2 fuzzy expert system for prediction of carbon monoxide concentration in mega-cities. *Appl. Soft Comput.* **2012**, *12*, 291–301. [CrossRef]
22. Kwok, L.K.; Lam, Y.F.; Tam, C.Y. Developing a statistical based approach for predicting local air quality in complex terrain area. *Atmos. Pollut. Res.* **2017**, *8*, 114–126. [CrossRef]
23. Quaassdorff, C.; Borge, R.; Pérez, J.; Lumbreras, J.; de la Paz, D.; de Andrés, J.M. Microscale traffic simulation and emission estimation in a heavily trafficked roundabout in Madrid (Spain). *Sci. Total Environ.* **2016**, *566*, 416–427. [CrossRef]
24. Shakerkhatibi, M.; Mohammadi, N.; Zoroufchi Benis, K.; Behrooz Sarand, A.; Fatehifar, E.; Asl Hashemi, A. Using ANN and EPR models to predict carbon monoxide concentrations in urban area of Tabriz. *Environ. Health Eng. Manag. J.* **2015**, *2*, 117–122.
25. Cai, M.; Yin, Y.; Xie, M. Prediction of hourly air pollutant concentrations near urban arterials using artificial neural network approach. *Transp. Res. Part D Transp. Environ.* **2009**, *14*, 32–41. [CrossRef]
26. Zheng, Y.; Liu, F.; Hsieh, H.P. U-Air: When urban air quality inference meets big data. In Proceedings of the 19th ACM SIGKDD International Conference on Knowledge Discovery and Data Mining, Chicago, IL, USA, 11–14 August 2013; pp. 1436–1444.
27. Namdeo, A.; Mitchell, G.; Dixon, R. TEMMS: An integrated package for modelling and mapping urban traffic emissions and air quality. *Environ. Model Softw.* **2002**, *17*, 177–188. [CrossRef]
28. Kho, F.W.L.; Law, P.L.; Ibrahim, S.H.; Sentian, J. Carbon monoxide levels along roadway. *Int. J. Environ. Sci. Technol.* **2007**, *4*, 27–34. [CrossRef]
29. Ranjbar, H.R.; Gharagozlou, A.R.; Nejad, A.R.V. 3D analysis and investigation of traffic noise impact from Hemmat highway located in Tehran on buildings and surrounding areas. *J. Geogr. Inf. Syst.* **2012**, *4*, 322. [CrossRef]
30. Li, F.; Liao, S.S.; Cai, M. A new probability statistical model for traffic noise prediction on free flow roads and control flow roads. *Transp. Res. Part D Transp. Environ.* **2016**, *49*, 313–322. [CrossRef]
31. Ragettli, M.S.; Goudreau, S.; Plante, C.; Fournier, M.; Hatzopoulou, M.; Perron, S.; Smargiassi, A. Statistical modeling of the spatial variability of environmental noise levels in Montreal, Canada, using noise measurements and land use characteristics. *J. Expos. Sci. Environ. Epidemiol.* **2016**, *26*, 597. [CrossRef]
32. Kirchstetter, T.W.; Singer, B.C.; Harley, R.A.; Kendall, G.R.; Chan, W. Impact of oxygenated gasoline use on California light-duty vehicle emissions. *Environ. Sci. Technol.* **1996**, *30*, 661–670. [CrossRef]
33. Goyal, P. Present scenario of air quality in Delhi: A case study of CNG implementation. *Atmos. Environ.* **2003**, *37*, 5423–5431. [CrossRef]
34. Chen, K.S.; Wang, W.C.; Chen, H.M.; Lin, C.F.; Hsu, H.C.; Kao, J.H.; Hu, M.T. Motorcycle emissions and fuel consumption in urban and rural driving conditions. *Sci. Total Environ.* **2003**, *312*, 113–122. [CrossRef]
35. Henderson, S.B.; Beckerman, B.; Jerrett, M.; Brauer, M. Application of land use regression to estimate long-term concentrations of traffic-related nitrogen oxides and fine particulate matter. *Environ. Sci. Technol.* **2007**, *41*, 2422–2428. [CrossRef]
36. Shi, Y.; Lau, K.K.L.; Ng, E. Developing street-level PM2.5 and PM10 land use regression models in high-density Hong Kong with urban morphological factors. *Environ. Sci. Technol.* **2016**, *50*, 8178–8187. [CrossRef]
37. Hall, M.A. Correlation-Based Feature Selection for Machine Learning. Ph.D. Thesis, The University of Waikato, Hamilton, New Zealand, 1999.

38. Boznar, M.; Lesjak, M.; Mlakar, P. A neural network-based method for short-term predictions of ambient SO2 concentrations in highly polluted industrial areas of complex terrain. *Atmos. Environ. Part B Urban Atmos.* **1993**, *27*, 221–230. [CrossRef]
39. Chaloulakou, A.; Saisana, M.; Spyrellis, N. Comparative assessment of neural networks and regression models for forecasting summertime ozone in Athens. *Sci. Total Environ.* **2003**, *313*, 1–13. [CrossRef]
40. De Cos Juez, F.J.; Nieto, P.G.; Torres, J.M.; Castro, J.T. Analysis of lead times of metallic components in the aerospace industry through a supported vector machine model. *Math. Comput. Model.* **2010**, *52*, 1177–1184. [CrossRef]
41. Gardner, M.W.; Dorling, S.R. Neural network modelling and prediction of hourly NOx and NO2 concentrations in urban air in London. *Atmos. Environ.* **1999**, *33*, 709–719. [CrossRef]
42. Hooyberghs, J.; Mensink, C.; Dumont, G.; Fierens, F.; Brasseur, O. A neural network forecast for daily average PM10 concentrations in Belgium. *Atmos. Environ.* **2005**, *39*, 3279–3289. [CrossRef]
43. Kukkonen, J.; Partanen, L.; Karppinen, A.; Ruuskanen, J.; Junninen, H.; Kolehmainen, M.; Cawley, G. Extensive evaluation of neural network models for the prediction of NO2 and PM10 concentrations, compared with a deterministic modelling system and measurements in central Helsinki. *Atmos. Environ.* **2003**, *37*, 4539–4550. [CrossRef]
44. Nieto, P.G.; Lasheras, F.S.; García-Gonzalo, E.; de Cos Juez, F.J. PM 10 concentration forecasting in the metropolitan area of Oviedo (Northern Spain) using models based on SVM, MLP, VARMA and ARIMA: A case study. *Sci. Total Environ.* **2018**, *621*, 753–761. [CrossRef]
45. Friedman, J.; Hastie, T.; Tibshirani, R. *The Elements of Statistical Learning*; Springer Series in Statistics: New York, NY, USA, 2001; Volume 1.
46. Rojek, I. Technological process planning by the use of neural networks. *AI EDAM* **2017**, *31*, 1–5. [CrossRef]
47. Mousavi, S.Z.; Kavian, A.; Soleimani, K.; Mousavi, S.R.; Shirzadi, A. GIS-based spatial prediction of landslide susceptibility using logistic regression model. *Geomat. Nat. Hazards Risk* **2011**, *2*, 33–50. [CrossRef]
48. Sharma, N.; Gulia, S.; Dhyani, R.; Singh, A. Performance evaluation of CALINE 4 dispersion model for an urban highway corridor in Delhi. *J. Sci. Ind. Res.* **2013**, *72*, 521–530.
49. Pao, Y.-H.; Phillips, S.M.; Sobajic, D.J. Neural-net computing and the intelligent control of systems. *Int. J. Control* **1992**, *56*, 263–289. [CrossRef]
50. Pao, H.-T.; Fu, H.-C.; Tseng, C.-L. Forecasting of CO_2 emissions, energy consumption and economic growth in China using an improved grey model. *Energy* **2012**, *40*, 400–409.

4

Study of the Relationship between Urban Expansion and PM_{10} Concentration using Multi-Temporal Spatial Datasets and the Machine Learning Technique: Case Study for Daegu, South Korea

Yun-Jae Choung [1,2] and Jin-Man Kim [3,*]

1 GIS Research Center, Geo C&I Co., Ltd., Daegu 41165, Korea; choung12osu@gmail.com
2 Global Land Satellite Information Center, Kyungpook National University, Daegu 41566, Korea
3 Geotechnical Engineering Research Institute, Korea Institute of Civil Engineering and Building Technology, Goyang-si, Gyeonggi-do 10223, Korea
* Correspondence: jmkim@kict.re.kr

Abstract: To protect the population from respiratory diseases and to prevent the damages due to air pollution, the main cause of air pollution should be identified. This research assessed the relationship between the airborne particulate concentrations (PM_{10}) and the urban expansion in Daegu City in South Korea from 2007 to 2017 using multi-temporal spatial datasets (Landsat images, measured PM_{10} data) and the machine learning technique in the following steps. First, the expanded urban areas were detected from the multiple Landsat images using support vector machine (SVM), a widely used machine learning technique. Next, the annual PM_{10} concentrations were calculated using the long-term measured PM_{10} data. Finally, the degrees of increase of the expanded urban areas and of the PM_{10} concentrations in Daegu from 2007 to 2017 were calculated by counting the pixels representing the expanded urban areas and computing variation of the annual PM_{10} concentrations, respectively. The experiment results showed that there is a minimal or even no relationship at all between the urban expansion and the PM_{10} concentrations because the urban areas expanded by 55.27 km^2 but the annual PM_{10} concentrations decreased by 17.37 $\mu g/m^3$ in Daegu from 2007 to 2017.

Keywords: coarse particle; particulate matter 10 (PM_{10}); landsat image; machine learning; support vector machine

---

1. Introduction

Urban expansion, also called "urban sprawl," is defined as "the spreading of urban development (e.g., houses, shopping centers) on undeveloped lands near a city" or "the rapid expansion of the geographic extent of cities and towns, often characterized by low-density residential housing, single-use zoning, and increased reliance on private automobiles for transportation" [1,2]. In general, urban expansion has a close relationship with urban development, infrastructure improvement, population growth, etc. [3].

Coarse particle, defined as particulate matter 10 (PM_{10}), consists of particles with a diameter of 10 μm or less [4,5]. PM_{10} is one of the main components of air pollution, and it also results in various environmental impacts (e.g., atmospheric pollution) and human health impacts (e.g., chronic respiratory diseases) [6,7]. In particular, exposure to a high PM_{10} concentration can cause a number of significant health impacts, ranging from coughing to high blood pressure, heart attack, stroke, and lung cancer [8].

Previous studies found that urban expansion generally has a significant impact on air pollution because more human activities that can cause air pollution (e.g., vehicular traffic) are expected in urban areas [9]. Stone (2007) assessed the relationship between urban expansion and air quality [10]. Cho and Choi (2014) investigated the effect of compact urban development on air quality [11]. Liu et al. (2018) assessed the relationship between urban air pollution and urban form, seasonality, and city size [12].

To protect the population from respiratory diseases and to prevent public health disasters due to PM_{10} concentration, the main causes of PM_{10} concentration in the city should be identified. Limited research has been conducted, however, to identify the main causes of PM_{10} concentration in each city in South Korea. In general, urban expansion is regarded as the main cause of urban air pollution owing to the urban development activities accompanying it. This research aims to assess the relationship between urban expansion and PM_{10} concentration by monitoring the 10 years (from 2007 to 2017) of urban expansion and the annual PM_{10} concentrations using multi-temporal spatial datasets acquired in Daegu, South Korea and the machine learning technique. First, the expanded urban areas were detected from the multi-temporal Landsat satellite images using the machine learning technique. Then the annual PM10 concentrations were calculated using the long-term measured PM10 data. Finally, the relationship between urban expansion and PM_{10} concentration was assessed by counting the pixels representing the expanded urban areas and computing variation of the annual PM10 concentrations, respectively.

2. Study Area and Datasets

Daegu Metropolitan City in South Korea was selected as the study area in this research for the following reasons. First, the urban areas of Daegu have been significantly expanded of late [13]. Second, there are long-term measured coarse particle (PM_{10}) data acquired by the 11 air quality monitoring stations (AQMSs) in Daegu, which can be used for the study [14]. Daegu has operated these AQMSs since 1973 for sustainably monitoring the air quality condition of the city. Figure 1 shows the locations of the 11 AQMSs in Daegu, South Korea.

The multi-temporal Landsat satellite images acquired on 13 May 2007 ("first Landsat image") and on April 29, 2017 ("second Landsat image") were used in this study for the following reasons. First, the urban areas in Daegu significantly expanded during such periods due to the city's urban development policy. Second, both Landsat images were less affected by the prevailing atmospheric conditions then. Figure 2 shows one section each of the first and second Landsat images.

The first Landsat image was acquired by the Landsat-5 thematic mapper (TM) sensor, and it consists of seven bands (blue: 450–520 nm; green: 520–600 nm; red: 630–690 nm; near-infrared: 770–900 nm; short-wave infrared 1: 1550–1750 nm; short-wave infrared 2: 2080–2350 nm; and thermal: 10,400–12,500 nm) [15]. The second Landsat image was acquired by the Landsat-8 operational land imager (OLI) and the thermal infrared sensor (TIRS), and it consists of nine bands (coastal aerosol: 435–451 nm; blue: 452–512 nm; green: 533–590 nm; red: 636–673 nm; near-infrared: 851–879 nm; short-wave infrared 1: 1566–1651 nm; short-wave infrared 2: 2107–2294 nm; thermal infrared 1: 10,600–11,190 nm; and thermal infrared 2: 11,500–12,510 nm) [15]. Both Landsat images were georeferenced to the coordinate system universal transverse mercator (UTM), zone 52 N, based on the 1984 datum world geodetic system (WGS).

To measure the annual PM_{10} concentrations during the same period for the monitoring of the urban expansion in Daegu using the Landsat 1 and 2 images, the measured PM_{10} data between 2007 and 2017 acquired through the Daegu atmospheric information system (https://air.daegu.go.kr/) were also utilized.

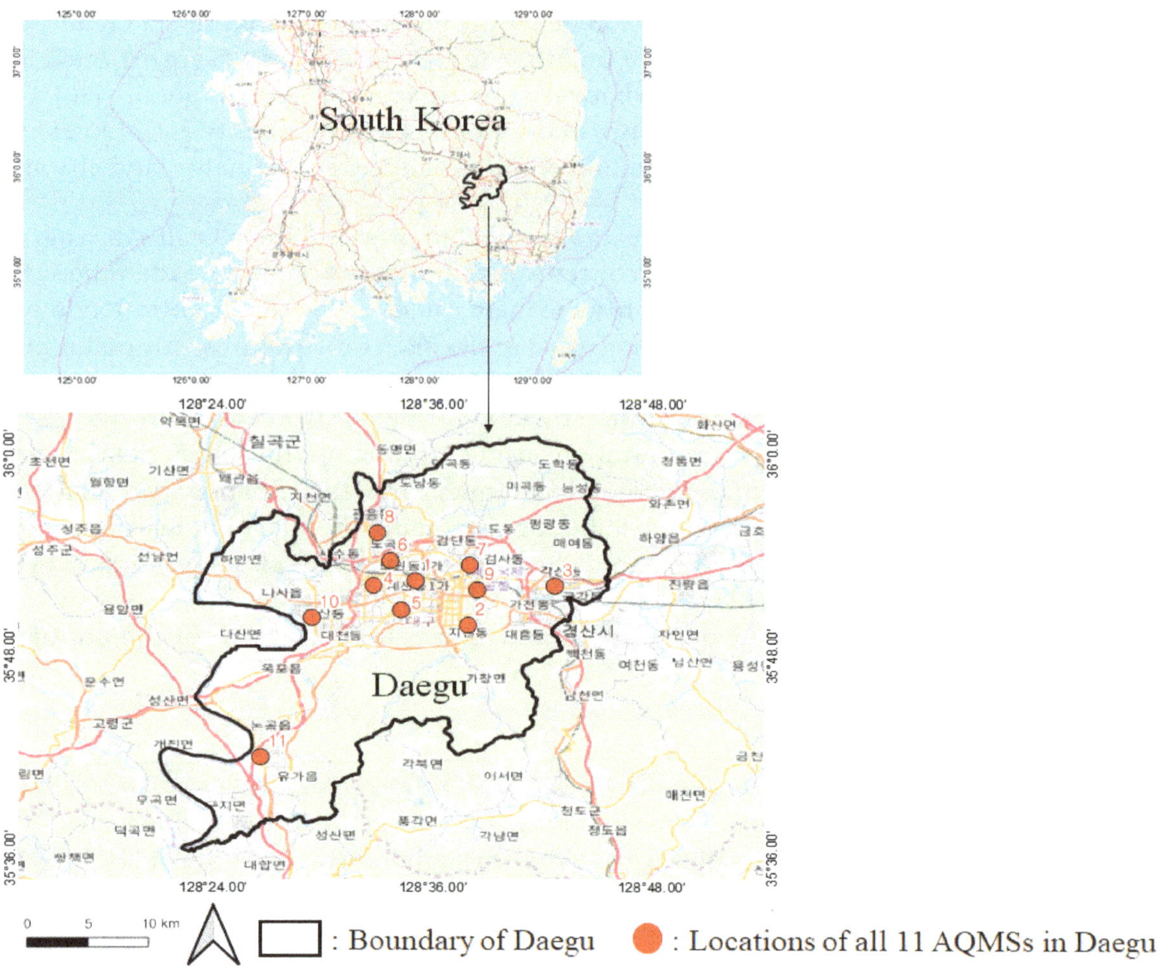

Figure 1. Locations of the 11 air quality monitoring stations (AQMSs) in Daegu, South Korea.

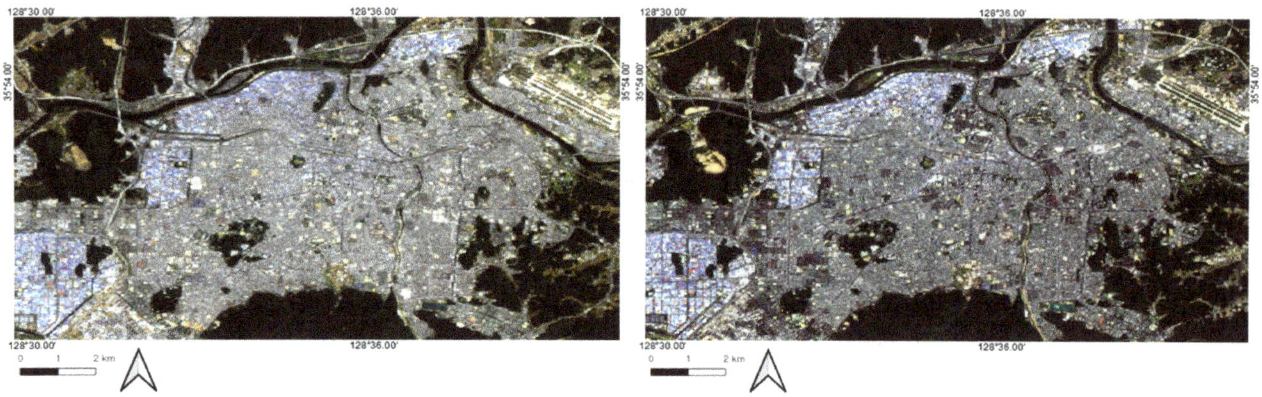

(**a**) (**b**)

Figure 2. One section each of the multi-temporal Landsat images utilized in this research: (**a**) a section of the first Landsat image (acquired on 13 May 2007); and (**b**) a section of the second Landsat image (acquired on 29 April 2017).

3. Methodology

Figure 3 presents a flowchart of the procedure that was employed to assess the relationship between urban expansion and PM_{10} concentration in Daegu from 2007 to 2017 using the given datasets.

As can be seen in Figure 3, in the first step of the proposed methodology, two urban maps were generated, respectively, from the first and second Landsat images, using the support vector machine

(SVM) technique, a widely used machine learning technique. Then the extent of urban expansion was detected using the generated first and second urban maps. In the next step, the annual PM_{10} concentrations were calculated using the measured PM_{10} data acquired by each AQMS of Daegu. Finally, the relationship between the urban expansion and the PM_{10} concentration rate in Daegu was assessed using the two calculated statistics: the increase of the expanded urban areas and the increase of the annual PM_{10} concentrations in Daegu from 2007 to 2017.

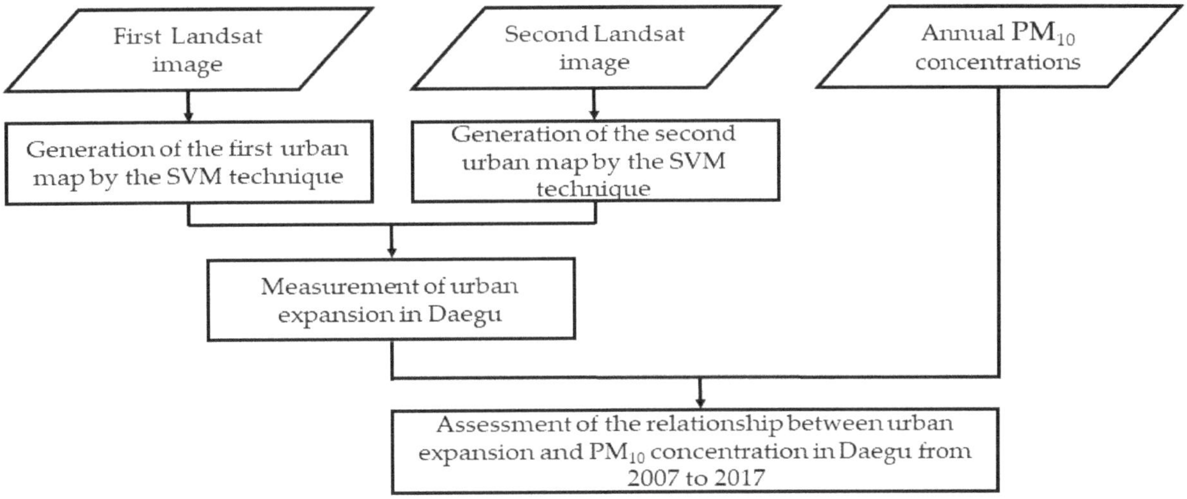

Figure 3. Flowchart showing the procedure that was employed to assess the relationship between urban expansion and PM_{10} concentration in Daegu from 2007 to 2017 using the given datasets.

3.1. Generation of the Urban Maps by the SVM Technique

This section illustrates the procedure for generating urban maps through the SVM technique, a widely used machine learning algorithm. Machine learning is defined as ""the ability of a machine to improve its performance based on previous results" [16]. The machine learning technique has been widely used of late in remote sensing applications for classifying land uses and detecting the significant features from the remote sensing datasets, due to its advantages for high-value classification [17,18]. SVM, a widely used machine learning technique for finding the linear hyperplane that maximizes the margins between the two clusters in n-dimensional spaces, has been widely used in remote sensing applications due to its superior advantages over the other machine learning techniques for classifying land uses, detecting significant features, and avoiding classification errors [19]. Hence, in this research, the SVM technique was used to generate urban maps, which distinguish the urban areas from the non-urban areas (water, soil, vegetation, etc.). Figure 4 shows the first and second urban maps separately generated from the first and second Landsat images, respectively, through the SVM technique.

3.2. Detection of the Expaned Urban Areas in Daegu from 2007 to 2017

In the next step of the proposed methodology, the expanded urban areas in Daegu were detected using the two urban maps that had been generated. The expanded urban areas from 2007 to 2017 were detected by intersecting the pixels representing the non-urban areas in the first urban map and the pixels representing the urban areas in the second urban map. Figure 5 shows the locations of the AQMSs and expanded urban areas in Daegu from 2007 to 2017 detected using the two generated urban maps.

As can be seen in Figure 5, all the AQMSs in Daegu were located near the expanded urban areas detected using the two generated urban maps.

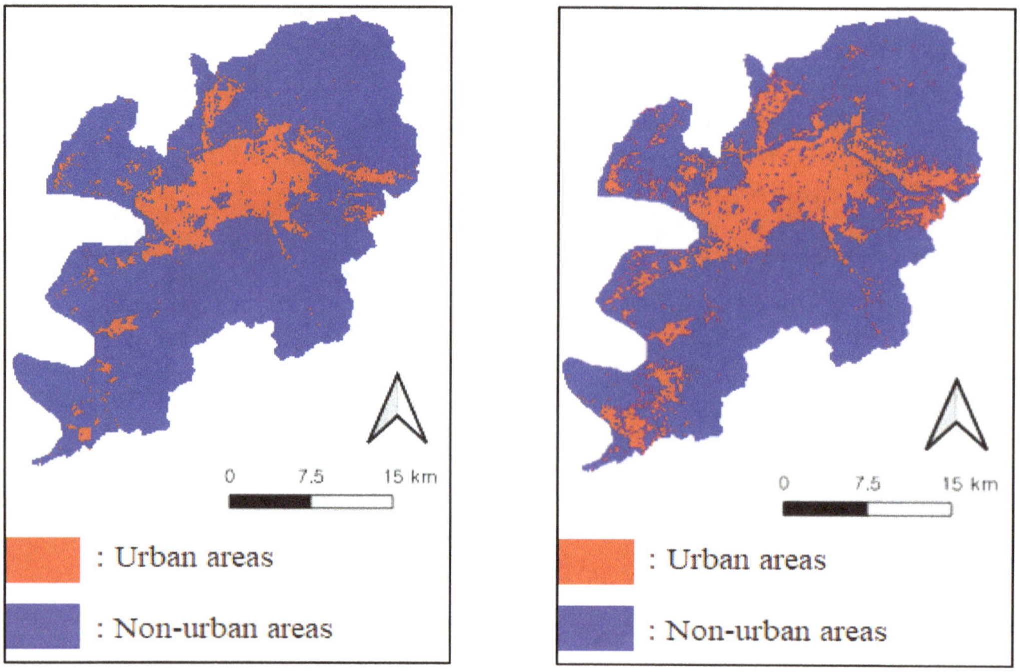

(a) (b)

Figure 4. First and second urban maps: (**a**) first urban map generated from the first Landsat image through the SVM technique; and (**b**) second urban map generated from the second Landsat image through the SVM technique.

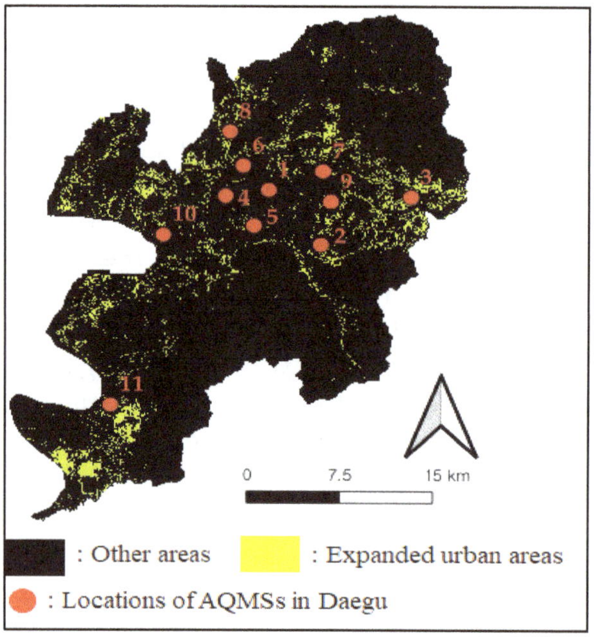

Figure 5. Locations of the AQMSs and expanded urban areas in Daegu from 2007 to 2017 detected using the two generated urban maps.

3.3. Calculation of the Statistics for the Annual PM_{10} Concentrations

In this section, the calculation of the statistics for the annual PM_{10} concentrations in each AQMS in Daegu from 2007 to 2017 is described. Figure 6 presents time series graphs showing the annual PM_{10} concentrations in each AQMS in Daegu from 2007 to 2017.

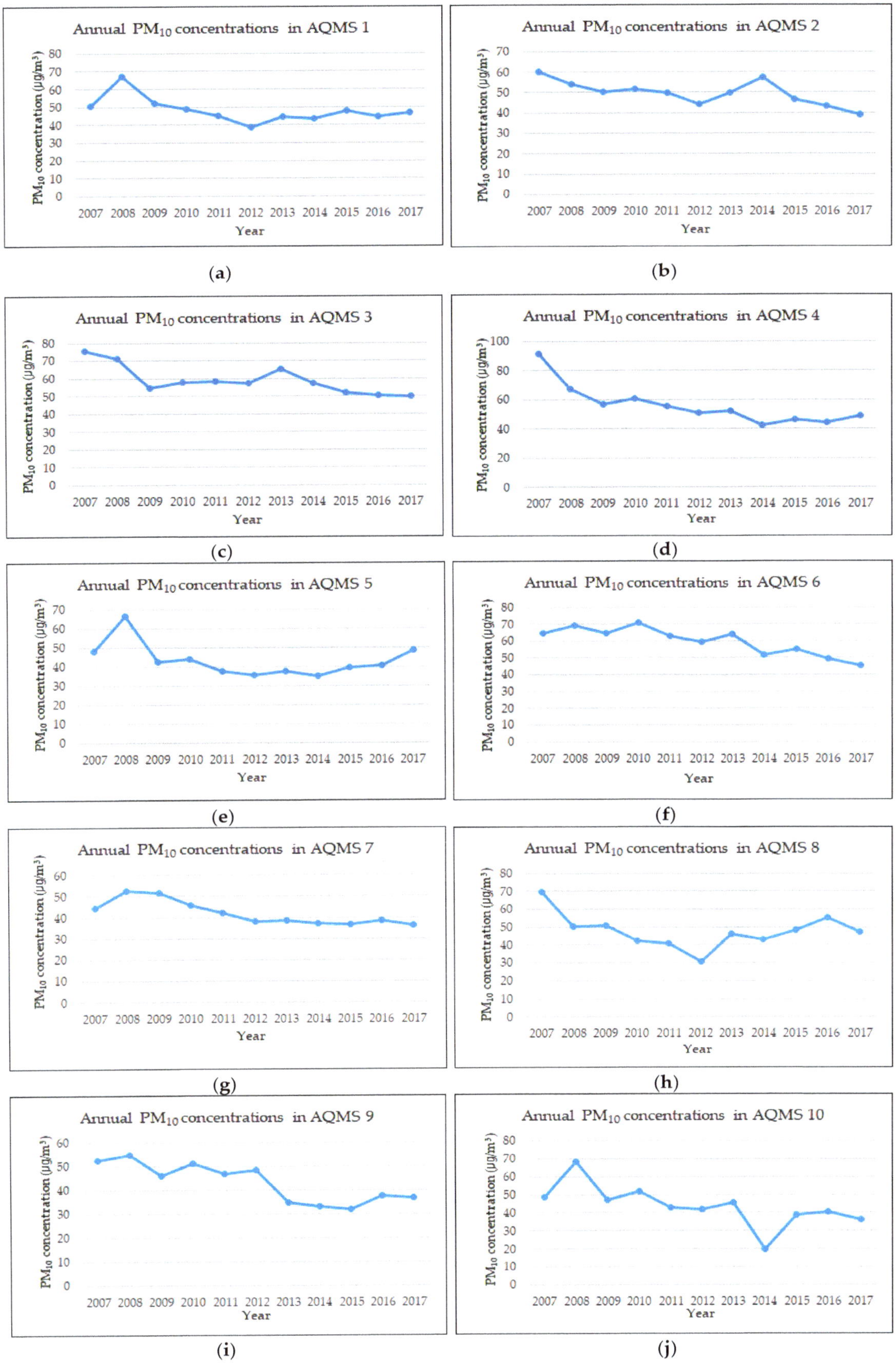

Figure 6. *Cont.*

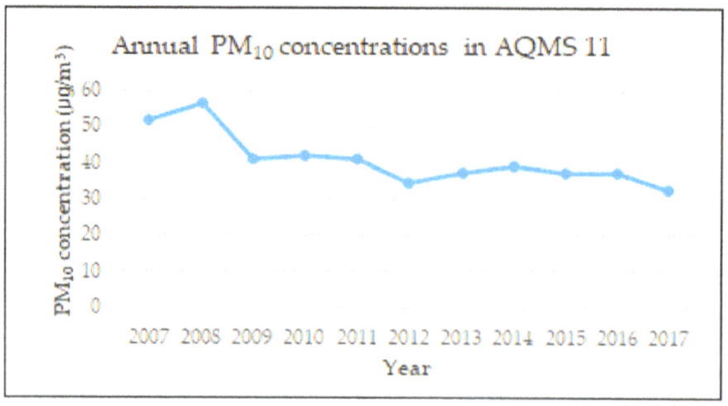

(**k**)

Figure 6. Time series graphs showing the annual PM_{10} concentrations in each AQMS in Daegu from 2007 to 2017. AQMS 1(**a**) to AQMS 11(**k**).

4. Results and Discussions

4.1. Accuracies of the Generated Urban Maps

In this section, the degrees of accuracy of the first and second urban maps separately generated from the first and second Landsat images, respectively, are assessed using the 100 checkpoints generated through manual digitization. Figure 7 shows examples of the checkpoints generated through manual digitization for measuring the degrees of accuracy of the first and second urban maps.

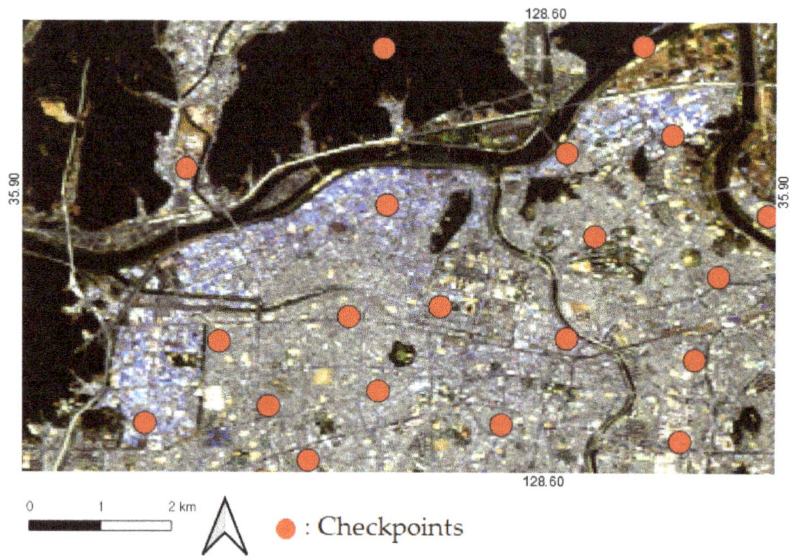

Figure 7. Examples of the checkpoints generated through manual digitization for measuring the degrees of accuracy of the first and second urban maps.

Table 1 shows the statistical results showing the degrees of accuracy of the first and second urban maps separately generated from the first and second Landsat images, respectively.

As can be seen in Table 1, the first and second urban maps separately generated from the first and second Landsat images, respectively, through the SVM technique had high accuracy in identifying the urban areas in the entire Daegu area. There were a few misclassification errors, however, in both urban maps because some urban features (e.g., man-made features) were misclassified as non-urban features (e.g., soil, water, vegetation), or vice versa. These misclassification errors were generally caused by the similar reflectance characteristics of the different materials owing to the shades, etc. [18].

Table 1. Statistical results showing the degrees of accuracy of the first and second urban maps separately generated from the first and second Landsat images, respectively: (**a**) degree of accuracy of the first urban map generated from the first Landsat image; and (**b**) degree of accuracy of the second urban map generated from the second Landsat image.

(a)			
Overall Accuracy		**97%**	
Producer's Accuracy (Error of Omission)		**User's Accuracy (Error of Commission)**	
Urban areas	94%	Urban areas	100%
Non-urban areas	100%	Non-urban areas	94%

(b)			
Overall Accuracy		**99%**	
Producer's Accuracy (Error of Omission)		**User's Accuracy (Error of Commission)**	
Urban areas	100%	Urban areas	98%
Non-urban areas	98%	Non-urban areas	100%

4.2. Relationship between the Urban Expansions and the PM_{10} Concentrations in Daegu from 2007 to 2017

Discussed in this section is the relationship between the urban expansion and the annual PM_{10} concentration rate in Daegu from 2007 to 2017 determined by calculating the following statistics: those showing the increase of the expanded urban areas in Daegu from 2007 to 2017 and those showing the annual PM_{10} concentration changes in each AQMS, each year and each season in Daegu from 2007 to 2017 (see Table 2).

As can be seen in Figure 6, Table 2a,b, the urban areas expanded by 55.27 km^2 in the entire Daegu area from 2007 to 2017, but the annual PM_{10} concentrations in all the AQMSs in Daegu only slightly increased by 0.45 $\mu g/m^3$ in only one station or sharply decreased by 4.12~42.28 $\mu g/m^3$ in the other 10 stations, within the same period. Table 2c shows that the annual PM_{10} concentrations had decreased in Daegu City from 2007 to 2017 by 17.37 $\mu g/m^3$.

Table 2b,c show that the annual PM_{10} concentrations were measured high in AQMSs 3, 4, and 6, while they were measured low in AQMSs 5, 7, 9, 10 and 11. Table 2c also shows that the highest annual PM_{10} concentrations was most frequently measured in AQMS 6 from 2007 to 2017 while the lowest annual PM_{10} concentrations most frequently measured in AQMS 11 during the same period. Based on Table 2b,c results, we assume that there are the number of other facilities (e.g., industrial factories) that emit the PM_{10} particles near AQMSs 3, 4, and 6, while there are few facilities emitting the PM_{10} particles near AQMS 5, 7, 9, 10, and 11.

Table 2c,d show that the annual PM_{10} concentrations generally measured high in spring and winter compared to summer and autumn. In addition, the annual PM_{10} concentrations decreased in all the seasons by 31.11 $\mu g/m^3$ for spring, by 8 $\mu g/m^3$ for summer, by 8.78 $\mu g/m^3$ for autumn, and by 18.77 $\mu g/m^3$ for winter. Based on Table 2d, we assume that the climate factors (e.g., air temperature, air pressure, rainfall, humidity, wind speed, and wind direction) can be significant for the annual PM_{10} concentrations.

The above experiment results show that there is a minimal or no relationship at all between the urban expansion and the PM_{10} concentrations rate in Daegu, which means that the urban expansion that occurred in Daegu from 2007 to 2017 was not the main cause of the rise in the PM_{10} concentration rate in Daegu, South Korea during the same period. We assume that, however, the types of facilities and climate factors can influence on the annual PM_{10} concentrations.

Table 2. Statistics showing the increase of the expanded urban areas in Daegu from 2007 to 2017 and the annual PM_{10} concentration changes in each air quality monitoring stations (AQMS) in Daegu within the same period: (**a**) statistics showing the total number of urbanized areas in Daegu from 2007 to 2017; (**b**) statistics showing the annual PM_{10} concentration changes in each AQMS in Daegu from 2007 to 2017; (**c**) statistics showing the range of variability of the annual PM_{10} concentrations in each year from 2007 to 2017; and (**d**) statistics showing the annual PM_{10} concentrations in each season (spring: March, April, and May; summer: June, July, and August; autumn: September, October, and November; and winter: December, January, and February) from 2007 to 2017.

(a)

Total Areas of the Urban Areas in the First Urban Map (km^2)	Total Areas of the Urban Areas in the Second Urban Map (km^2)	Increase of the Expanded Urban Areas in Daegu from 2007 to 2017 (km^2)
148.08	203.35	+ 55.27

(b)

AQMS ID	Maximum (μg/m^3)	Minimum (μg/m^3)	Average (μg/m^3)	Standard Deviation	Variation of Annual PM_{10} Concentration (2017 vs 2007) (μg/m^3)
AQMS 1	67.28	38.79	48.07	7.35	−4.12
AQMS 2	60.21	38.80	49.61	6.30	−21.41
AQMS 3	75.32	49.30	58.80	8.40	−26.02
AQMS 4	91.14	41.99	55.85	13.79	−42.28
AQMS 5	66.16	34.64	42.95	8.97	+0.45
AQMS 6	70.65	45.23	59.54	8.34	−19.29
AQMS 7	52.22	36.29	41.85	5.78	−7.88
AQMS 8	56.75	32.78	47.61	9.72	−22.75
AQMS 9	54.86	31.78	43.03	8.43	−15.59
AQMS 10	68.53	19.66	43.78	11.85	-12.81
AQMS 11	56.75	32.78	41.05	7.26	−19.35

(c)

Year	Maximum (μg/m^3)	AQMS ID for Maximum	Minimum (μg/m^3)	AQMS ID for Minimum	Average(μg/m^3)	Standard Deviation
2007	91.14	AQMS 4	44.17	AQMS 7	59.72	14.31
2008	71.04	AQMS 3	50.36	AQMS 8	61.58	7.83
2009	64.37	AQMS 6	41.21	AQMS 11	50.52	6.59
2010	70.65	AQMS 6	42.27	AQMS 11	50.51	8.66
2011	62.63	AQMS 6	37.21	AQMS 5	47.42	8.15
2012	59.51	AQMS 6	30.74	AQMS 8	43.61	9.44
2013	65.03	AQMS 3	34.63	AQMS 9	46.73	10.34
2014	57.22	AQMS 2	19.66	AQMS 10	41.63	11.02
2015	54.55	AQMS 6	31.78	AQMS 9	43.48	7.13
2016	54.98	AQMS 8	31.32	AQMS 11	43.6	5.71
2017	49.30	AQMS 3	32.78	AQMS 11	42.35	6.21

(d)

Year	Spring (μg/m^3)	Summer (μg/m^3)	Autumn (μg/m^3)	Winter (μg/m^3)
2007	81.58	41.55	48.47	64.32
2008	73.63	52.11	55.88	64.10
2009	51.74	42.37	45.51	62.87
2010	57.39	39.86	50.21	58.59
2011	59.77	37.59	42.20	50.59
2012	50.63	33.98	41.76	48.14
2013	52.92	38.99	39.90	55.02
2014	50.01	33.82	35.30	47.49
2015	50.51	34.59	31.14	57.61
2016	53.33	32.71	40.02	48.49
2017	50.47	33.55	39.69	45.55

5. Conclusions and Future Works

In this research, an experiment was performed to assess the relationship between the urban expansion and the PM_{10} concentration rate in Daegu from 2007 to 2017 by calculating the expanded urban areas and the annual PM_{10} concentration changes in each AQMS, each year and each season. The experiment results showed that there is a minimal or no relationship at all between the urban expansion that occurred in Daegu from 2007 to 2017 and the rise in the PM_{10} concentration rate in the same city during the same period because the urban areas significantly expanded but the annual PM_{10} concentrations sharply decreased.

This research proved that the urban expansion that occurred in Daegu from 2007 to 2017 was not the main cause of the rise in the PM_{10} concentration rate in Daegu during the same period. This research, however, also suggested that the other factors, such as types of facilities or climate factors, can be significant for the annual PM_{10} concentrations. To protect the public health, it is necessary to identify the main causes of PM_{10} concentrations, hence, further research, will be carried out to identify the main causes (e.g., the climate factors and the type of facilities) of PM_{10} concentrations in general. In addition, research will be conducted to identify the main cause of $PM_{2.5}$ concentrations that also causing serious air pollution problems.

Author Contributions: Y.-J.C. performed the experiments, analyzed the data and wrote the manuscript; J.-M.K. proposed the method, designed the experiments and modified the manuscript.

References

1. Merriam-Webster. Urban Sprawl. Available online: https://www.merriam-webster.com/dictionary/urban%20sprawl (assessed on 22 November 2018).
2. Encyclopedia Britanica. Urban Sprawl. Available online: https://www.britannica.com/topic/urban-sprawl (assessed on 22 November 2018).
3. Conserve Energy Future. What Is Urban Sprawl. Available online: https://www.conserve-energy-future.com/causes-and-effects-of-urban-sprawl.php (assessed on 22 November 2018).
4. Kang, D.; Kim, J.-E. Fine, Ultrafine and Yellow Dust: Emerging Health Problems in Korea. *J. Korean Med. Sci.* **2014**, *29*, 621–622. [CrossRef] [PubMed]
5. United States Environmental Protection Agency. Particulate Matter (PM) Basics. Available online: https://www.epa.gov/pm-pollution/particulate-matter-pm-basics (assessed on 22 November 2018).
6. Ghorani-Azam, A.; Riahi-Zanjani, B.; Balali-Mood, M. Effects of air pollution on human health and practical measures for prevision in Iran. *J. Res. Med. Sci.* **2016**, *21*, 65. [CrossRef] [PubMed]
7. World Health Organization (WHO) Working Group. *Health Aspects of Air Pollution with Particulate Matter, Ozone and Nitrogen Dioxide*; World Health Organization: Bonn, Germany, 2003.
8. Yin, H.; Xu, L.; Cai, Y. Monetary Valuation of PM_{10}—Related Health Risks in Beijing China: The Necessity for PM_{10} Pollution Indemnity. *Int. J. Environ. Res. Public Health* **2015**, *12*, 9967–9987. [CrossRef] [PubMed]
9. Pourahmad, A.; Baghavand, A.; Shahraki, S.Z.; Givehchi, S. The Impact of Urban Sprawl up on Air Pollution. *Int. J. Environ. Res.* **2007**, *1*, 347–353. [CrossRef]
10. Stone, B. Urban Sprawl and Air Quality in Large US Cities. *J. Environ. Manag.* **2008**, *86*, 688–698. [CrossRef] [PubMed]
11. Cho, H.S.; Choi, M.J. Effects of Compact Urban Development on Air Pollution: Empirical Evidence from Korea. *Sustainability* **2014**, *6*, 5968–5982. [CrossRef]
12. Liu, Y.; Wu, J.; Yu, D.; Ma, Q. The relationship between urban form and air pollution depends on seasonality and city size. *Environ. Sci. Pollut. Res.* **2018**, *25*, 15554–15567. [CrossRef] [PubMed]
13. Daegu Metropolitan City. The Document for the Urban Planning Policy of Daegu. Available online: http://ebook.daegu.go.kr/Viewer/RBL5N2AFGRNW (accessed on 22 November 2018).
14. Daegu Atmospheric Information System. PM10 Statistics. Available online: https://air.daegu.go.kr/open_content/ko/index.do (assessed on 22 November 2018).
15. United States Geological Survey (USGS). Landsat Missions. Available online: https://www.usgs.gov/land-resources/nli/landsat (assessed on 22 November 2018).
16. DicionaryCom. Machine Learning. Available online: https://www.dictionary.com/browse/machine-learning (assessed on 22 November 2018).

17. Choung, Y.-J.; Jo, M.-H. Comparison between a Machine-learning-based Method and a Water-index-based Method for Shoreline Mapping Using a High-Resolution Satellite Image Acquired in Hwado Island, South Korea. *J. Sens.* **2017**, *2017*, 8245204. [CrossRef]
18. Lary, D.J.; Alavi, A.H.; Gandomi, A.H.; Walker, A.L. Machine learning in geosciences and remote sensing. *Geosci. Front.* **2016**, *7*, 3–10. [CrossRef]
19. Mountrakis, G.; Im, J.; Ogole, C. Support vector machines in remote sensing: A review. *ISPRS J. Photogramm. Remote Sens.* **2011**, *66*, 247–259. [CrossRef]

5

Convolutional Neural Network-Based Remote Sensing Images Segmentation Method for Extracting Winter Wheat Spatial Distribution

Chengming Zhang [1,2,*], Shuai Gao [3,*], Xiaoxia Yang [1,2], Feng Li [4], Maorui Yue [5], Yingjuan Han [6], Hui Zhao [6], Ya'nan Zhang [1] and Keqi Fan [1]

1 College of Information Science and Engineering, Shandong Agricultural University, 61 Daizong Road, Taian 271000, China; yangxx@sdau.edu.cn (X.Y.); zyn980113@hotmail.com (Y.Z.); fkq980810@hotmail.com (K.F.)
2 Shandong Technology and Engineering Center for Digital Agriculture, 61 Daizong Road, Taian 271000, China
3 Chinese Academy of Sciences, Institute of Remote Sensing and Digital Earth, 9 Dengzhuangnan Road, Beijing 100094, China
4 Shandong Climate Center, Mountain Road, Jinan 250001, China; lifengsd@outlook.com
5 Taian Agriculture Bureau, Naihe Road, Taian 271000, China; yuemaoruita@outlook.com
6 Key Laboratory for Meteorological Disaster Monitoring and Early Warning and Risk Management of Characteristic Agriculture in Arid Regions, CMA, 71 Xinchangxi Road, Yinchuan 750002, China; yjhan_nx@outlook.com (Y.H.); zhaohui_cau@outlook.com (H.Z.)
* Correspondence: chming@sdau.edu.cn (C.Z.); gaoshuai@radi.ac.cn (S.G.)

Featured Application: In Gaofen-2 images, it is difficult to accurately extract winter wheat spatial distribution using traditional methods. Because our approach can better solve this problem, it has played an important role in agricultural surveys and improved the efficiency of agricultural surveys. Our approach has been utilized by the Department of Agriculture and the Meteorological Bureau of Shandong Province, China.

Abstract: When extracting winter wheat spatial distribution by using convolutional neural network (CNN) from Gaofen-2 (GF-2) remote sensing images, accurate identification of edge pixel is the key to improving the result accuracy. In this paper, an approach for extracting accurate winter wheat spatial distribution based on CNN is proposed. A hybrid structure convolutional neural network (HSCNN) was first constructed, which consists of two independent sub-networks of different depths. The deeper sub-network was used to extract the pixels present in the interior of the winter wheat field, whereas the shallower sub-network extracts the pixels at the edge of the field. The model was trained by classification-based learning and used in image segmentation for obtaining the distribution of winter wheat. Experiments were performed on 39 GF-2 images of Shandong province captured during 2017–2018, with SegNet and DeepLab as comparison models. As shown by the results, the average accuracy of SegNet, DeepLab, and HSCNN was 0.765, 0.853, and 0.912, respectively. HSCNN was equally as accurate as DeepLab and superior to SegNet for identifying interior pixels, and its identification of the edge pixels was significantly better than the two comparison models, which showed the superiority of HSCNN in the identification of winter wheat spatial distribution.

Keywords: remote sensing image segmentation; convolutional neural networks; Gaofen-2; hybrid structure convolutional neural networks; winter wheat spatial distribution; classification-based learning

1. Introduction

Winter wheat is the most important food crop in China, comprising 21.38% of the gross cropped area of the domestic food crops in 2017 according to the data released by the National Bureau of Statistics, with its output accounting for 21.00% of the total food crop production [1]. For national food security, the Chinese government has assigned a minimum area of arable land in each region that needs to be safeguarded (the "red line") [2]. Timely and accurate acquisition of the size and spatial distribution of winter wheat fields assists the relevant government departments in guiding the farming activities, estimating the yield, and adjusting the agricultural structure for ensuring food security [3].

Remote sensing is capable of imaging and large-area monitoring, making it a good data source for rapid and accurate extraction of winter wheat planting information. Researchers have successfully extracted winter wheat spatial distribution information from MODIS (moderate-resolution imaging spectroradiometer) and ETM/TM (enhanced thematic mapper plus/thematic mapper), achieving accuracies of 85.5% and 89.1%, respectively [4,5]. This exhibits the advantage of remote sensing in this application. However, owing to limitations in the spatial resolution of the data source, the spatial resolution of the extraction results is also rather coarse and unable to satisfy the requirement of the application [6–10]. With the development of high-resolution remote sensing satellites, a crop planting area can be monitored more accurately using the corresponding images as the data source [11,12]. The winter wheat cultivation information is extracted from the remote-sensing images captured by Gaofen-1 of the Gaofen series of Chinese satellites, yielding satisfactory results, with maximum accuracy reaching about 89% [13–18]. Most researchers still use traditional methods, such as decision trees and textures features. These methods can only take advantage of low-level features, which make it easy to make mistakes in identifying pixels at the edge of winter wheat planting area.

Image segmentation has been successfully used in the processing of camera images and applied by researchers to high-resolution remote sensing images, achieving significantly more accurate classification by a pixel-by-pixel segmentation [19–21]. Feature extraction is the key step in remote sensing image segmentation. In high-resolution remote sensing images, as the spectral difference between the same type of objects is increased, and between different types of objects is diminished, the former has more probability of exhibiting different spectral properties, whereas the latter tends to be spectrally similar, which makes feature extraction increasingly difficult [22,23]. Traditional methods including k-nearest neighbors and maximum entropy can only identify low-level image features such as color, shape, and texture. They are not capable of visually providing a semantic description. This hinders the extraction of higher-level features and limits the use of these methods in the segmentation of high-resolution remote sensing images [24,25].

With the development of machine learning, algorithms such as neural networks (NNs) [26] and support vector machine (SVM) [27,28] are being used in the segmentation of high-resolution images [29–31]. In some studies, when compared with traditional statistical methods and object-oriented methods, machine learning algorithms yielded better image segmentation results [32, 33]. Both SVM and NNs are shallow-learning algorithms [34–36], which do not express complex functions well owing to the limitations in their network structure. Therefore, these models cannot adapt to the continuously increasing complexity caused by the increasing sample size and diversity [37,38].

Progress in deep learning has facilitated solving these problems by using deep neural networks (DNNs) [39–42]. As an important branch of deep learning, a convolutional NN (CNN) is widely used with visual data because of its excellent feature learning ability [43–45]. A CNN is a deep learning network, composed of several layers, capable of nonlinear mapping. Its strength in learning is exemplified by the good image segmentation results achieved [46–52]. Further, the capacity of many large CNNs can be scaled according to the size of the training data and complexity and processing ability of the model, and their performance in image segmentation has improved significantly [53–60].

A fully convolutional network (FCN) is a deep learning network for image segmentation, which was proposed in 2015. Taking advantage of convolution computation in its feature organization and extraction abilities, an FCN realizes pixel-by-pixel segmentation of camera images by constructing a multi-layer convolutional structure and setting appropriate deconvolutional layers [61–63]. Accordingly, a series of convolution-based segmentation models has been developed including SegNet [64], UNet [65], DeepLab [66], multi-scale FCN [67], and ReSeg [68]. Of these models, SegNet and UNet are clearly structured, and it is easy to understand the convolution structure of the model. The processing speed is fast. DeepLab uses a method called "Atrous Convolution", which has a strong advantage in processing detailed images. multi-scale FCN is designed to address the huge scale gap between different classes of targets, i.e., sea/land and ships. ReSeg exploits the local generic features extracted by Convolutional Neural Networks and the capacity of Recurrent Neural Networks (RNN) to retrieve distant dependencies. Each model has its own strengths and is adept at dealing with certain image types.

In the work of extracting the spatial distribution of crops with high GF-1 as the data source, in addition to methods such as decision trees, textures features, and maximum entropy, research has also been carried out using deep learning. However, most of these studies directly use the existing deep learning model as a tool, and seldom consider the influence of characteristics difference of edge pixels and inner pixels in the crop planting area are large.

On board the Gaofen-2 satellite is a panchromatic camera with a spatial resolution of 1 m, and a multi-spectral camera with a spatial resolution of 4 m, which provides ideal data for extracting winter wheat plantation information. Before the application of a CNN to GF-2 remote-sensing images for this purpose, trial extraction is performed with classical network architectures (such as SegNet) where misidentified pixels are categorized, of which approximately 90% are found at the edge of the crop field. Further analysis indicates the structure of the convolutional layer as the source of this problem. The outcome produced by operating the convolution kernel in the pixel block is treated as the eigenvalue of the central pixel of the pixel block. As such, for the pixels at the edge, 50% of the pixels involved in each convolution are from negative samples, whereas, for the pixels at the corner, this number is 75% or higher. This results in a significant difference between the eigenvalues of the pixels at these locations and those at the center of the image, and an increase in the probability of the recognition results being placed in a wrong category. To avoid these problems, a new method is herein proposed for the extraction of the winter wheat field information from the GF-2 remote sensing images. The main procedures are as follows.

1. First, a CNN consisting of two independent sub-networks of different depths is established. The deep and shallow sub-networks are trained to be sensitive only to the pixels at the interior and edge of a winter wheat planting field, respectively, and only these pixels are extracted. This model is named as a Hybrid Structure Convolutional Neural Network (HSCNN).
2. A classification algorithm is adopted in the model training. For initial training of the sub-network used for the edge pixel extraction, edge pixels are considered as positive samples, with the pixels at other locations being treated as negative samples. The inner pixels are then designated as positive samples, with the pixels at other locations as negative samples, for training the sub-network used for the inner pixel extraction. After the successful completion of the training, the neural network is able to extract the winter wheat field from the GF-2 images accurately.
3. Finally, a GF-2 image is segmented by the trained model. Because SegNet and DeepLab are classic semantic segmentation models of images, and, the working principles of these two models are very similar to our work, we choose these two models as the comparison model, and a comparison is performed with them to evaluate the accuracy of the segmentation results.

2. Data Sources and Methods

2.1. Data Sources

2.1.1. Study Region

The whole study region is Shandong province, China. Shandong is located along the eastern coast of China (in the lower stream of the Yellow river), within 34°22′ N–38°24′ N and 114°47.5′ E–122°42′ E. It measures 721.03 km from east to west, and 437.28 km from north to south. The land area of the province is 155,800 km^2, of which 14.59% is mountainous, 5.56% is water (such as lakes), 15.98% is forest, and 53.82% is cultivated land. The annual total planting area of crops in the province is approximately 162 million mu. The main food crops of this region are wheat and maize. In 2016, the wheat planting area was 57.45405 million mu, and in 2017 it was 57.6435 million mu [69].

In this paper, we used the ground data and remote sensing data of Feicheng county, Ningyang county and Zhangqiu county, Shandong province. The three counties are similar in topography, all relatively flat, which can eliminate the influence of topographic fluctuations on the experimental results.

2.1.2. Ground-Based Data

For manufacturing sample to train our model, we conducted a field survey in Feicheng county, Ningyang county and Zhangqiu county in 2017 and 2018, and obtained the land use data of 369 sample points, among which 257 were winter wheat sample points and 112 were bare land. The survey results include the time, location and type of land use.

2.1.3. Remote Sensing Data

We selected 39 GF-2 remote sensing image, size of each image is 7300 × 6900. Of these images, 15 were captured on 17 February 2017, 11 were captured on 21 March 2018 and 13 were captured on 12 April 2018. We select images from different periods to increase the anti-interference abilities of the HSCNN. These remote sensing data cover Feicheng county, Ningyang county and Zhangqiu county, and are matched with ground investigation time. At the same time, the selected remote sensing data have fewer clouds and better clarity.

The Environment for Visualizing Images (ENVI) software was used for preprocessing the tasks, including fusion of panchromatic spectrum and multispectral band to obtain 1-m spatial resolution multispectral data, and the contrast stretch to generate a color-enhanced color composite image.

2.2. Network Architecture of Our Method

The HSCNN model is divided into five functional groups of components, input (a), inner-CNN (b), edge-CNN (f), vote function (j), and output (k), as shown in Figure 1. Both the edge-CNN and inner-CNN have convolution layers, an encoder layer, and a classifier layer. In the training stage, the inputs are original images and artificial classification labels. In the classification stage, the inputs are original GF-2 images, output is a single-band file, and content of each pixel in the output is the category number of the corresponding original image pixel. The HSCNN indicates the winter wheat area using category number 100, and category number 200 distinguishes other land use. The reason for adopting the two numbers is to fit with the coding value table we are working on to obtain detailed land use information.

2.2.1. Inner-Layers and Edge-Layers

The operational characteristics of the pixel block-based convolution for image segmentation are described in Section 1, in addition to the effect of the pixel block location on the convolution results. Based on this analysis, two convolution sub-structures of different depths are setup for the feature extraction of the winter wheat field. The deep convolution sub-network is used to extract the features of the pixels in the interior of the winter wheat plantation, shown as inner-layers (c) in Figure 1.

The shallower sub-network is used to extract the features of the pixels at the edge of the winter wheat plantation, shown as edge-layers (g) in Figure 1. The benefits of this design are discussed in Section 4 based on the experimental results.

In our approach, an inner pixel refers to the pixel that only contains winter wheat pixels in the pixel block when convolution operation is carried out with the pixel as the center pixel. An edge pixel refers to the pixel that contains winter wheat pixels and other pixels when computing the feature of the pixel.

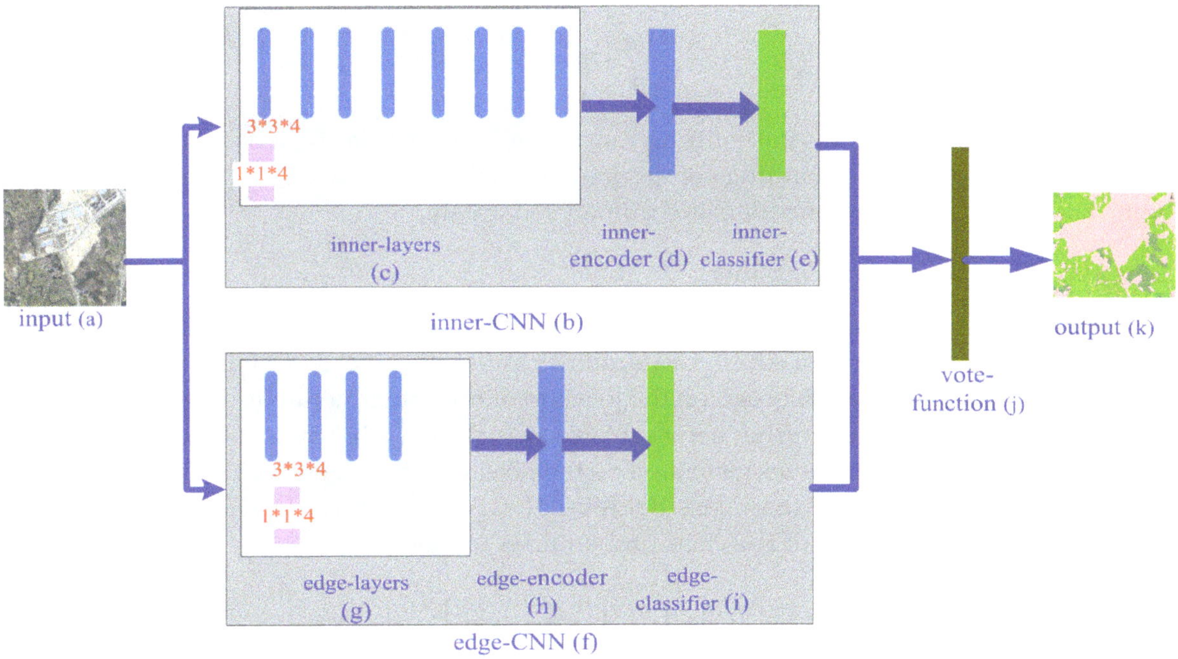

Figure 1. Network architecture of the Hybrid Structure Convolutional Neural Network (HSCNN): (**a**) input; (**b**) inner-CNN; (**c**) inner-layers; (**d**) inner-encoder; (**e**) inner-classifier; (**f**) edge-CNN; (**g**) edge-layers; (**h**) edge-encoder; (**i**) edge-classifier; (**j**) vote function; (**k**) output.

All kernels of the HSCNN take the form, $w \times h \times c$, where w is the width, h is the height, and c is the number of channels of a kernel. Two types of kernels are used in the first convolutional layers of inner-layers (c) and edge-layers (g). For one type w and h are set to 1, and for the other type the values are set to 3. In both cases, c is set to 4 because the data in the four multi-spectral bands of GF-2 are used. Kernels of the form $1 \times 1 \times 4$ are used to extract the features of the pixels. The generated feature map is used instantaneously as the input of the encoder, and does not participate in the subsequent convolution. Convolution kernels of the form $3 \times 3 \times 4$ are used to extract the spatial relation between the pixels and generate the spatial semantics by multi-level convolution.

After the operation of first convolution layer on the original image, we obtain a feature map which has only one channel. Because the input of convolution layer is the feature map calculated by the previous convolution layer, so the w and h values of the kernels used in all other convolutional layers are set to 3, and c is 1 from the second layer. To extract more features from the edge pixels of the crop field, the number of kernels used in each convolutional layer of edge-layers (g) is twice that used in the corresponding layer of inner-layers (c).

In the HSCNN, each convolution layer has only one activation layer attached, and there is no pool layer. Accordingly, the convolution result of each pixel block can be used directly as the feature of its central pixel, without the need to determine the position of the pixel that the feature corresponds to through deconvolution. As such, the HSCNN does not utilize a deconvolutional layer. This reduces the extent of computation and positioning error of the deconvolution, thereby improving the accuracy of the segmentation.

2.2.2. Inner-Encoder and Edge-Encoder

The inner-encoder and edge-encoder are used to encode the eigenvector extracted by the convolution layers on the pixel, ensuring that the classifier can establish the relationship between the eigenvector and pixel type. In the HSCNN model, the inner- and edge-encoders are both $2 \times n$ matrices, where n is the length of the eigenvector.

Let X denote the eigenvector of the pixel, W denote the encoder matrix, and R the encoded vector result. The encoding calculation is displayed in Equation (1).

$$\begin{bmatrix} r_1 \\ r_2 \end{bmatrix} = \begin{bmatrix} w_{11} & w_{12} & \cdots & w_{1n} \\ w_{21} & w_{22} & \cdots & w_{2n} \end{bmatrix} \times \begin{bmatrix} x_1 & x_2 & \cdots & x_n \end{bmatrix}^T + \begin{bmatrix} b_1 \\ b_2 \end{bmatrix} \quad (1)$$

where each row of matrix w represents a fitting function for a particular type of pixel, b_1 and b_2 are the respective biases, and the corresponding component of r is the encoded value of eigenvector x on that pixel type. The inner- and edge-encoders are trained separately.

2.2.3. Inner-Classifier and Edge-Classifier

For each pixel, the inner-classifier converts its vector of the encoded values given by the inner-encoder into a probability distribution over a set of classes, and classifies the pixel as an inner pixel or a non-inner pixel of the winter wheat plantation based on the location of the component with the highest probability. Similarly, the edge-classifier distinguishes between the edge and non-edge pixels of the winter wheat field using the vectors of the encoded values generated on the pixels by the edge-encoder.

In reference to the classic softmax classifier [60–65], Equation (2) is used here to convert vector r of the encoded values to vector p of the class probabilities for each pixel.

$$\begin{bmatrix} p_1 \\ p_2 \end{bmatrix} = \begin{bmatrix} \dfrac{e^{r_1}}{e^{r_1} + e^{r_2}} \\ \dfrac{e^{r_2}}{e^{r_1} + e^{r_2}} \end{bmatrix} \quad (2)$$

After the transformation, the index of max(p_1 and p_2) is taken as the predicted category of the pixel. For the inner-classifier, index numbers of 1 and 0 are assigned to the inner pixels of the winter wheat field and other pixels, respectively. Accordingly, index numbers of 1 and 0 are assigned to the edge pixels of the winter wheat field and other pixels, respectively.

2.2.4. Vote-Function

The vote-function determines the category number of a pixel given by the inner-classifier and edge-classifier and writes it to the output file. As described in the beginning of Section 2.2, the HSCNN indicates the winter wheat area using category number 100 and other land uses using category number 200. The category number of a pixel is calculated in Equation (3).

$$o = \begin{cases} 100 & p_{inner} = 1 \text{ or } p_{edge} = 1 \\ 200 & p_{inner} = 0 \text{ and } p_{edge} = 0 \end{cases} \quad (3)$$

where o represents the final category number of a pixel and p_{inner} and p_{edge} are the outputs of the inner-classifier and edge-classifier, respectively.

2.3. HSCNN Training

We manually labeled all images at the pixel level as ground truth (GT) label data. In other words, for each image, there exists a 7300×6900 label map, having a pixel-class (row-col indexed) correspondence with it. We used 36 images for training, and the remaining 3 images for testing. The GF-2 images and their corresponding artificial classification labels will be input to the HSCNN as training samples.

The training process includes error calculation, error back propagation and weight update. This process is iterated until the difference becomes smaller than the predetermined threshold.

We calculated the errors between the predicted classification label and manual classification label by the chain rule. The chain rule, the derivative rule in calculus, is used to find the derivative of a complex function, which is a common way to do the derivative calculation of calculus. The derivative of a composite function is the product of the derivatives of this finite number of functions at the corresponding point, as a chain. Then, the errors are back-propagated through the network. The backward propagation algorithm is a kind of training and learning method in deep learning, which can spread the error of the output layer backward to realize weight adjustment, adjust the weight between each node in the deep network, and achieve the goal that the sample tag output from the network is consistent with the actual tag. We use gradient descent method to update HSCNN parameters. Gradient descent method is the most commonly used optimization method. The idea is to use the negative gradient direction of the current position as the search direction, because that direction is the fastest descending direction of the current position.

2.3.1. Sample Labeling

We use the ENVI software for labeling and designing a preprocessor to build the labels. The process of artificial labeling is as follows:

1. The region-of-interest (RoI) tool in the ENVI software is used to select the winter wheat regions and other regions in the image. Then, the map locations of the pixels in each region are output to different files based on the category.
2. A band is added to the image file by the preprocessor as a mask band. The spatial resolution, size, and other parameters of the mask band are the same as the original image. Then, the category number of each pixel is written to the mask band according to the map location of the pixel previously output. We manually label all the images at the pixel level. Thus, for each image, there exists a 7300 × 6900 label map, with a row-column-indexed pixel-class correspondence.
3. The pixels marked as winter wheat are further categorized as edge pixels and inner pixels. Based on the parameters given above, the inner-layers comprise of eight convolutional layers, each with a 3 × 3 (length × width) convolution kernel. Therefore, the feature extraction from pixel s involves a 9 × 9 pixel block centered at s in the calculation. As defined in Section 2.2.1, the winter wheat pixels are divided sequentially into edge pixels and inner pixels. For training class by class, we use temporary code 160 to denote edge pixels and 170 to denote inner pixels in the mask band.

Figure 2 shows an example of an image-label pair.

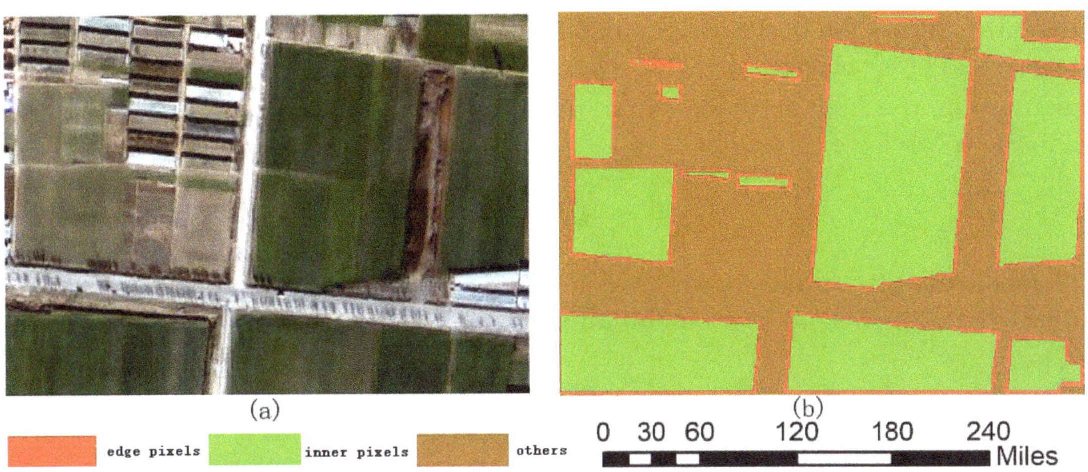

Figure 2. Image-label pair example: (**a**) original image; and (**b**) labels.

2.3.2. Loss Function

In our method, new loss functions are defined for the inner-CNN and edge-CNN, which still use the cross entropy as the basic element for the calculation, as expressed in Equation (4).

$$H(p,q) = -\sum_{i=1}^{2} q_i \log(p_i) \tag{4}$$

where p and q are, respectively the predicted and actual probability distribution, and i is index of a component in the probability distribution. On this basis, the loss function of the inner-CNN is defined as

$$\text{loss} = -\frac{1}{m}\sum_{m}\sum_{i=1}^{2} q_i \log(p_i) \tag{5}$$

When computing loss of inner-CNN, m is obtained by subtracting the number of edge pixels of the winter wheat field from the total number of samples, and when computing loss of edge-CNN, m is obtained by subtracting the number of inner pixels of the winter wheat field from the total number of samples.

2.3.3. Model Training

Images from two different periods were selected as the training data. We selected images from different periods for increasing the anti-interference abilities of the HSCNN and mitigating the complications such as the change in the seasons, and thus enhancing applicability. The training stage proceeded through the following steps:

1. Image-label pairs are input into the HSCNN as training samples. Network parameters are initialized.
2. Forward propagation is performed on the sample images.
3. The [loss]_inner is calculated and back propagated to the inner-CNN, whereas the [loss]_edge is calculated and back propagated to the edge-CNN.
4. The network parameters are updated using the stochastic gradient descent (SGD) [41,48] with momentum.
5. Steps (2)–(4) are iterated until both [loss]_inner and [loss]_edge are less than the predetermined threshold values.

The training yields two sub-networks, an inner-CNN and edge-CNN. The former can accurately extract the inner pixels of the winter wheat plantation from the sweet GF-2 remote sensing images, whereas the latter allows the best possible distinction between the edge pixels of the winter wheat planting region and other pixels.

In our training, the SGD method with momentum was used for parameter updates, which is illustrated in the following expression:

$$W^{(n+1)} = W^{(n)} - \Delta W^{(n+1)} \tag{6}$$

where $W^{(n)}$ denote the old parameters, $W^{(n+1)}$ denote the new parameters, and $\Delta W^{(n+1)}$ is the increment in the current iteration, which is a combination of the old parameters, gradient, and historical increment, i.e.,

$$\Delta W^{(n+1)} = \vartheta\left(d_w \cdot W^{(n)} + \frac{\partial J(W)}{\partial W^{(n)}}\right) + m \cdot \Delta W^{(n)} \tag{7}$$

where J(W) is the loss function, ϑ is the learning rate for step length control, d_w denotes the weight decay, and m denotes the momentum.

2.4. Segmentation Using the Trained Network

After successful training, the HSCNN can be used to segment an input imagery pixel-by-pixel. According to our design, the output is written in a new band. The benefit of this design avoids damaging the original file.

3. Experiments and Results

The data used in the experiment are presented in Section 2.1. In this section, the models used for comparison are described in Section 3.1, and the experimental results and assessment of accuracy are given in Section 3.2.

3.1. Comparison Model

Feature selection is the basis of remote sensing image segmentation. At present, there are mainly two methods based on artificial feature selection and machine learning. Haralick et al. (1973) put forward the method of gray level co-occurrence matrix, which is a classical artificial selection feature method. This method is mainly used to select image texture features. Since the texture is formed by repeated alternating changes of gray distribution in image space, so there is a certain gray-scale relationship between two separate pixels away certain distance, Haralick et al. described this correlation through a matrix [70]. Based on the artificial selection feature, only limited, shallow features can generally be selected. The feature selection based on machine learning can fully explore the deep feature and spatial semantic feature of the image. SegNet and DeepLab are classic semantic segmentation models of images, which have achieved very good results in the processing of images. Moreover, the working principles of these two models are very similar to our work, so we choose these two models as the comparison model, which can better reflect the advantages of our model in feature extraction. A comparative experiment was conducted using the methods established in the published literature.

3.1.1. SegNet

For the SegNet model, we directly employed the structure proposed by Badrinarayanan et al. [64], which consists of an encoder, a decoder, and a classifier. The encoder uses the first 13 convolutional layers of the VGG16 network, each having its corresponding decoder layer, totaling 13 decoder layers. The last decoder generates a multi-channel feature map as the input to the classifier, which outputs a probability vector of length K, where K is the number of classes. The final predicted category corresponds to the class having maximum probability at each pixel. In terms of training, SegNet can be trained end-to-end using SGD.

3.1.2. DeepLab

For DeepLab, we directly employed the DeepLab v3 model proposed by Liang-Chieh Chen et al. [66]. DeepLab was also developed based on the VGG network. To ensure that the output size would not be not too small without excessive padding, DeepLab changes the stride of the pool4 and pool5 layers of the VGG network from the original 2 to 1, plus 1 padding. To compensate for the effect of the stride change on the receptive field, DeepLab uses a convolution method called "atrous convolution" to ensure that the receptive field after pooling remains unchanged and the output is more refined. Finally, DeepLab incorporates a fully connected conditional random field (CRF) model to refine the segmentation boundary.

3.2. Results and Result Comparison

In the comparative experiment, we applied our trained model to three GF-2 images for segmentation. These images were only used for testing and not involved in training. Figure 3 illustrates the results obtained from the comparison methods and proposed method. In Figure 3, the first column illustrates the results of Experiment 1, the second column illustrates the results of Experiment 2 and the third column illustrates the results of Experiment 3.

Tables 1–3 are confusion matrices c for the segmentation results of SegNet model, DeepLab model, and HSCNN model, respectively. Each row of the confusion matrix represents the proportion taken by the actual category, and each column represents the proportion taken by the predicted category. As can be seen from the tables, our method achieves better classification results. In the example above, the proportion of "winter wheat" wrongly categorized as "background" is on average 0.069, and the proportion of "background" wrongly classified as "winter wheat" is on average 0.019, resulting in an overall accuracy of 91.2%.

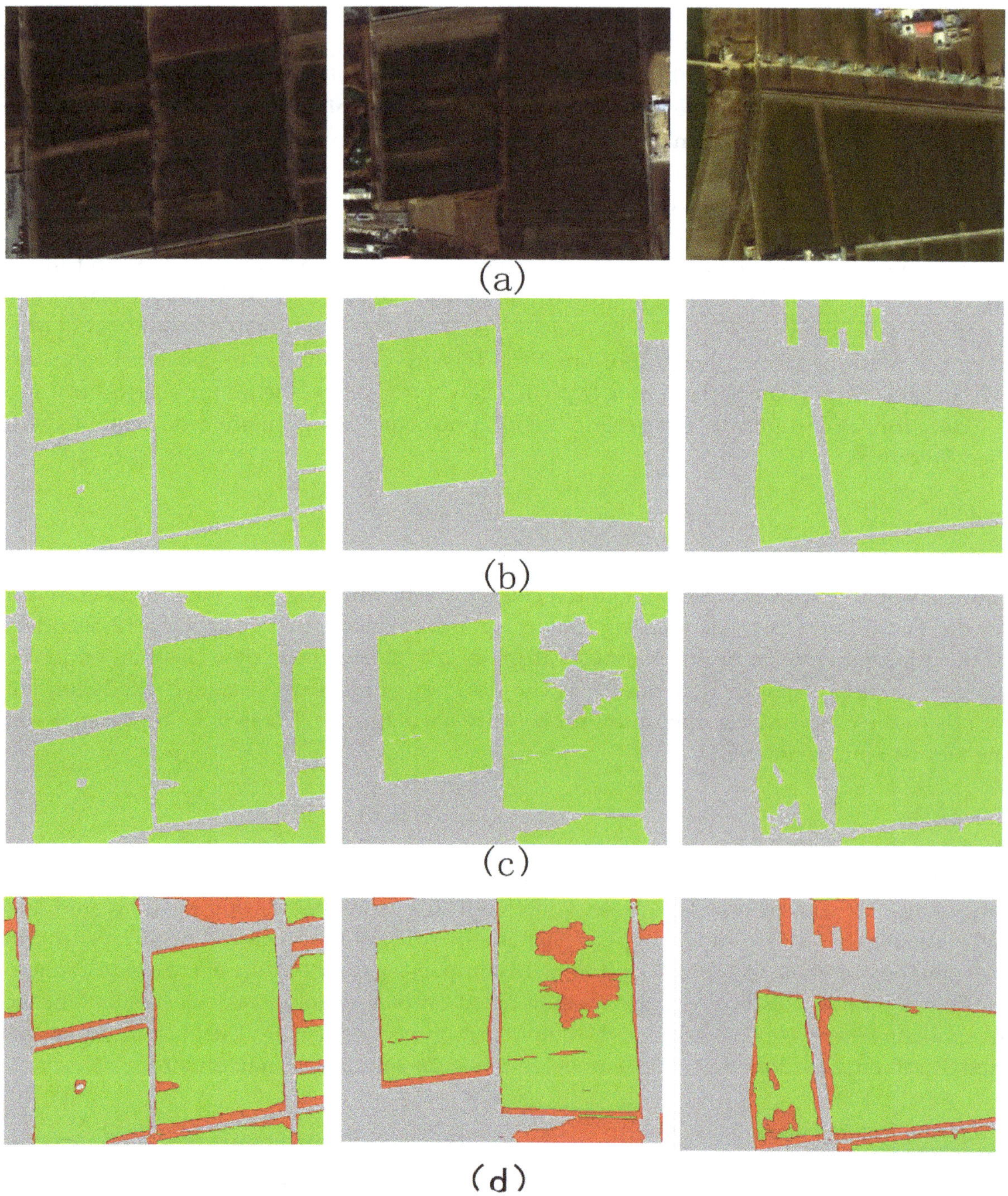

Figure 3. *Cont.*

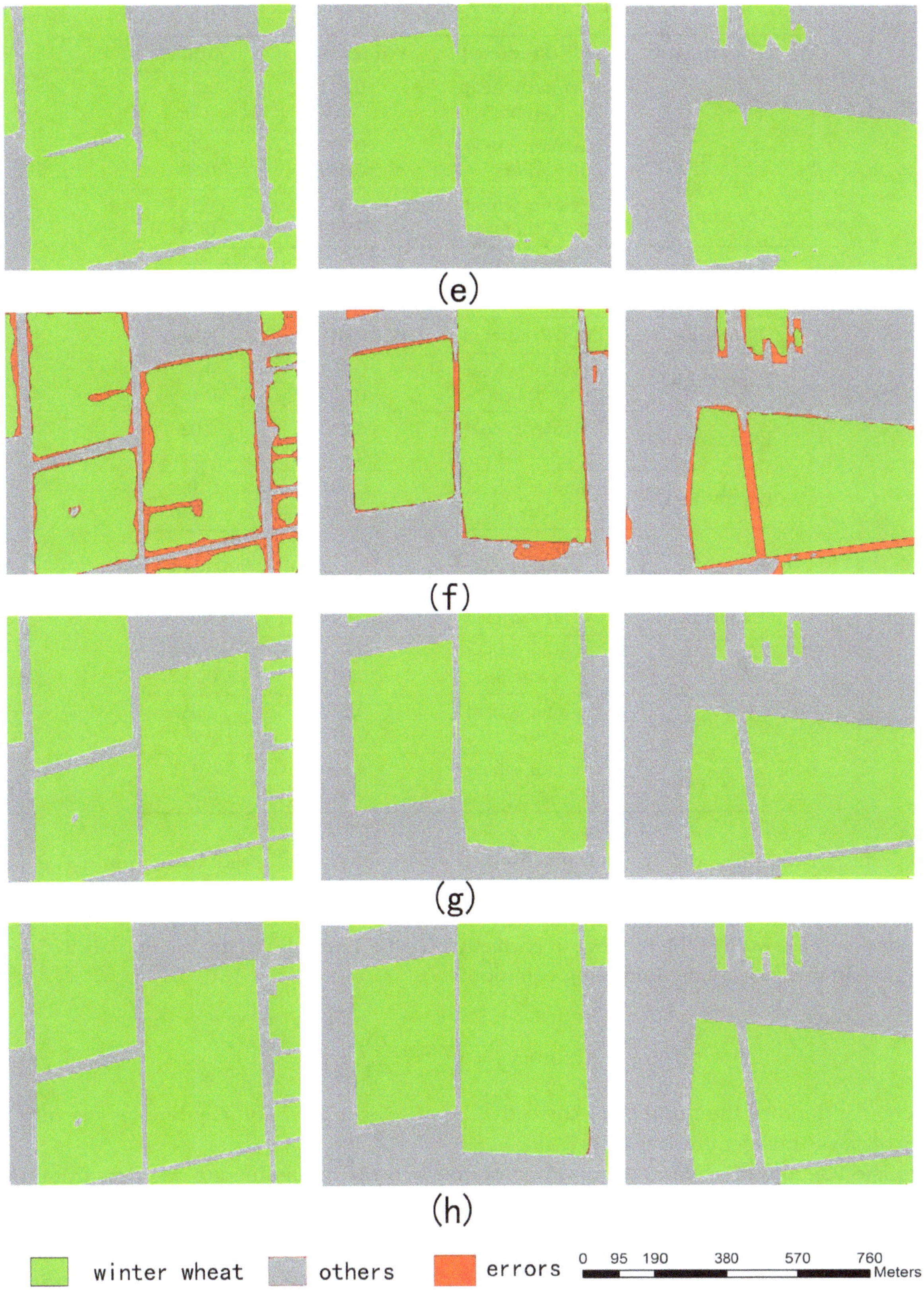

Figure 3. Segmentation results for Gaofen-2 (GF-2) images: (**a**) original images; (**b**) ground truth, (**c**) results of SegNet corresponding to the images in (**a**); (**d**) errors of SegNet; (**e**) results of DeepLab; (**f**) errors of DeepLab; (**g**) results of HSCNN; and (**h**) errors of HSCNN.

Table 1. Confusion matrix of the SegNet approach for Figure 4.

Experiment	GT/Predicted	Winter Wheat	Others
Experiment-1	winter wheat	0.621	0.129
	Others	0.087	0.163
Experiment-2	winter wheat	0.387	0.153
	Others	0.084	0.376
Experiment-3	winter wheat	0.217	0.123
	Others	0.129	0.531

Table 2. Confusion matrix of the DeepLab approach for Figure 4.

Experiment	GT/Predicted	Winter Wheat	Others
Experiment-1	winter wheat	0.653	0.107
	Others	0.055	0.185
Experiment-2	winter wheat	0.432	0.086
	Others	0.039	0.443
Experiment-3	winter wheat	0.301	0.108
	Others	0.045	0.546

Table 3. Confusion matrix of our HSCNN approach for Figure 4.

Experiment	GT/Predicted	Winter Wheat	Others
Experiment-1	winter wheat	0.681	0.075
	Others	0.027	0.217
Experiment-2	winter wheat	0.458	0.062
	Others	0.013	0.467
Experiment-3	winter wheat	0.329	0.071
	Others	0.017	0.583

Accuracy, precision, recall, and the Kappa coefficient were used to evaluate the models. These indices are calculated via mixing matrix c.

Accuracy is the ratio of the number of correctly classified samples to the total number of samples, and is given in this case by the following equation:

$$\text{Accuracy} = \frac{\sum_{i=1}^{2} C_{ii}}{\sum_{i=1}^{2}\sum_{j=1}^{2} C_{ij}} \tag{8}$$

Here, C_{ii} denotes the number of correctly classified samples, and C_{ij} denotes the number of samples of class i misidentified as class j.

Precision denotes the average proportion of pixels correctly classified to one class from the total retrieved pixels. Precision is calculated as:

$$\text{Precision} = \frac{1}{2}\sum_{i} C_{ii} / \sum_{j} C_{ij} \tag{9}$$

Recall represents the average proportion of pixels that are correctly classified in relation to the actual total pixels of a given class. Recall is computed as:

$$\text{Recall} = \frac{1}{2}\sum_{i} C_{ii} / \sum_{i} C_{ij} \tag{10}$$

The Kappa coefficient measures the consistency of the predicted classes with artificial labels. The Kappa coefficient is computed as:

$$\text{Kappa} = \frac{\frac{\sum_{i=1}^{2} C_{ii}}{\sum_{i=1}^{2}\sum_{j=1}^{2} C_{ij}} - \frac{\sum_{i=1}^{2} C_{ii}\sum_{j=1}^{2} C_{ij}}{(\sum_{i=1}^{2}\sum_{j=1}^{2} C_{ij})^2}}{1 - \frac{\sum_{i=1}^{2} C_{ii}\sum_{j=1}^{2} C_{ij}}{(\sum_{i=1}^{2}\sum_{j=1}^{2} C_{ij})^2}} \tag{11}$$

Equations (8)–(11) use the definitions given in Reference [18] and are modified according to our actual situation. The minimum accepted precision is 89% according to practical application.

The indicator values are listed in Table 4.

Table 4. Comparison of the approaches using SegNet, DeepLab, and HSCNN.

Approach	Index	Experiment-1	Experiment-2	Experiment-3	Average
SegNet	Accuracy	0.784	0.763	0.748	0.765
	Precision	0.740	0.767	0.721	0.743
	Recall	0.718	0.766	0.720	0.734
	Kappa	0.579	0.617	0.564	0.586
DeepLab	Accuracy	0.838	0.875	0.847	0.853
	Precision	0.815	0.877	0.830	0.840
	Recall	0.778	0.877	0.852	0.836
	Kappa	0.665	0.778	0.716	0.720
HSCNN	Accuracy	0.898	0.925	0.912	0.912
	Precision	0.895	0.927	0.897	0.906
	Recall	0.853	0.927	0.921	0.900
	Kappa	0.776	0.860	0.826	0.821

4. Analysis and Discussion

From the experimental results in Section 3, it is clear that our method significantly improves the accuracy of winter wheat extraction. In this section, the advantages of our model are discussed in terms of the differences between the remote sensing images and camera images. This is followed by more specific comparisons with SegNet and DeepLab. Finally, the role of our model in the classification of land uses by remote sensing is discussed briefly.

4.1. Advantages of the HSCNN Model

CNNs have achieved significant success in camera image segmentation, which has motivated researchers to apply them to remote sensing images. The HSCNN model proposed here is developed based on a previous work followed by a further in-depth analysis of the fundamental difference between camera images and remote sensing images. Thus, it possesses clear advantages compared with the traditional practice of the straightforward application of camera image segmentation model to remote sensing images.

Camera images and remote sensing images essentially differ in information representation. Owing to their advantages in shooting distance and the pixel quality of the camera, camera images are superior in terms of the rich details they contain, such that one object is formed by multiple pixels. Thus, the color of a pixel reflects the information at a certain point on an object but not the spatial relation between the pixels, which is found and expressed only by convolution. The nature of convolution is to represent the spatial correlation between the pixels by constructing a complex fitting function by operating on the pixel value of a pixel block. Particularly because it makes good use of the essential characteristics of camera images, deep convolution is extremely successful in camera image processing.

In remote sensing images, particularly for crop fields, a pixel generally contains multiple objects. For example, generally in GF-2 images 1 m^2 of ground is covered by a pixel, which contains 600–700 winter wheat plants. A pixel embodies the color information of the plants and the spatial information between them. However, at the edge of a winter wheat field, the region covered by a pixel is often a mixture of the winter wheat and bare land or winter wheat and other geographical objects, with varying percentages of winter wheat in the space. In this view, the information contained in a pixel at the edge region is significantly different from that at the interior. These two types of pixels can even be regarded as two different types of objects.

Based on the above analyses, the HSCNN network architecture is designed with a complete consideration of the properties of the remote sensing image and extraction target, and it makes good use of the characteristics of the winter wheat field captured in the GF-2 remote sensing images. The strengths of this model are exhibited in the following three aspects:

1. Considering the significant difference between the pixels of the interior and edge region of the winter wheat plantation (during extraction), these two regions are treated as two subclasses. Accordingly, the features of the inner pixels are more focused, which facilitates the model training. Two sub-networks with different depths are then designed with respect to the characteristics of the two subclasses. The deep sub-network extracts the pixels at the interior, whereas the shallower sub-network extracts those at the edge. This scheme reduces the effect of the non-winter wheat pixels on the features and improves the stability of the model for edge pixel extraction.
2. Two types of kernels are used in the first convolutional layer. The $1 \times 1 \times 4$ kernels are used to extract the feature of the pixels, and the $3 \times 3 \times 4$ kernels are used to extract the spatial relation between pixels. This design takes advantage of the ability of convolution for extracting higher-level spatial semantics and for obtaining the rich pixel information contained in the remote sensing images.
3. Our model does not utilize pooling, instead the convolution result is taken as the eigenvalue of the central pixel of the pixel block. In the application of the convolutional network to image classification, the basic target (sample) for the classification is the entire image. Thus, pooling can produce the main features of the feature map and reduce the amount of subsequent computation. Although the information on the accurate position of the features is lost during this process, their relative positions are nevertheless retained, which ensures the normal operation of the subsequent computational steps. However, in image segmentation, the basic target (sample) for the classification is an individual pixel, whose exact location must be mapped by the eigenvalue. Therefore, the major advantage of our model is its ability to preserve the spatial location of the eigenvalue, which makes it possible to remove the deconvolution adopted by the traditional FCN. Accordingly, the amount of computation is reduced. Further, the loss in precision due to positioning error is reduced, as the accurate position of the eigenvalue is kept.

4.2. Comparison with SegNet and Analysis

SegNet is founded on the FCN model. Its main strength lies in the search and extraction of the rich details of an image by deep convolution, and it is very distinct when extracting target objects with relatively few pixels. If the target objects contain only a few pixels or even one pixel, the deep convolution does not generate more details and may introduce more noise owing to the expanded field of view, affecting the determination of the pixel type.

In the remote sensing images of GF-2, the edge and interior of the winter wheat plantation are very different in composition, which makes it more difficult for SegNet to locate the common features, because of its structure containing a single convolutional network. In comparison, the HSCNN is equipped with two sub-networks of different depths and is adaptable to the characteristics of the edge and interior. It also uses two different sizes of kernel, which are capable of uncovering the spatial relation between the pixels, and the information embedded in the pixels.

As shown in Figure 3, the segmentation results of HSCNN and SegNet are nearly identical for the interior of the winter wheat field. SegNet, however, produces prominent errors at the edge of the field, while HSCNN does not.

Both HSCNN and SegNet use classifiers to generate the probability distribution of the classes, and consider the class with the maximum probability (max) in the distribution as the type to which the pixel belongs. Clearly, a larger difference between the max and background implies a higher separability of the pixels and more reliable results. The probability differences given by the HSCNN and SegNet model for the inner wheat and edge wheat classes are presented in Figures 4 and 5, respectively. It is clear in Figure 4 that HSCNN and SegNet lead to significant probability differences for many pixels in the interior, which demonstrates the high separability of this region and the strength of CNN. In the probability distribution in Figure 5, fewer pixels are noted as having large probability differences in both the HSCNN and SegNet; nevertheless, the number is maintained at a quite high level for the HSCNN, whereas SegNet exhibits a reduced performance.

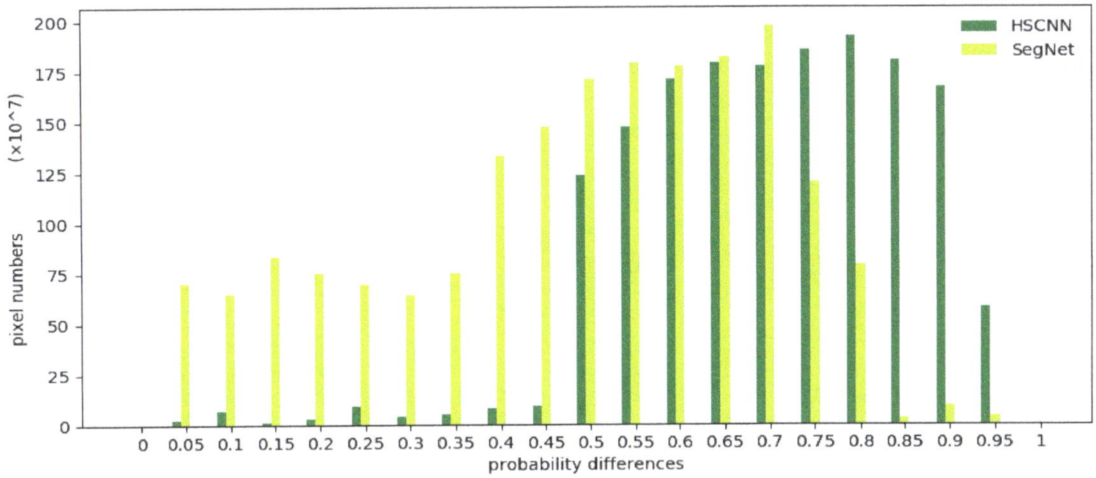

Figure 4. Distribution of the probability differences for the inner wheat pixels.

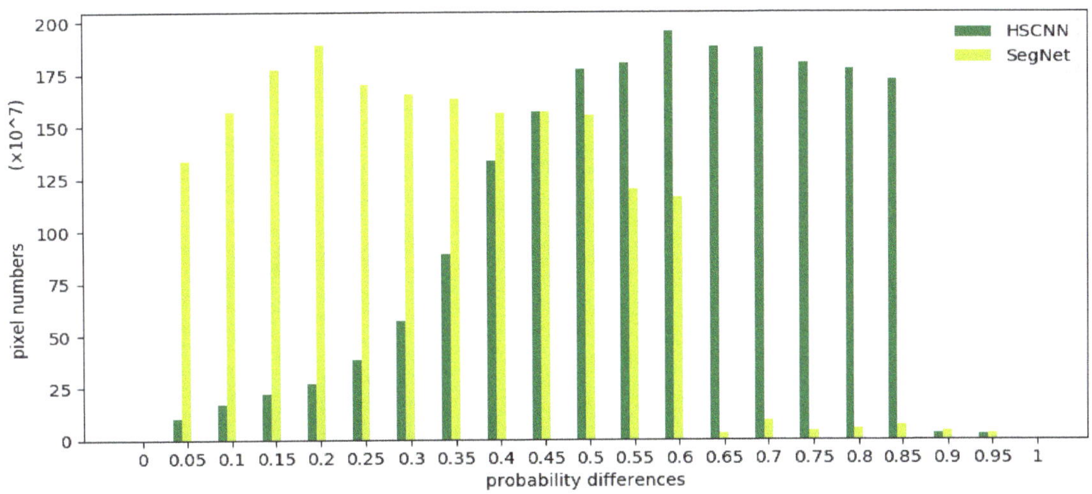

Figure 5. Distribution of the probability differences for the edge wheat pixels.

4.3. Comparison with DeepLab and Analysis

Compared with the FCN and SegNet, DeepLab has significant improvements in two aspects: (1) the deconvolution; and (2) the refinement of the boundary of the segmentation result by fully connected CRFs. These two improvements are beneficial for the segmentation of individual objects covering numerous pixels. Based on the published literature, DeepLab displays a higher segmentation accuracy at the boundary than the FCN and SegNet, because it better utilizes the detailed information contained in the image and the large-scale spatial correlation between the pixels. However, in its application to winter wheat identification, the strength of DeepLab is not fully realized, because the details within a pixel block of the winter wheat plantation do not change significantly. Therefore, less information is available to the model, and the spatial correlation within the farmlands and woods is not strong over large regions.

As mentioned in Section 4.2, the HSCNN completely utilizes the characteristics of the pixels and the spatial relation between them. Therefore, it is well adapted to the data characteristics of the winter wheat plantation. Further, it effectively avoids the deficiencies of DeepLab and ensures the accuracy of segmentation.

As in Section 4.2, the probability differences between the HSCNN and DeepLab models in the inner wheat and edge wheat class are displayed in Figures 6 and 7, respectively. It is clear in Figure 6 that both the models produce large probability differences for many pixels in the interior. In the probability distribution of Figure 7, a considerable number of pixels still display large probability differences after the HSCNN processing, whereas DeepLab shows a much poorer performance (even lower than SegNet), proving again the notion that the atrous convolution is not suitable for farmlands.

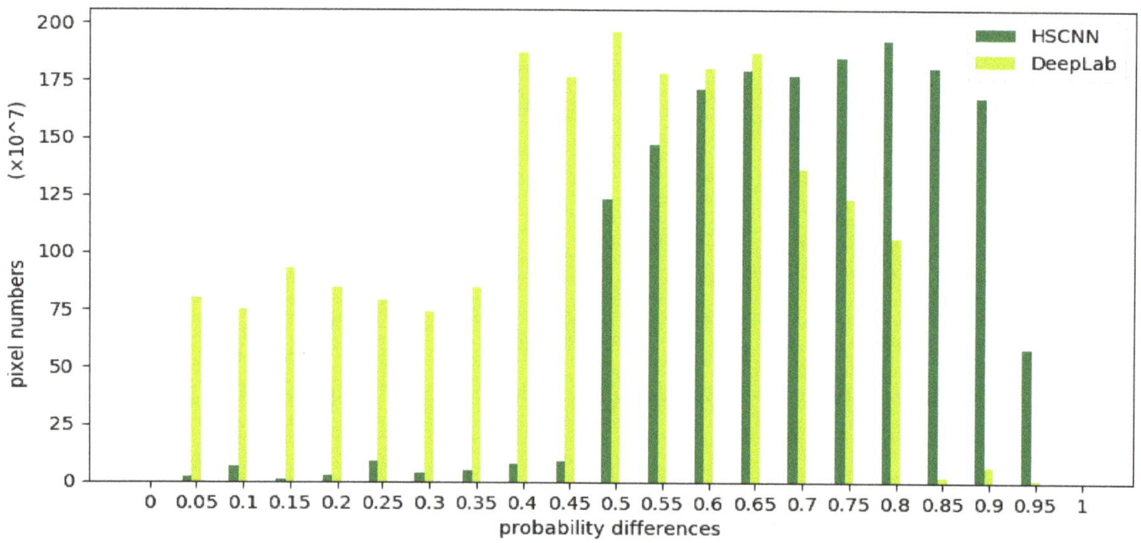

Figure 6. Distribution of the probability differences for the inner wheat pixels.

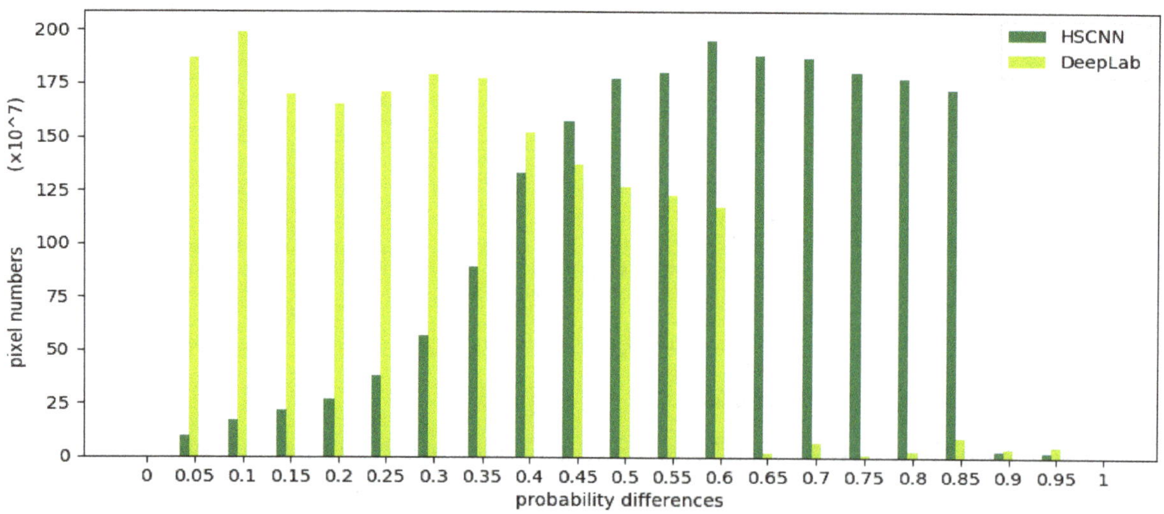

Figure 7. Distribution of the probability differences for the edge wheat pixels.

4.4. *Benefits of Using the Proposed Approach to Classify Land Use*

Accurate land use classification is of tremendous importance in scientific research and agriculture with the use of remote sensing data as an increasingly common practice for this purpose. Based on a CNN and taking complete advantage of the convolution in feature extraction, the design of the CNN architecture adapting to the features of the remote-sensing images is the key in land use classification by this method.

We have taken the feature difference between the edge pixel and the inner pixel in the white wheat planting area into full consideration, this significantly improve the extraction accuracy of the edge pixel. Compared with earlier research, the model presented in this paper has the following advantages.

Firstly, two types of kernels were used in the convolution of the model, which allowed the full utilization of the strength of the convolution in the extraction of spatial semantics and made appropriate use of the rich information contained in the pixels of the remote sensing images, thus achieving a more accurate segmentation.

Secondly, pooling layers were not used in the model. Although the speed of the feature aggregation was consequently reduced, the information of the exact location to which an eigenvalue corresponds was retained, thereby effectively mitigating the loss in the accuracy due to the positioning error of the deconvolution and improving the overall effect of the segmentation.

The model presented in this paper provides a solution for the edge extraction problem or the segmentation of the winter wheat plantation using GF-2 images. It has an important role to play and enhances the efficiency of the agricultural survey. This model has been utilized by the Department of Agriculture and the Meteorological Bureau of Shandong Province, China.

5. Conclusions

This paper presents a novel approach for the extraction of the winter wheat distribution from GF-2 remote-sensing images. Compared with the two typical deep learning-based approaches, the extraction accuracy is obviously improved. Our approach combines the segmentation and classification stages, taking the accuracy as the only constraint, and achieves high quality classification in an end-to-end way. The GT classes of ground objects are taken as the supervised information that guides both the feature extraction and the region generation. Taking into account the significant differences between pixel and edge pixel in the planting area, different convolution structures were used to extract the feature of edge and interior pixels, focusing on the common features in the two subclasses for more effective model training, and obtained a high resolution class prediction.

Our model is still limited in many aspects, and further improvements could be made in the following two areas: (1) The current encoder uses a relatively simple regression algorithm for encoding; thus, a regression that can express the complex relationship between the eigenvalues needs to be explored. (2) A new pooling method, which allows for expedited feature aggregation without the loss of the spatial information of the eigenvalues, should be established. We will continue our work in the future to improve the current model and obtain better classification performance.

Author Contributions: C.Z. wrote the manuscript; C.Z. and S.G. presented the direction of this study and designed the experiments; X.Y. and F.L. carried out the experiments and analyzed the results; M.Y. and Y.Z. carried out ground investigation and preprocessing; and Y.H., H.Z. and K.F. carried out sample labeling.

References

1. Announcement of the National Statistics Bureau on Grain Output in 2017. Available online: http://www.gov.cn/xinwen/2017-12/08/content_5245284.htm (accessed on 8 December 2017).
2. Wang, L.M.; Liu, J.; Yao, B.M.; Ji, F.H.; Yang, F.G. Area change monitoring of winter wheat based on relationship analysis of GF-1 NDVI among different years. *Trans. CSAE* **2018**, *34*, 184–191. [CrossRef]
3. He, H.; Zhu, X.F.; Pan, Y.Z.; Zhu, W.Q.; Zhang, J.S.; Jia, B. Study on scale issues in measurement of winter wheat plant area by remote sensing. *J. Remote Sens.* **2008**, *1*, 168–176.

4. Zhang, J.H.; Feng, L.L.; Yao, F.M. Improved maize cultivated area estimation over a large scale combining MODIS-EVI time series data and crop phonological information. *ISPRS J. Photogramm. Remote Sens.* **2014**, *94*, 102–113. [CrossRef]
5. Wu, M.Q.; Wang, C.Y.; Niu, Z. Mapping paddy field in large areas, based on time series multi-sensors data. *Trans. CSAE* **2010**, *26*, 240–244.
6. Xu, Q.Y.; Yang, G.J.; Long, H.L.; Wang, C.C.; Li, X.C.; Huang, D.C. Crop information identification based on MODIS NDVI time-series data. *Trans. CSAE* **2014**, *30*, 134–144.
7. Becker-Reshef, I.; Vermote, E.; Lindeman, M.; Justice, C. A′generalized regression-based model for forecasting winter wheat yields in Kansas and Ukraine using MODIS data. *Remote Sens. Environ.* **2010**, *114*, 1312–1323. [CrossRef]
8. Zhang, J.G.; Li, X.W.; Wu, Y.L. Object oriented estimation of winter wheat planting area using remote sensing data. *Trans. CSAE* **2008**, *24*. [CrossRef]
9. Zhu, C.M.; Luo, J.C.; Shen, Z.F.; Chen, X. Winter wheat planting area extraction using multi-temporal remote sensing data based on filed parcel characteristic. *Trans. CSAE* **2011**, *27*, 94–99.
10. Lu, L.L.; Guo, H.D. Extraction method of winter wheat phenology from time series of SPOT/VEGETATION data. *Trans. CSAE* **2009**, *25*, 174–179.
11. Jha, A.; Nain, A.S.; Ranjan, R. Wheat acreage estimation using remote sensing in tarai region of Uttarakhand. *Vegetos* **2013**, *26*, 105–111. [CrossRef]
12. Wu, M.Q.; Yang, L.C.; Yu, B.; Wang, Y.; Zhao, X.; Niu, Z.; Wang, C.Y. Mapping crops acreages based on remote sensing and sampling investigation by multivariate probability proportional to size. *Trans. CSAE* **2014**, *30*, 146–152.
13. You, J.; Pei, Z.Y.; Wang, F.; Wu, Q.; Guo, L. Area extraction of winter wheat at county scale based on modified multivariate texture and GF-1 satellite images. *Trans. CSAE* **2016**, *32*, 131–139. [CrossRef]
14. Wang, L.M.; Liu, J.; Yang, F.G.; Fu, C.H.; Teng, F.; Gao, J. Early recognition of winter wheat area based on GF-1 satellite. *Trans. CSAE* **2015**, *31*, 194–201. [CrossRef]
15. Ma, S.J.; Yi, X.S.; You, J.; Guo, L.; Lou, J. Winter wheat cultivated area estimation and implementation evaluation of grain direct subsidy policy based on GF-1 imagery. *Trans. CSAE* **2016**, *32*, 169–174. [CrossRef]
16. Wang, L.M.; Liu, J.; Yang, L.B.; Yang, F.G.; Teng, F.; Wang, X.L. Remote sensing monitoring winter wheat area based on weighted NDVI index. *Trans. CSAE* **2016**, *32*, 127–135. [CrossRef]
17. Wu, M.Q.; Huang, W.J.; Niu, Z.; Wang, Y.; Wang, C.Y.; Li, W.; Hao, P.Y.; Yu, B. Fine crop mapping by combining high spectral and high spatial resolution remote sensing data in complex heterogeneous areas. *Comput. Electron. Agric.* **2017**, *139*, 1–9. [CrossRef]
18. Fu, G.; Liu, C.J.; Zhou, R.; Sun, T.; Zhang, Q.J. Classification for high resolution remote sensing imagery using a fully convolutional network. *Remote Sens.* **2017**, *9*, 498. [CrossRef]
19. Liu, Y.D.; Cui, R.X. Segmentation of Winter Wheat Canopy Image Based on Visual Spectral and Random Forest Algorithm. *Spectrosc. Spect. Anal.* **2015**, *35*, 3480–3485.
20. Dong, Z.P.; Wang, M.; Li, D.R. A High Resolution Remote Sensing Image Segmentation Method by Combining Superpixels with Minimun Spanning Tree. *Acta Geod. Cartogr. Sin.* **2017**, *46*, 734–742. [CrossRef]
21. Basaeed, E.; Bhaskar, H.; Al-Mualla, M. Supervised remote sensing image segmentation using boosted convolutional neural networks. *Knowl.-Based Syst.* **2016**, *99*, 19–27. [CrossRef]
22. Liu, D.W.; Han, L.; Han, X.Y. High Spatial Resolution Remote Sensing Image Classification Based on Deep Learning. *Acta Opt. Sin.* **2016**, *36*, 0428001. [CrossRef]
23. Luo, B.; Zhang, L.P. Robust autodual morphological profiles for the classification of high-resolution satellite images. *IEEE Trans. Geosci. Remote* **2014**, *52*, 1451–1462. [CrossRef]
24. Li, D.R.; Zhang, L.P.; Xia, G.S. Automatic Analysis and Mining of Remote Sensing Big Data. *Acta Geod. Cartogr. Sin.* **2014**, *43*, 1211–1216. [CrossRef]
25. Chan, T.H.; Jia, K.; Guo, S.H.; Lu, J.; Zeng, Z.; Ma, Y. PCANet: A Simple Deep Learniing Baseline for Image Classification. *IEEE Trans. Image Process.* **2015**, *24*, 5017–5032. [CrossRef] [PubMed]
26. Mas, J.F.; Flores, J.J. The application of artificial neural networks to the analysis of remotely sensed data. *Int. J. Remote Sens.* **2008**, *29*, 617–663. [CrossRef]

27. Gustavo, C.V.; Bruzzone, L. Kernel-based methods for hyperspectral image classification. *IEEE Trans. Geosci. Remote Sens.* **2005**, *43*, 1351–1362.
28. Mountrakis, G.; Im, J.; Ogole, C. Support vector machines in remote sensing: A review. *ISPRS J. Photogramm. Remote Sens.* **2011**, *66*, 247–259. [CrossRef]
29. Pacifici, F.; Chini, M.; Emery, W.J. A neural network approach using multi-scale textural metrics from very high-resolution panchromatic imagery for urban land-use classification. *Remote Sens. Environ.* **2009**, *113*, 1276–1292. [CrossRef]
30. Huang, X.; Zhang, L. An SVM ensemble approach combining spectral, structural, and semantic features for the classification of high resolution remotely sensed imagery. *IEEE Trans. Geosci. Remote Sens.* **2013**, *51*, 257–272. [CrossRef]
31. Liu, C.; Hong, L.; Chen, J.; Chun, S.S.; Deng, M. Fusion of pixel-based and multi-scale region-based features for the classification of high-resolution remote sensing image. *J. Remote Sens.* **2015**, *19*, 228–239. [CrossRef]
32. Yuan, Y.; Lin, J.; Wang, Q. Hyperspectral Image Classification via Multitask Joint Sparse Representation and Stepwise MRF Optimization. *IEEE Trans. Cybern.* **2016**, *46*, 2966–2977. [CrossRef] [PubMed]
33. Xie, F.D.; Li, F.F.; Lei, C.K.; Ke, L.N. Representative Band Selection for Hyperspectral Image Classification. *ISPRS Int. J. Geo-Inf.* **2018**, *7*, 338. [CrossRef]
34. Bengio, Y. Learning deep architectures for AI. *Found. Trends Mach. Learn.* **2009**, *2*, 1–127. [CrossRef]
35. Larochelle, H.; Bengio, Y.; Louradour, J.; Lamblin, P. Exploring strategies for training deep neural networks. *J. Mach. Learn. Res.* **2009**, *10*, 1–40.
36. Jones, N. The learning machines. *Nature* **2014**, *505*, 146–148. [CrossRef] [PubMed]
37. Gao, Q.S.; Lim, S.S.; Jia, X.P. Hyperspectral Image Classification Using Convolutional Neural Networks and Multiple Feature Learning. *Remote Sens.* **2018**, *10*, 299. [CrossRef]
38. Dong, Y.; Liu, Y.N.; Lian, S.G. Automatic age estimation based on deep learning algorithm. *Neurocomputing* **2016**, *187*, 4–10. [CrossRef]
39. Taormina, R.; Chau, K.W. Data-driven input variable selection for rainfall–runoff modeling using binary-coded particle swarm optimization and Extreme Learning Machines. *J. Hydrol.* **2015**, *529*, 1617–1632. [CrossRef]
40. Liang, Z.; Shan, S.; Liu, X.; Wen, Y. Fuzzy prediction of AWJ turbulence characteristics by using typical multi-phase flow models. *Eng. Appl. Comput. Fluid Mech.* **2017**, *11*, 225–257. [CrossRef]
41. Bellary, S.A.I.; Adhav, R.; Siddique, M.H.; Chon, B.H.; Kenyery, F.; Samad, A. Application of computational fluid dynamics and surrogate-coupled evolutionary computing to enhance centrifugal-pump performance. *Eng. Appl. Comput. Fluid Mech.* **2016**, *10*, 171–181. [CrossRef]
42. Zhang, J.; Chau, K.W. Multilayer Ensemble Pruning via Novel Multi-sub-swarm Particle Swarm Optimization. *J. Univ. Comput. Sci.* **2009**, *15*, 840–858.
43. Wang, W.C.; Chau, K.W.; Xu, D.M.; Chen, X.Y. Improving forecasting accuracy of annual runoff time series using ARIMA based on EEMD decomposition. *Water Resour. Manag.* **2015**, *29*, 2655–2675. [CrossRef]
44. Zhang, S.; Chau, K.W. Dimension reduction using semi-supervised locally linear embedding for vegetation leaf classification. *Emerg. Intell. Comput. Technol. Appl.* **2009**, *5754*, 948–955.
45. Wu, C.; Chau, K.; Fan, C. Prediction of rainfall time series using modular artificial neural networks coupled with data-preprocessing techniques. *J. Hydrol.* **2010**, *389*, 146–167. [CrossRef]
46. Castelluccio, M.; Poggi, G.; Sansone, C.; Verdoliva, L. Land Use Classification in Remote Sensing Images by Convolutional Neural Networks. *arXiv*, 2015; arXiv:1508.00092.
47. Hu, F.; Xia, G.S.; Hu, J.; Zhang, L. Transferring deep convolutional neural networks for the scene classification of high-resolution remote sensing imagery. *Remote Sens.* **2015**, *7*, 14680–14707. [CrossRef]
48. Zhu, C.; Cheng, T. Research on geological hazard identification based on deep learning. In Proceedings of the 6th International Conference on Computer-Aided Design, Manufacturing, Modeling and Simulation, Busan, Korea, 14–15 April 2018. [CrossRef]
49. Wu, Z.; Zhang, Q. On combining spectral, textural and shape features for remote sensing image segmentation. *Acta Geod. Cartogr. Sin.* **2013**, *42*, 44–50.
50. Noh, H.; Hong, S.; Han, B. Learning deconvolution network for semantic segmentation. In Proceedings of the IEEE International Conference on Computer Vision, Santiago, Chile, 7–13 December 2015; pp. 1520–1528.

51. Paisitkriangkrai, S.; Sherrah, J.; Janney, P.; Hengel, V.D. Effective semantic pixel labelling with convolutional networks and conditional random fields. In Proceedings of the IEEE Conference on Computer Vision and Pattern Recognition Workshops, Boston, MA, USA, 7–12 June 2015; pp. 36–43.
52. Papandreou, G.; Kokkinos, I.; Savalle, P.A. Untangling local and global deformations in deep convolutional networks for image classification and sliding window detection. *arXiv*, 2014; arXiv:1412.0296.
53. Badrinarayanan, V.; Handa, A.; Cipolla, R. SegNet: A deep convolutional encoder-decoder architecture for robust semantic pixel-wise labelling. *arXiv*, 2015; arXiv:1505.07293.
54. Ioffe, S.; Szegedy, C. Batch normalization: Accelerating deep network training by reducing internal covariate shift. *arXiv*, 2015; arXiv:1502.03167.
55. Liu, J.; Liu, B.; Lu, H. Detection guided deconvolutional network for hierarchical feature learning. *Pattern Recognit.* **2015**, *48*, 2645–2655. [CrossRef]
56. Volpi, M.; Ferrari, V. Semantic segmentation of urban scenes by learning local class interactions. In Proceedings of the IEEE Conference on Computer Vision and Pattern Recognition Workshops, Boston, MA, USA, 7–12 June 2015; pp. 1–9.
57. Simonyan, K.; Zisserman, A. Very deep convolutional networks for large-scale image recognition. *arXiv*, 2014; arXiv:1409.1556.
58. Mittal, A.; Hooda, R.; Sofat, S. LF-SegNet: A Fully Convolutional Encoder–Decoder Network for Segmenting Lung Fields from Chest Radiographs. *Wirel. Pers. Commun.* **2018**, *101*, 511–529. [CrossRef]
59. Kendall, A.; Badrinarayanan, V.; Cipolla, R. Bayesian SegNet: Model uncertainty in deep convolutional encoder-decoder architectures for scene understanding. *arXiv*, 2015; arXiv:1511.02680.
60. Chen, L.C.; Papandreou, G.; Kokkinos, I.; Murphy, K.; Yuille, A.L. Semantic Image Segmentation with Deep Convolutional Nets and Fully Connected Crfs. *arXiv*, 2014; arXiv:1412.7062.
61. Long, J.; Shelhamer, E.; Darrell, T. Fully convolutional networks for semantic segmentation. In Proceedings of the IEEE Conference on Computer Vision and Pattern Recognition, Boston, MA, USA, 7–12 June 2015; pp. 3431–3440.
62. Längkvist, M.; Kiselev, A.; Alirezaie, M.; Loutfi, A. Classification and segmentation of satellite orthoimagery using convolutional neural networks. *Remote Sens.* **2016**, *8*, 239. [CrossRef]
63. Dolz, J.; Desrosiers, C.; Ben, A. 3D fully convolutional networks for subcortical segmentation in MRI: A large-scale study. *NeuroImage* **2018**, *170*, 456–470. [CrossRef] [PubMed]
64. Badrinarayanan, V.; Kendall, A.; Cipolla, R. SegNet: A deep convolutional encoderdecoderarchitecture for image segmentation. *IEEE Trans. Pattern Anal. Mach. Intell.* **2017**, *39*, 2481–2495. [CrossRef] [PubMed]
65. Ronneberger, O.; Fischer, P.; Brox, T. U-Net: Convolutional Networks for Biomedical Image Segmentation. *arXiv*, 2015; arXiv:1505.04597.
66. Chen, L.C.; Papandreou, G.; Kokkinos, I.; Murphy, K.; Yuille, A.L. Deeplab: Semantic image segmentation with deep convolutional nets, atrous convolution, and fully connected crfs. *arXiv*, 2016; arXiv:1606.00915.
67. Lin, H.N.; Shi, Z.W.; Zou, Z.X. Maritime Semantic Labeling of Optical Remote Sensing Images with Multi-Scale Fully Convolutional Network. *Remote Sens.* **2017**, *9*, 480–501. [CrossRef]
68. Visin, F.; Ciccone, M.; Romero, A.; Kastner, K.; Cho, K.; Bengio, Y.; Matteucci, M.; Courville, A. Reseg: A recurrent neural network-based model for semantic segmentation. *arXiv*, 2016; arXiv:1511.07053.
69. Statistical Yearbook of Shandong Province. Available online: http://www.stats-sd.gov.cn/col/col6279/index.html (accessed on 27 October 2017).
70. Haralick, R.M.; Shanmugam, K. Texture features for image classification. *IEEE Trans. Syst. Man Cybern.* **1973**, *SMC-3*, 610–621. [CrossRef]

6

Comparison of Machine Learning Regression Algorithms for Cotton Leaf Area Index Retrieval using Sentinel-2 Spectral Bands

Huihui Mao [1,2], Jihua Meng [1,*], Fujiang Ji [1,2], Qiankun Zhang [1,3] and Huiting Fang [1,2]

1 Key Laboratory of Digital Earth Sciences, Institute of Remote Sensing and Digital Earth, Chinese Academy of Sciences, Beijing 100101, China; maohh@radi.ac.cn (H.M.); jifj@radi.ac.cn (F.J.); zhang_q_k@163.com (Q.Z.); fanght@yeah.net (H.F.)
2 University of Chinese Academy of Sciences, Beijing 100049, China
3 College of Geomatics and Geoinformation, Guilin University of Technology, Guilin 541004, China
* Correspondence: mengjh@radi.ac.cn

Abstract: Leaf area index (LAI) is a crucial crop biophysical parameter that has been widely used in a variety of fields. Five state-of-the-art machine learning regression algorithms (MLRAs), namely, artificial neural network (ANN), support vector regression (SVR), Gaussian process regression (GPR), random forest (RF) and gradient boosting regression tree (GBRT), have been used in the retrieval of cotton LAI with Sentinel-2 spectral bands. The performances of the five machine learning models are compared for better applications of MLRAs in remote sensing, since challenging problems remain in the selection of MLRAs for crop LAI retrieval, as well as the decision as to the optimal number for the training sample size and spectral bands to different MLRAs. A comprehensive evaluation was employed with respect to model accuracy, computational efficiency, sensitivity to training sample size and sensitivity to spectral bands. We conducted the comparison of five MLRAs in an agricultural area of Northwest China over three cotton seasons with the corresponding field campaigns for modeling and validation. Results show that the GBRT model outperforms the other models with respect to model accuracy in average ($\overline{R^2}$ = 0.854, $\overline{RMSE}$ = 0.674 and $\overline{MAE}$ = 0.456). SVR achieves the best performance in computational efficiency, which means it is fast to train, and to validate that it has great potentials to deliver near-real-time operational products for crop management. As for sensitivity to training sample size, GBRT behaves as the most robust model, and provides the best model accuracy on the average among the variations of training sample size, compared with other models ($\overline{R^2}$ = 0.884, $\overline{RMSE}$ = 0.615 and $\overline{MAE}$ = 0.452). Spectral bands sensitivity analysis with dCor (distance correlation), combined with the backward elimination approach, indicates that SVR, GPR and RF provide relatively robust performance to the spectral bands, while ANN outperforms the other models in terms of model accuracy on the average among the reduction of spectral bands ($\overline{R^2}$ = 0.881, $\overline{RMSE}$ = 0.625 and $\overline{MAE}$ = 0.480). A comprehensive evaluation indicates that GBRT is an appealing alternative for cotton LAI retrieval, except for its computational efficiency. Despite the different performance of the ML models, all models exhibited considerable potential for cotton LAI retrieval, which could offer accurate crop parameters information timely and accurately for crop fields management and agricultural production decisions.

Keywords: leaf area index (LAI); machine learning; Sentinel-2; sensitivity analysis; training sample size; spectral bands

---

1. Introduction

Leaf area index (LAI), which characterizes the structure and functioning of vegetation, is usually defined as half of the total green leaf area per unit horizontal ground surface area [1,2]. LAI is one of the most important vegetation biophysical parameters, and a key variable for climate modeling, evapotranspiration modeling and crop modeling, and it is recognized as an Essential Climate Variable (ECV) by the Global Climate Observing System [3–8]. LAI has a wide range of applications regarding agricultural fields, and it has been demonstrated to be an essential indicator for crop growth monitoring and key variables for crop yield forecasting [9–11]. Therefore, it is of special relevance to retrieve LAI in a timely and accurate manner.

Remote sensing techniques provide promising alternatives to obtaining crop biophysical parameters by high temporally and spatially continuous means over large areas. To date, there are mainly three categories of methods developed to retrieve LAI based on optical remote sensing data, which are statistical methods, physically based methods, and hybrid methods [12–14]. Statistical methods can be further divided into parametric and non-parametric regression methods. Parametric regression methods usually consist of an explicit relationship between biophysical parameters and vegetation indices, while non-parametric regression methods define regression models learnt from the training dataset [15]. Non-parametric algorithms can be split into linear and non-linear regression methods; the latter is also commonly referred to as machine learning regression algorithms (MLRAs). While physically-based methods are applications of physical laws establishing cause-effect relationships, a hybrid method combines elements of non-parametric statistics and physically-based methods [13], whereas these two methods are both sophisticated models that demand a large number of parameters, which are usually difficult to obtain in practice. Empirical parametric models typically make use of a limited number of spectral bands [13,16]. However, nonparametric models can make full use of spectral information, and directly learn the input-output relationships from a given training dataset, which makes these models attractive alternatives for crop LAI retrieval.

With the development of remote sensing, more and more optical remote sensing satellites have been launched (e.g., Landsat 8, Sentinel-2, and Chinese GF-1, GF-2 and newly launched GF-6), which ensures the availability of high spatial, high temporal resolution satellite remote sensing data, and correspondingly, high dimensional (spatial, temporal and spectral) of remote sensing data amounts to large data volume, which also poses great challenges for more efficient, robust and accurate algorithms in a wide variety of applications with remote sensing.

Recently, machine learning (ML), a broad subfield of artificial intelligence, has attracted considerable attention in remote sensing applications for classification and regression problems, and encouraging results have been obtained [17–23]. With advances in computer technology and associated techniques, ML has drawn tremendous interest in a variety of fields to address complex problems. ML can be broadly defined as computational methods using experience to improve performance or to make accurate predictions [24]. ML has been extensively applied to biophysical parameter retrievals due to the ability to accurately approximate robust relationships between input-output data, which provides tremendous opportunities for remote sensing-based applications. Considering ML regression algorithms, a more efficient, robust, and accurate model for crop LAI retrieval should be established. Despite the considerable advances in ML for remote sensing applications, challenging problems remain in the selection of MLRAs for crop LAI retrieval among the variety of ML algorithms available, as well as the optimal number of training sample size and spectral bands to different MLRAs.

As for the versatile ML algorithms, artificial neural network (ANN), support vector regression (SVR), Gaussian process regression (GPR) and random forest (RF) are reportedly effective for crop LAI retrieval [25–28]. However, gradient boosting regression tree (GBRT), a highly robust ML algorithm for a wide range of applications, is capable of achieving high levels of accuracy for regression problems [29–31] and to our knowledge, has not been investigated for LAI retrieval. Further studies should be conducted to assess the performance of the GBRT model in crop LAI retrievals.

Many studies have been dedicated to crop LAI retrieval using MLRAs. However, there are a limited number of academic studies involving comparisons of different MLRAs for crop LAI retrievals using remote sensing. Apparently, none of these studies have focused on multispectral remote sensing data, and none of these studies have evaluated the different impact factor together, to conduct a comprehensive comparison.

In addition, the validation of global LAI products are important procedures to ensure the application of the products in a wide range of fields [32]. Regional high-resolution LAI maps can be used as a reference LAI map to validate the global LAI products, which calls for efficient, robust and accurate algorithms for LAI retrieval.

The objective of this study is to compare the performance of five advanced MLRAs (ANN, SVR, GPR, RF and GBRT) for cotton LAI retrieval in a relatively comprehensive manner. We conducted the study over the entire growth period of cotton using Sentinel-2 spectral bands and corresponding ground data. Specifically, the following research questions are addressed:

(1) Which of the five MLRAs perform best with regard to model accuracy?
(2) Which is the fastest model during the training and validation processes?
(3) How does the number of training sample size influence the performance of the five MLRAs?
(4) How does the number of spectral bands influence the performance of the five MLRAs?
(5) Which is the best model in consideration of model accuracy, computational efficiency, sensitivity to training sample size, and sensitivity to spectral bands together?
(6) How accurate are the global LAI products in Northwest China?

2. Materials

2.1. Study Area and Field Campaigns

China is one of the largest cotton producers and importers in the world, and Xinjiang is the primary cotton-growing region in China [33,34]. The chosen study area was on a large agriculture region (6118.08 km^2 or 1,511,810 acres) in Shihezi (44°37′ N, 85°42′ E), Xinjiang Province, China. The region is located in a temperate continental climate zone. The average field size in the study area is 73139.7 m^2 (18.07 acres), and the majority of fields have a flat topography, which is preferable for decametric remote sensing applications (e.g., Sentinel-2). The annual mean temperature of the study area is 7.39 °C, the annual total precipitation is 206 mm, and the average altitude is 450.8 m.

The field campaigns were conducted in early June 2018, mid-July 2018, and mid-August 2018 (Table 1), with 117 total quadrats obtained. Each quadrat was assigned one leaf area index (LAI) value, obtained as the average leaf area index (LAI) of the three sample points that matches the corresponding Sentinel-2 pixel (Table 2). Each sample point datum was collected using an LAI-2200C Plant Canopy Analyzer (Li-Cor, Inc., Lincoln, NE, USA). The main planted crops in the study area are cotton, grape, spring maize and winter wheat, and cotton holds the largest planting proportion, which is 71.7%. The cotton season is from mid-April to early October. The spring maize season is from mid-April to late September. The winter wheat season is from late September of the previous year to late June. Notably, the study area has completely achieved mechanization for agricultural production and management. Figure 1 shows the location of the study area and field observation sites with Sentinel-2A imagery (20 August 2018). A descriptive statistic of the measured cotton LAI at three observation dates is shown in Figure 2.

Table 1. Sentinel-2 imagery and the corresponding field campaign data.

Satellite	Date (dd/mm/yy)	Field Campaign Date	Quadrats
Sentinel-2B	6 June 2018	1–3 June 2018	23
Sentinel-2A	11 July 2018	8–11 July 2018	58
Sentinel-2A	20 August 2018	18–21 August 2018	36

Table 2. Sentinel-2 satellite imagery spectral band characteristics [35].

Band	Central Wavelength (nm)	Bandwidth (nm)	Spatial Resolution (m)	Purpose
B1	443	20	60	Atmospheric correction (aerosol scattering).
B2 [1]	490	65	10	Sensitive to vegetation senescing, carotenoid, browning and soil background; atmospheric correction (aerosol scattering).
B3	560	35	10	Green peak, sensitive to total chlorophyll in vegetation.
B4	665	30	10	Maximum chlorophyll absorption.
B5	705	15	20	Position of red edge; consolidation of atmospheric corrections/fluorescence baseline.
B6	740	15	20	Position of red edge; atmospheric correction, retrieval of aerosol load.
B7	783	20	20	Leaf Area Index (LAI), edge of the Near Infrared plateau.
B8	842	115	10	LAI.
B8A	865	20	20	NIR plateau, sensitive to total chlorophyll, biomass, LAI and protein; water vapor absorption reference; retrieval of aerosol load and type.
B9	945	20	60	Water vapor absorption, atmospheric correction.
B10	1380	30	60	Detection of thin cirrus for atmospheric correction.
B11	1610	90	20	Sensitive to lignin, starch and forest above ground biomass; snow/ice/cloud separation.
B12	2190	180	20	Assessment of Mediterranean vegetation conditions; distinction of clay soils for the monitoring of soil erosion; distinction between live biomass, dead biomass and soil, e.g., for burn scars mapping.

[1] The spectral bands in bold are ones used in this study.

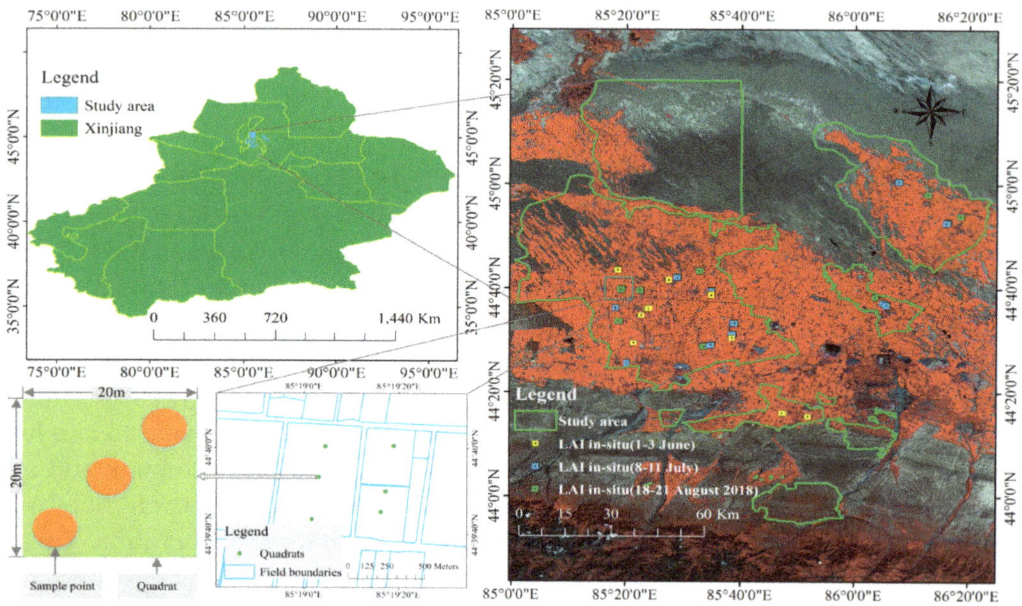

Figure 1. Location of study area and in-situ leaf area index (LAI) quadrats from three field campaigns. The background was the Sentinel-2A image acquired on August 20, 2018 and was shown in a false color band composition of R (8) G (4) B (3), with standard deviation stretch (**right**). The red color in the Sentinel image represents the vegetation (mainly crops and a few trees), the grey color represents the desert and bare land (**right**). The green dots represent the quadrats (including three sample points), and the blue polygons represent agricultural field boundaries (**lower left**).

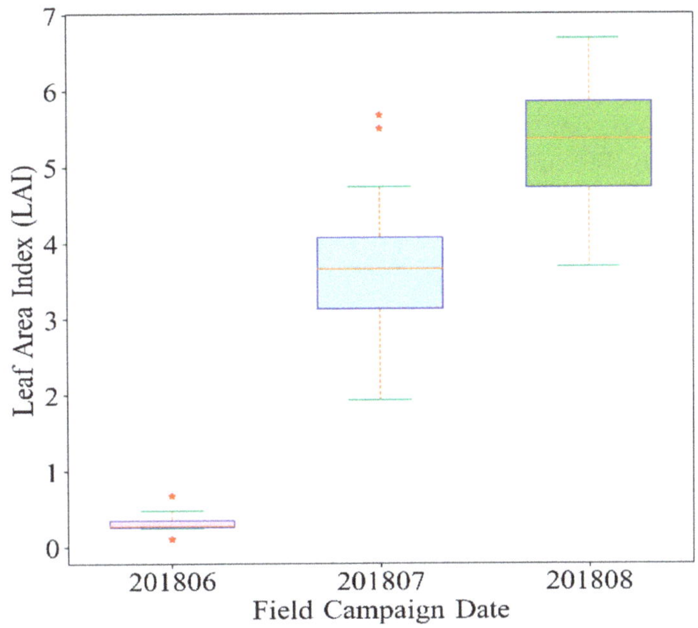

Figure 2. Field campaign cotton LAI descriptive statistics on three observation dates, shown with boxplots. The box extends from the lower to upper quantile values of the field observation data, with a line at the median, the whiskers extend from the box to show the range of the data. Flier points are those past the end of the whiskers, named as outliers (**red dots**).

2.2. Sentinel-2 Data and Preprocessing

The remote sensing data used in this study were Sentinel-2 imagery. Sentinel-2 is a constellation of satellites, Sentinel-2A and Sentinel-2B, which were launched by the European Space Agency (ESA) on June 2015 and March 2017, respectively. Each satellite carries a MultiSpectral Instrument (MSI) that provides a variety of spectral bands covering the visible, near infrared and shortwave infrared bands. The MSI contains four bands at 10 m, six bands at 20 m and three bands at 60 m [36]. It is of great importance that the MSI incorporates three bands in the red-edge region, centered at 705, 740 and 783 nm, and two Shortwave Infrared (SWIR) bands centered at 1610 and 2190 nm at 20 m (S2-20 m). Many studies have revealed that the red-edge bands and SWIR bands have the potential to improve the accuracy of LAI retrievals [37–39], which open great opportunities for crop LAI retrievals considering Sentinel-2's high revisit frequency. The Sentinel-2 spectral band characteristics are shown by Table 2.

The acquired Sentinel-2 imagery products are Level-1C products, which are top of atmosphere (TOA) reflectances [40]. Atmospheric correction was conducted using the Sen2Cor (2.5.5, ESA, Frascati, Italy, 2018) and Sentinel Application Platform (SNAP) toolbox (6.0.1, ESA, Frascati, Italy, 2018) provided by the ESA to produce Level-2A bottom of atmosphere (BOA) products [41,42]. A flowchart of Sentinel-2 preprocessing is presented in Figure 3.

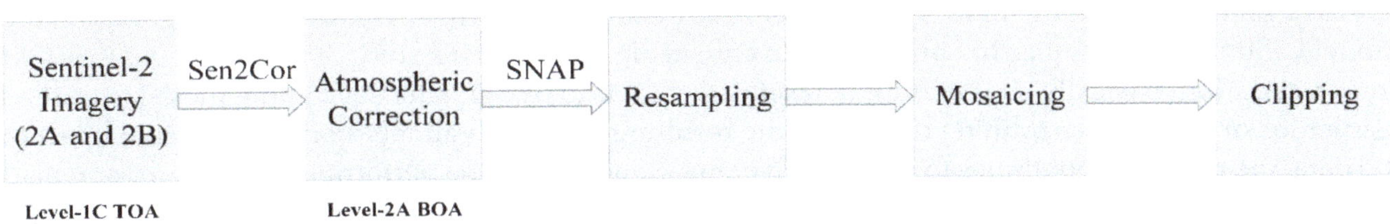

Figure 3. Flowchart of the preprocessing of Sentinel-2 imagery.

In this study, aiming at making full use of the spectral bands of Sentinel-2 (especially the red-edge and SWIR bands), we focus on S2-20 m (the same size as our quadrats) with 10 spectral bands considering the red-edge and SWIR spectral bands of Sentinel-2 imagery. A resampling process was performed using the Nearest Neighbor method with 4 spectral bands (B2, B3, B4 and B8) from 10 m to 20 m in the SNAP toolbox. Three of the atmospheric spectral bands at 60 m were not used in this study because these bands contributed to atmospheric applications, such as aerosol correction (B1), water vapor correction (B9) and cirrus detection (B10) [42]. The reflectance data collected on the Sentinel-2 images were using Extract by Points on the ArcToolbox of ArcGIS Desktop software (10.5, Environmental Systems Research Institute, Redlands, CA, USA, 2016), the points used to extract data on the Sentinel-2 images are the center GPS coordinate of the three sample points that represent one quadrat.

2.3. Global LAI Products

Many global LAI products with different spatial resolution and temporal characteristics has been produced, in which MODIS LAI products retrieved from Terra and Aqua platforms are one of the most famous global LAI products [2,43]. All MODIS products are available at [44]. In addition, GEOV1 LAI products have also been widely used for variety of applications. The GEOV LAI product was downloaded from the Copernicus Global Land Service [45]. More recently, new versions of these two products have been delivered with great improvement in spatial and temporal resolution [46–48]. Table 3 presents the main characteristics of the latest version of these two global LAI products.

Table 3. Main characteristics of two global LAI products under study.

Products	Version	Spatial Resolution	Temporal Resolution	Algorithms	Temporal Coverage	References
MODIS	MCD15 C6	500 m	4-day	RTM 3D [1] (LUT)	2002-present	Myneni, et al. 2015 [46]
GEOV3	V1.0.1	1/3 km	10-day	Neural network (red, NIR)	2014-present	Baret, et al. 2013, 2016 [47,48]

[1] RTM and LUT stands for "Radiative Transfer Model" and "Look Up Table", respectively.

3. Methods

ML algorithms can automatically learn the relationships in any given data between input (reflectances) and output (LAI). To identify the performance of five popular ML algorithms for cotton LAI retrieval, regression models were established based on artificial neural network (ANN), SVR, Gaussian process regression (GPR), random forest (RF) and gradient boosting regression tree (GBRT), at S2-20 m for the whole growth period of cotton (using all 117 quadrats data over the three observation dates).

All the ML models were implemented using the Scikit-learn package [49], which is an open-source Python [50] module project that integrates a wide range of prevalent ML algorithms [51]. All hyperparameter tuning of the models is based on GridSearchCV in the Scikit-learn package, which can evaluate all possible combinations of hyperparameter values using five-fold cross-validation to determine the best combination of hyperparameter values (the hyperparameter combination that has the best accuracy of the model in terms of root-mean-square error (RMSE)). Cross-validation are model validation techniques to obtain reliable and stable models. This study implemented a five-fold cross-validation, basically, the training datasets are split into five smaller sets, and a model is trained using four of the folds as training data, then the resulting model is validated on the remaining part of the data, the processes continues to circulate five times, and finally, the performance measure estimated by five-fold cross-validation is the average of the values computed in the loop. Training and testing sampling distribution has a great impact on machine learning regression algorithms (MLRAs), and some previous studies demonstrated that 70%/30% split option is appropriate for model training and validation [52–54], nonetheless, other studies argue that the 80%/20% split option is preferable [27,55]. In our study, to validate the performance of the ML models, all the datasets were randomly split into

75% (n = 87) for model training, and 25% (n = 30) for model validation. Regression models that achieve satisfactory performance in training datasets may fail to predict unseen datasets, and therefore, a model that performs well at both training and unseen testing datasets is referred to as having excellent generalization ability. This type of model could be used for crop LAI retrieval applications.

3.1. Artificial Neural Network (ANN)

The ANN has been one of the most widely used ML algorithms for wide range of remote sensing applications. ANN is a computational model that is inspired by the human brain. ANN is formed by a collection of interconnected units (neurons) that learn from experience by modifying connections (weights) [56,57]. ANN usually consists of an input layer, hidden layer, and output layer. The backpropagation (BP) ANN used in this study implements a multilayer perceptron (MLP) algorithm that trains with a BP algorithm, while an MLP refers to a feedforward network that generalizes the standard perceptron by having a hidden layer that resides between the input and output layers. It has been demonstrated that MLP can approximate any continuous function to an arbitrary degree of accuracy, given a sufficiently large but finite number of hidden neurons [56,58]. In this study, the input layer and output layer are referred to as the Sentinel-2 spectral bands (reflectances) and the cotton LAI, respectively.

The major tuning hyperparameters for ANN are the number of hidden layers and the number of neurons in the hidden layer. Here, we used the one-hidden-layer network because it was demonstrated to be powerful enough to approximate any measurable function to any desired degree of accuracy [59]. Many studies have been dedicated to the investigation of the optimal number of neurons in the hidden layer, and several empirical equations have been proposed [60–62]. In this study, the number of neurons in the hidden layer is determined by the following equation [62]:

$$N_h = \sqrt{(m+2)N} + 2\sqrt{\frac{N}{m+2}} \tag{1}$$

where N_h is the number of neurons in the hidden layer, m specifies the number of layers, and N denotes the number of input neurons. The number of neurons in the hidden layer was set to ten in this study according to Equation (1). In the input layer, the input variables include 10 spectral bands. Other important hyperparameters were optimized using GridSearch and 5-fold cross-validation. The remainder of hyperparameters for the ANN model remain defaults. The structure of the neural network used in this study is presented in Figure 4. The hyperparameter values adopted in this study are listed in Table 4.

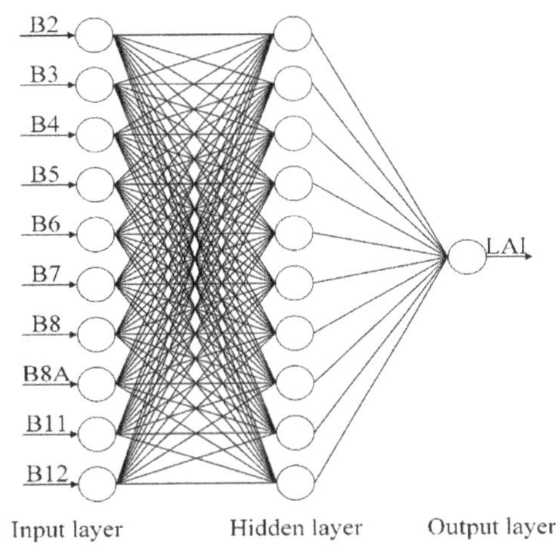

Figure 4. The architecture of backpropagation (BP) neural networks used in this study for cotton LAI retrieval.

Table 4. Parameter settings to determine the optimal hyperparameters for the artificial neural network (ANN) model.

Parameters	Description	GridSearch Values	Searching Results
activation	Activation function for the hidden layer.	'identity', 'logistic', 'tanh', 'relu'	'logistic'
solver	The solver for weight optimization.	'lbfgs', 'sgd', 'adam'	'lbfgs'
alpha	L2 penalty (regularization term) parameter.	1e-4, 1e-2, 0.1, 1	0.1
learning_rate	Learning rate schedule for weight updaters.	'constant', 'invscaling', 'adaptive'	'constant'

3.2. Support Vector Regression (SVR)

SVR is a significant application form of support vector machine (SVM), which was first introduced by Corinna Cortes (b. 1961 in Denmark) and Vapnik [63,64]. SVM is based on the idea of mapping the input space into a new feature space with higher dimensions using the kernel function, after which a hyperplane, known as the decision boundary, is constructed with the maximum margin [65,66]. SVM extension to SVR is realized by introducing an ε-insensitive region around the function, which is referred to as the ε-tube that best approximates the regression function [67]. Given training vectors $x_i \in \mathbb{R}^p$, I = 1, ⋯, n, and a vector $y \in \mathbb{R}^n$, ε-SVR solves the following original problem:

$$\min_{w,\,b,\,\xi,\,\xi^*} \frac{1}{2}\|w\|^2 + C\sum_{i=1}^{n}(\xi_i + \xi_i^*) \quad (2)$$

$$subject\ to \begin{cases} y_i - \langle w, x_i\rangle - b \le \varepsilon + \xi_i \\ \langle w, x_i\rangle + b - y_i \le \varepsilon + \xi_i^* \\ \xi_i, \xi_i^* \ge 0,\ i = 1, \cdots, n \end{cases} \quad (3)$$

where C > 0 is a regularization parameter that gives more weight to minimizing the flatness or error.

The principal hyperparameter of SVR is the kernel, as it defines the kernel functions of the model. The radial basis function (RBF) was selected as the kernel function because it has been found to be efficient and accurate for regression problems [68,69]. The RBF kernel is described as follows:

$$k(x_i, x_j) = exp\left(-\frac{\|x_i - x_j\|^2}{2\sigma^2}\right) \quad (4)$$

The SVR model is easy to establish because only two hyperparameters must be tuned: Penalty parameter C and the kernel coefficient gamma. The optimal values of the two hyperparameters that were optimized using GridSearch and 5-fold cross-validation are presented in Table 5. The remainder of hyperparameters for the SVR model remain defaults.

Table 5. Parameter settings to determine the optimal hyperparameters for the support vector regression (SVR) model.

Parameters	Description	GridSearch Values	Searching Results
C	Penalty parameter C of the term.	0.1, 0.5, 1, 5, 10, 15, 20, 50, 100, 500	50
gamma	Kernel coefficient for 'rbf', 'poly' and 'sigmoid'.	0.01, 0.1, 0.2, 0.4, 0.8, 1, 1.5, 2, 5, 10	0.2

3.3. Gaussian Processes Regression (GPR)

A Gaussian process is a stochastic process that is formed by a collection of random variables and has a Gaussian probability distribution [70]. The major factor in the Gaussian process is the covariance function known as the kernel function. The learning problem in the Gaussian process amounts to adjusting the covariance hyperparameters. The mean function $m(x)$ and covariance function $k(x, x')$ characterize the Gaussian process $f(x)$, which can be described as follows:

$$m(x) = \mathbb{E}[f(x)] \tag{5}$$

$$k(x, x') = \mathbb{E}[(f(x) - m(x))(f(x') - m(x'))] \tag{6}$$

$$f(x) \sim GP(m(x), k(x, x')) \tag{7}$$

For simplicity, we usually consider the mean function to be zero.

Similar to SVR, GPR has been advanced to solve complex nonlinear problems through projection of inputs into high dimensional space by applying highly flexible kernels, where such a technique is referred to as kernel trick. Additionally, GPR models have a great ability to provide the most informative feature (spectral band) from the input dataset. In this study, we used the sum of an RBF and a noise component function kernel:

$$k(x_i, x_j) = \theta_0{}^2 exp\left(-\frac{(x_i - x_j)^2}{2\sigma^2}\right) + \theta_1{}^2\delta_{ij} \tag{8}$$

where θ_0 is the scaling factor, σ is the length-scale, and θ_1 corresponds to the independent noise component.

Beyond the kernel functions, alpha and n_restarts_optimizer, are significant for the GPR model. The results of hyperparameter tuning using GridSearch and 5-fold cross-validation are shown in Table 6 as follows. The remainder of hyperparameters for the GPR model are set as defaults.

Table 6. Parameter settings to determine the optimal hyperparameters for the Gaussian process regression (GPR) model.

Parameters	Description	GridSearch Values	Searching Results
alpha	Value added to the diagonal of the kernel matrix during fitting; larger values correspond to increased noise level in the observations; this can also prevent a potential numerical issue during fitting, by ensuring that the calculated values form a positive definite matrix.	1e-2, 1e-1, 1	1
n_restarts_optimizer	The number of restarts of the optimizer for finding the kernel's parameters which maximize the log-marginal likelihood; the first run of the optimizer is performed from the kernel's initial parameters, the remaining ones (if any) from thetas sampled log-uniform randomly from the space of allowed thetas-values.	0, 1, 2, 4, 8, 10, 12, 16, 20, 32, 64	16

3.4. Random Forest (RF)

RF has been a prevalent ML algorithm for a wide range of fields for classification, regression and other complicated problems. As one of the ensemble learning methods, RF grows a multitude of decision trees as base learners, and combines these trees together to obtain a better performance by averaging the predictions [71,72]. Each tree grows independently with training samples obtained using bootstrap sampling from the original data. Then, m variables out of M input variables are chosen, after which the best of m is used for splitting the node (note that m $\ll$ M). The final prediction comes from the averaging predictions of each independent tree. This kind of technique is referred to as bagging. RF models can provide feature importance estimates, which enables insight into feature selection processes.

Key hyperparameters include n_estimators, max_depth, min_samples_split and min_samples_leaf. Hyperparameter tuning results using GridSearch and 5-fold cross-validation are displayed in Table 7. The remainder of the RF model hyperparameters are set as defaults.

Table 7. Parameter settings to determine the optimal hyperparameters for the random forest (RF) model.

Parameters	Description	GridSearch Values	Searching Results
n_estimators	The number of trees in the forest.	1, 5, 10, 20, 40, 60, 80, 120	40
max_depth	The maximum depth of the tree.	2, 5, 8, 15, 25, 30, 50, None	15
min_samples_split	The minimum number of samples required to split an internal node.	2, 5, 10, 15, 100	2
min_samples_leaf	The minimum number of samples required to be at a leaf node; a split point at any depth will only be considered if it leaves at least min_samples_leaf training samples in each of the left and right branches; this may have the effect of smoothing the model, especially in regression.	1, 2, 5, 10, 50	1

3.5. Gradient Boosting Regression Tree (GBRT)

GBRT, also known as gradient boosting decision tree (GBDT) or multiple additive regression tree (MART), is one of the most widely used ML algorithms for the model's great generalization ability and highly robust performance in practical applications, and this GBRT was introduced by Friedman [73,74]. As one of the ensemble learning methods that combines different weak learners to generate strong learners, GBRT uses boosting techniques that aim to reduce bias, rather than bagging (e.g., RF algorithm), which aims to reduce variance. To evaluate the accuracy of the model, a variety of loss functions can be used during boosting, including least squares, least absolute deviation, Huber and quantile for regression, binomial deviance, multinomial deviance, and exponential loss for classification. GBRT considers additive models of the following form:

$$F(x) = \sum_{m=1}^{M} \gamma_m h_m(x) \tag{9}$$

where $h_m(x)$ represents the basis learners in boosting. Then, GBRT builds the additive model in a forward stagewise fashion:

$$F_m(x) = F_{m-1}(x) + \gamma_m h_m(x) \tag{10}$$

At each stage, the decision tree $h_m(x)$ is chosen to minimize the loss function L, given the current model F_{m-1} and its fitting of $F_{m-1}(x_i)$:

$$F_m(x) = F_{m-1}(x) + arg\ \min_{h} \sum_{i=1}^{n} L(y_i, F_{m-1}(x_i) + h(x)) \tag{11}$$

The initial model F_0 is problem specific. Gradient boosting attempts to solve this minimization problem numerically via the steepest descent. The steepest descent direction is the negative gradient of the loss function evaluated at the current model F_{m-1}, which can be calculated for any differentiable loss function as follows:

$$F_m(x) = F_{m-1}(x) - \gamma_m \sum_{i=1}^{n} \nabla_F L(y_i, F_{m-1}(x_i)) \tag{12}$$

where the step length γ_m is chosen using a line search:

$$\gamma_m = arg\ \min_{\gamma} \sum_{i=1}^{n} L\left(y_i, F_{m-1}(x_i) - \gamma \frac{\partial L(y_i, F_{m-1}(x_i))}{\partial F_{m-1}(x_i)}\right) \tag{13}$$

The GBRT algorithm can provide feature relevance information and partial dependence, where this partial dependence shows the dependence among the target response and the most important features (not shown in this study). However, despite the GBRT's demonstrated satisfactory accuracy for versatile domains of regression problems [29,75,76], to our knowledge, this algorithm has not been previously applied to crop LAI retrievals with remote sensing. GBRT should be studied as it might be a promising alternative in crop LAI retrieval with remote sensing.

With regard to GBRT model hyperparameters, loss, n_estimators, max_depth, min_samples_split and min_samples_leaf are selected for hyperparameter tuning using GridSearch and 5-fold cross-validation, the results are exhibited in Table 8. The remainder of the GBRT model hyperparameters are set as defaults.

Table 8. Parameter settings to determine the optimal hyperparameters for the gradient boosting regression tree (GBRT) model.

Parameters	Description	GridSearch Values	Searching Results
loss	Loss function to be optimized.	'ls', 'lad', 'huber', 'quantile'	'lad'
n_estimators	The number of boosting stages to perform; gradient boosting is fairly robust to over-fitting so a large number usually results in a better performance.	1, 5, 10, 20, 40, 60, 80, 120, 300	300
max_depth	Maximum depth of the individual regression estimators; the maximum depth limits the number of nodes in the tree; tune this parameter for best performance; the best value depends on the interaction of the input variables.	2, 3, 5, 8, 15, 25, 30, None	25
min_samples_split	The minimum number of samples required to split an internal node.	2, 5, 10, 15	10
min_samples_leaf	The minimum number of samples required to be at a leaf node; a split point at any depth will only be considered if it leaves at least min_samples_leaf training samples in each of the left and right branches; this may have the effect of smoothing the model, especially in regression.	1, 2, 5, 10, 50	1

3.6. Performance Evaluation

To evaluate the performance of the ML regression models, the root mean square error (RMSE), mean absolute error (MAE) and coefficient of determination (R^2) between the measured and predicted values were used to assess the performance of the models. The RMSE, MAE and R^2 are calculated as follows:

$$RMSE(y,\ \hat{y}) = \sqrt{\frac{1}{n_{samples}} \sum_{i=0}^{n_{samples}-1} (y_i - \hat{y}_i)^2} \tag{14}$$

$$MAE(y,\ \hat{y}) = \frac{1}{n_{samples}} \sum_{i=0}^{n_{samples}-1} |y_i - \hat{y}_i| \tag{15}$$

$$R^2(y,\ \hat{y}) = 1 - \frac{\sum_{i=0}^{n_{samples}-1} (y_i - \hat{y}_i)^2}{\sum_{i=0}^{n_{samples}-1} (y_i - \overline{y})^2} \tag{16}$$

where $\overline{y} = \frac{1}{n_{samples}} \sum_{i=0}^{n_{samples}-1} y_i$, $\hat{y}_i$ is the estimated cotton LAI value, y_i is the measured cotton LAI value and $n_{samples}$ is the number of validation datasets. The higher the R^2, the smaller the RMSE and MAE, and thus the higher the model precision and accuracy.

4. Results

4.1. Performance Evaluation

To avoid skew results caused by the random sampling of training and testing datasets, we performed 20 random repetitions of the five ML regression models (87 samples for training and 30 for testing). The performance of five ML models with respect to R^2, RMSE and MAE is displayed in Figure 5. GBRT surpasses the other models on the average ($\overline{R^2}$ = 0.854, $\overline{RMSE}$ = 0.674 and $\overline{MAE}$ = 0.456), however, GBRT acts less robust to the training/testing random split according to the distribution of R^2, RMSE and MAE. RF delivers the worst accuracy on the average ($\overline{R^2}$ = 0.807, $\overline{RMSE}$ = 0.781 and $\overline{MAE}$ = 0.545), and it also acts less robust to the training/testing random split. Nonetheless, SVR achieves a desirable result while it also acts reasonably robust to the training/testing random split ($\overline{R^2}$ = 0.835, $\overline{RMSE}$ = 0.730 and $\overline{MAE}$ = 0.550), which indicates that SVR is highly stable to the random sampling processes. Overall, all models achieved satisfactory performances, which indicates that ML algorithms are appealing methods for cotton LAI estimation.

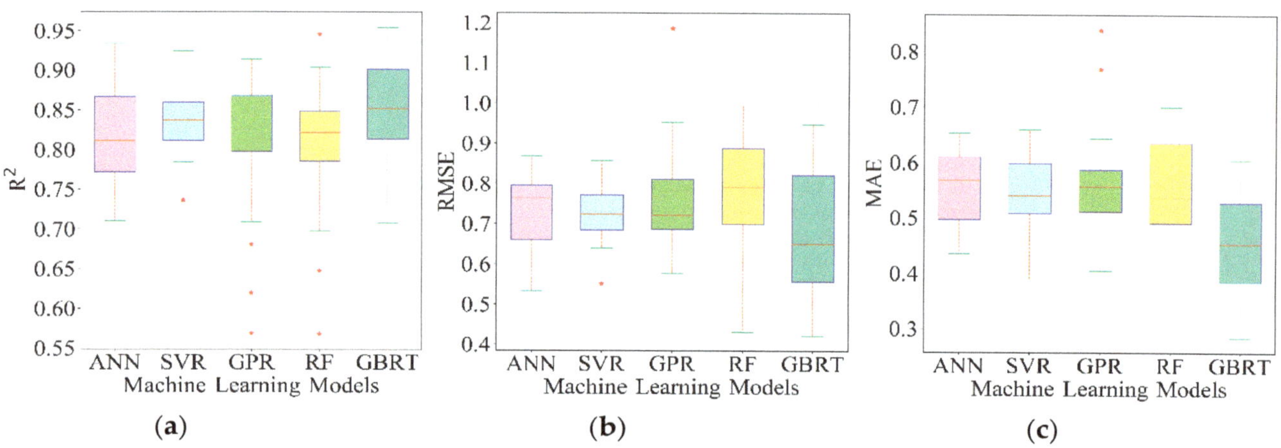

Figure 5. The R^2, RMSE and MAE distribution of 20 repetitions between the predicted LAI values and corresponding measured LAI. (**a**) R^2; (**b**) RMSE; (**c**) MAE.

4.2. Computational Efficiency

Beyond the model accuracy of the five models for the training and testing random split, it is of particular relevance to compare the computational efficiency (time required to the model during training and validation processes) of the models, as it is an important criterion for operational algorithms. All models were implemented in a Python environment on an Intel(R) Xeon(R) CPU E5-2620 v2 @ 2.10 GHz processor and installed memory (RAM) of 32.0 GB. The computational efficiency is recorded from the 20 repetitions in Section 4.1. The averaged results of 20 repetitions are illustrated in Figure 6. Large differences among the five ML models are clearly found. SVR performs incredibly fast (less than 1 s). However, GBRT is frustrating, owing to the large amounts of its hyperparameter tuning processes. In general, SVR is able to deliver near real time operational products with a highly efficient processing speed, whereas the GBRT model is not recommended for this kind of application, because the GBRT model is computationally more demanding. RF, GPR and ANN showed moderate performances in terms of computational efficiency.

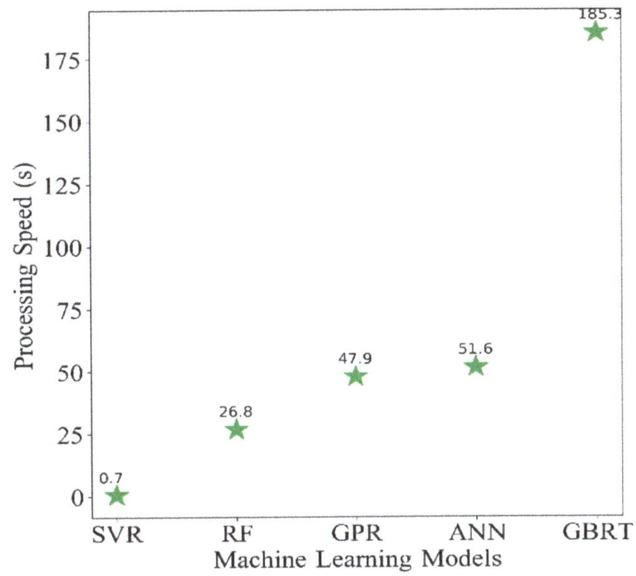

Figure 6. The average processing speed for five ML models.

4.3. Sensitivity to Training Sample Size

We use R^2, RMSE and MAE between the predicted LAI values and measured LAI values to assess the sensitivity of five ML methods to the training sample size. Eight datasets with different sample sizes (17-87) were generated by randomly sampling from the total training datasets (87 samples) at intervals of 10. Table 9 shows the number of training and testing samples of the eight datasets, while the testing datasets keep as the same datasets with 30 samples.

Table 9. Number of training and testing samples of the eight datasets.

Sample Name	Set 1	Set 2	Set 3	Set 4	Set 5	Set 6	Set 7	Set 8
Training Datasets	17	27	37	47	57	67	77	87
Testing Datasets				30				

To evaluate the robustness of five ML models for the training samples, we conducted 10 random repetitions of each model and provided the corresponding standard error. The standard error is given as follows:

$$\sigma_{\overline{x}} = \frac{\sigma}{\sqrt{n}} \tag{17}$$

where σ denotes the standard deviation, n represents the number of repetitions.

Figure 7 shows the changes in R^2, RMSE and MAE as the training sample sizes varied, the standard error is given at each training sample size (filling area of each model), lower values of the filling area corresponds to smaller standard error, and consequently, more robust model performance. According to Figure 7, different sensitivity performances of the sample size variation are found among the ML models. GPR and GBRT show robust performances for the training samples according to the standard error of different models. GBRT produces a remarkable performance for the training samples that is more robust than the other models, GPR behaves suboptimally. ANN and SVR are very sensitive to the training samples. Moreover, all models behave more robust with the growth of training sample size according to the variation of standard error. In summary, the model accuracy improves overall with an increasing training sample size, while GBRT provides the most robust performance for the training samples, and the best model accuracy among the variations in training sample size on average ($\overline{R^2}$ = 0.884, $\overline{RMSE}$ = 0.615 and $\overline{MAE}$ = 0.452, calculated by all the 8 groups of training sample size).

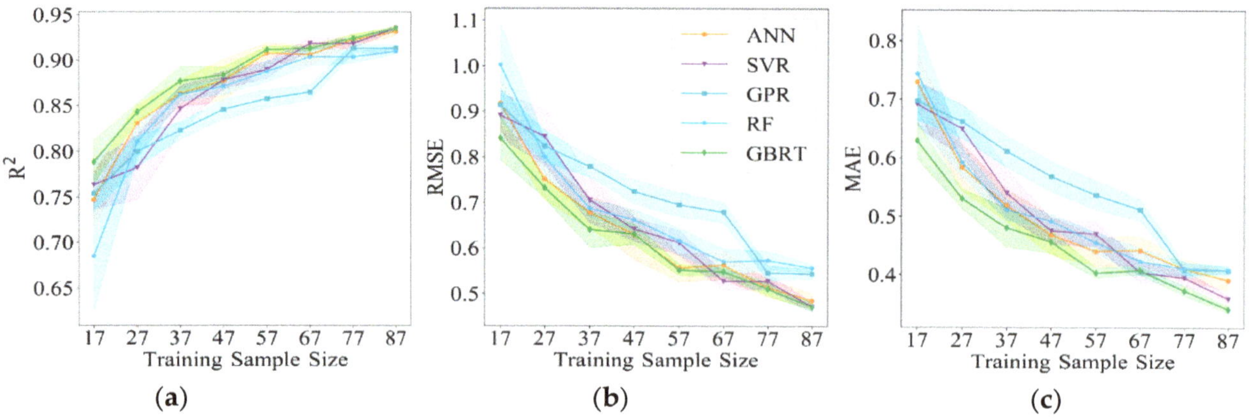

Figure 7. Sensitivity of the ML models to the training sample size (with intervals of 10) in terms of R^2, RMSE and MAE with standard error. (**a**) R^2; (**b**) RMSE; (**c**) MAE.

4.4. Sensitivity to Spectral Bands

Variable and feature selection have been one of the most significant processes in ML, with the aim of improving model accuracy and accelerating model processing. With 10 spectral bands of Sentinel-2 available (B2, B3, B4, B5, B6, B7, B8, B8A, B11, B12), it may not be practical to evaluate the performance of all the possible band combinations. This process could be challenging, as the characteristics of different ML models are distinct, and thus it is challenging to conduct a comparison among the different models under a unified standard. For example, GPR, RF and GBRT could provide feature relevance (spectral bands relevance) information that enables insight into feature selection processes, however, the feature relevance results are subject to the model itself, in other words, we could obtain different feature selection schemes, and such information is practical with single ML model based applications. In addition, ANN and SVR are not able to deliver feature relevance information.

Székely, Rizzo, and Bakirov [77] proposed the distance correlation (dCor) to measure the dependence of random vectors. dCor ranges between 0 and 1, and dCor (**X**, **Y**) = 0 only if **X** and **Y** are independent, which is effective in the feature selection processes [78–80]. To identify the optimal number of Sentinel-2 spectral bands required for cotton LAI retrieval using MLRAs, we used the distance correlation combined with the backward elimination method [81,82]. We also performed 10 random repetitions of each model to assess the model's robustness with the spectral bands. Notably, the distance correlation is subject to the training dataset adopted in this study. Figure 8 displays the distance correlation among the 10 spectral bands and LAI. The ranking of the dCor values between the 10 spectral bands and LAI are identified as B12 > B3 > B4 > B2 > B5 > B11 > B8A > B7 > B8 > B6 (B12: 0.88631, B3: 0.88630, B4: 0.88016, B2: 0.87721, B5: 0.87400, B11: 0.87338, B8A: 0.80219, B7: 0.76812, B8: 0.76585, B6: 0.68998). The visible and SWIR bands occupied the top rankings, while differences among the red-edge bands were distinct. The sensitivity results of different ML models to spectral bands using dCor and the backward elimination method (iteratively removing the spectral band with the lowest dCor value until only one spectral band is left) are shown in Figure 9, similarly, the standard error is given at each spectral bands combination (filling area of each model), lower values of the filling area corresponds to smaller standard error, and consequently, more robust model performance. Differences exist among the sensitivity of the models with the reduction in spectral bands. A turning point occurred at 6 for ANN, the accuracy begins to increase with the reduction in spectral bands before 6, though with some fluctuations, and then, the accuracy begins to decrease dramatically until reaching approximately 3. Similar trends are found among all the models except for GPR, as there is a relatively strong fluctuation at 7. GPR has a stable performance after 4. SVR, GPR and RF are robust for the spectral bands according to the standard error. Overall, model accuracy decreases as the spectral bands decrease, although there are some fluctuations in the models, whereas ANN provides the best model accuracy among the spectral band reductions on average ($\overline{R^2}$ = 0.881, $\overline{RMSE}$ = 0.625 and $\overline{MAE}$ = 0.480, calculated by all the 10 groups of spectral bands).

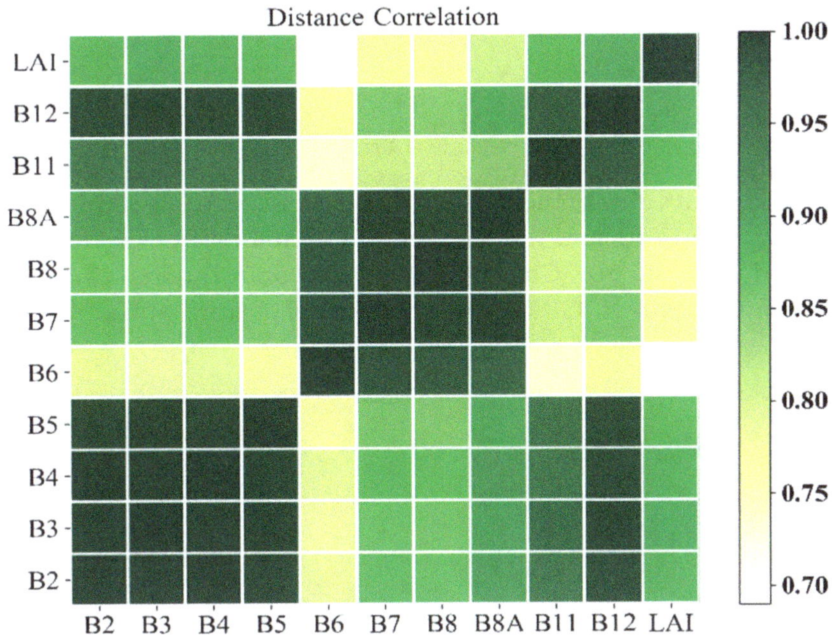

Figure 8. Distance correlation among the Sentinel-2 spectral bands and corresponding cotton LAI.

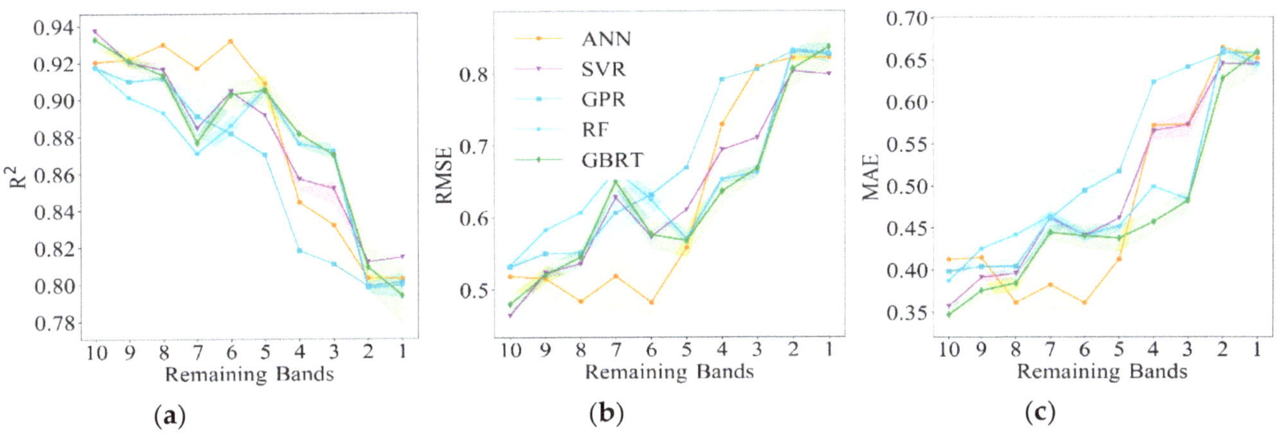

Figure 9. Sensitivity of the ML models to the spectral bands with respect to R^2, RMSE and MAE using the dCor and backward elimination method (the standard error is displayed). The horizontal axis (Remaining Bands) represents the number for the rest of spectral bands after each iterative removing process. (**a**) R^2; (**b**) RMSE; (**c**) MAE.

The minimum number of bands required for cotton LAI retrieval are recognized as 6 for ANN and SVR, 5 for RF and GBRT, and 8 for GPR, in other words, increasing of the spectral bands of Sentinel-2 does not significantly improves the model accuracy. Given the recognized number of spectral bands of each of the ML models, this is vital for associated applications with great amounts of samples, as it may reduce model processing time and desirable model accuracy could be acquired in the meantime.

4.5. Comprehensive Evaluation

In this study, we conducted a comparison of five universal MLRAs for cotton LAI retrieval with regard to model accuracy, computational efficiency, sensitivity to training sample size and sensitivity to spectral bands. In this section, we provide a comprehensive evaluation of five ML algorithms according to the obtained results, which were discussed earlier in this paper. To combine all the metrics, we performed a standardization process on the results. The mean values of the results in the sensitivity analysis (training sample size and spectral bands) were used, and we make RMSE, MAE and processing speed negative for the sake of comprehensive comparison. An alternative standardization is interval scaling, which is as follows:

$$x' = \frac{x - Min}{Max - Min} \tag{18}$$

where Max denotes the maximum value of the data, and Min represents the minimum value of the data. The results are displayed in Figure 10, and the metrics used in the radar chart and its corresponding implications are displayed in Table 10. Clearly, GBRT shows an outstanding performance from a comprehensive viewpoint, however, GPR shows the worst performance. ANN and SVR show moderate performances.

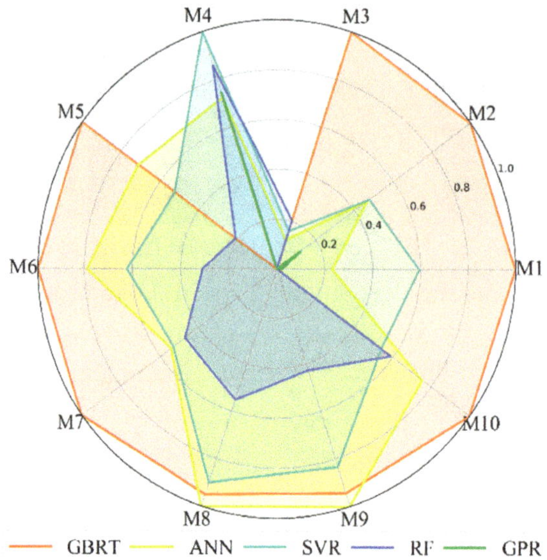

Figure 10. Comprehensive comparison of five machine learning regression algorithms (MLRAs), with different metrics using a radar chart.

Table 10. Metrics used in a radar chart and the corresponding descriptions.

Metrics in the Radar Chart	Corresponding Description
M1	Fitness ($\boldsymbol{R}^2$)
M2	Accuracy (RMSE)
M3	Accuracy (MAE)
M4	Computational efficiency
M5	Sensitivity to training sample size ($\boldsymbol{R}^2$)
M6	Sensitivity to training sample size (RMSE)
M7	Sensitivity to training sample size (MAE)
M8	Sensitivity to spectral bands ($\boldsymbol{R}^2$)
M9	Sensitivity to spectral bands (RMSE)
M10	Sensitivity to spectral bands (MAE)

4.6. Final LAI Maps

The best performing ML regression model has been applied to map cotton LAI with Sentinel-2 imagery, as analyzed in previous sections, we use the GBRT model to map cotton LAI in the study area. The fields in the study area are regular, we conducted a manual vectorization based on Google Earth image data to obtain accurate agricultural fields boundaries, and Landsat 8 image data were collected to update the fields, as Google Earth image data has poor temporal information. Totally, 40,211 fields were collected by this way in the study area. Figure 11 displayed the agricultural fields boundaries. A per-field crop classification was performed to extract cotton fields. The final cotton LAI map was obtained by masking the results using the extracted cotton fields. Figure 12 presents the final cotton

LAI map. From Figure 12, it is clear to find that there is an increasing trend of cotton LAI growth, as revealed by Figure 2.

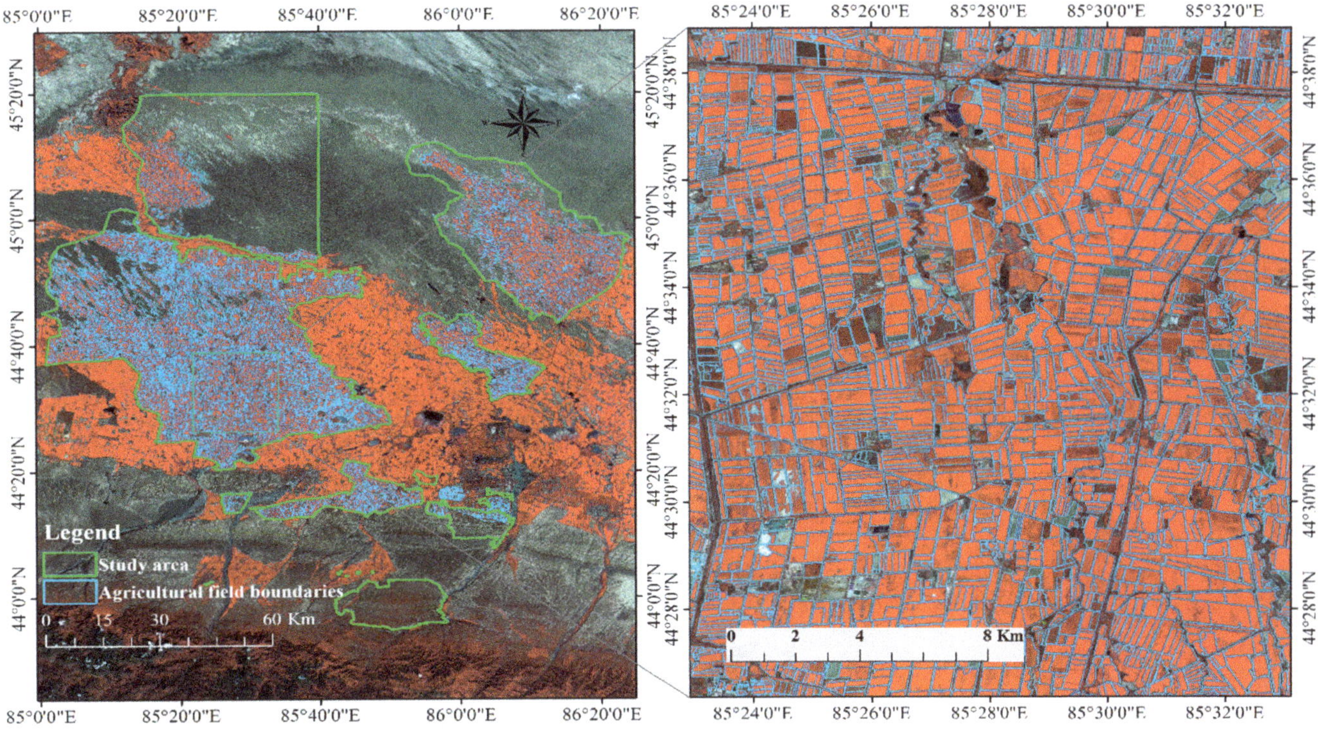

Figure 11. Agricultural fields boundaries collected by manual vectorization.

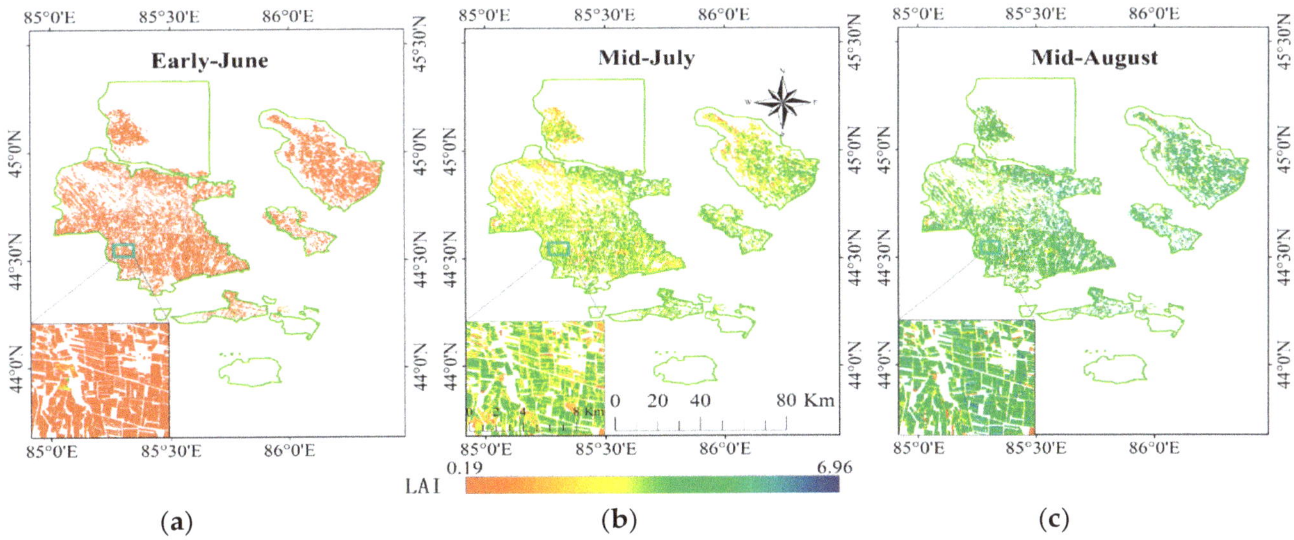

Figure 12. Final cotton LAI map obtained using GBRT model and masked by the extracted cotton fields. (**a**) Early June 2018; (**b**) Mid July 2018;(**c**) Mid August 2018.

In the early June cotton LAI map, some areas have LAI values much larger than the whole map (green color), as well as LAI values much smaller found in mid-July and mid-August (orange color), these fields are some other crops rather than cotton, and it can be ascribed to the misclassification.

4.7. Comparison with Global LAI Products

The validation of moderate-resolution products are key steps to assess the quality of the global LAI products. In this study, we performed a comparison among GEOV3, MODIS LAI products and upscaled Sentinel-2 LAI to assess the quality of these two products in Northwest China. To conduct the comparison of the cotton LAI results obtained using the GBRT model with global LAI products, we choose the closest dates of the products at a spatial resolution of the products to minimize the spatial and temporal differences [2].

All Sentinel-2 LAI maps were resampled to the same spatial resolution of MODIS and GEOV3 LAI products, and then GEOV3 LAI product were also resampled to the same spatial resolution of MODIS LAI product to make a straight comparison of these two products. Figure 13 shows the scatter plots among the comparison of GEOV3, MODIS and Sentinel-2 cotton LAI over three observation dates. Regarding the comparison of the MODIS LAI product and Sentinel-2 LAI maps, it was found that both the MODIS LAI product and the Sentinel-2 LAI suffered from the influence of the background, and are kind of overestimated in some areas in early June 2018. MODIS LAI product underestimated a little bit in mid-August 2018. In terms of GEOV3 LAI product and Sentinel-2 LAI maps, the GEOV3 LAI product overestimated at some areas both in early June and mid-July 2018, and underestimated in mid-August 2018. Finally, with respect to the comparison of the two global LAI products, relatively, large differences are found in early June and mid-July 2018, whereas a good relationship appeared in mid-August 2018.

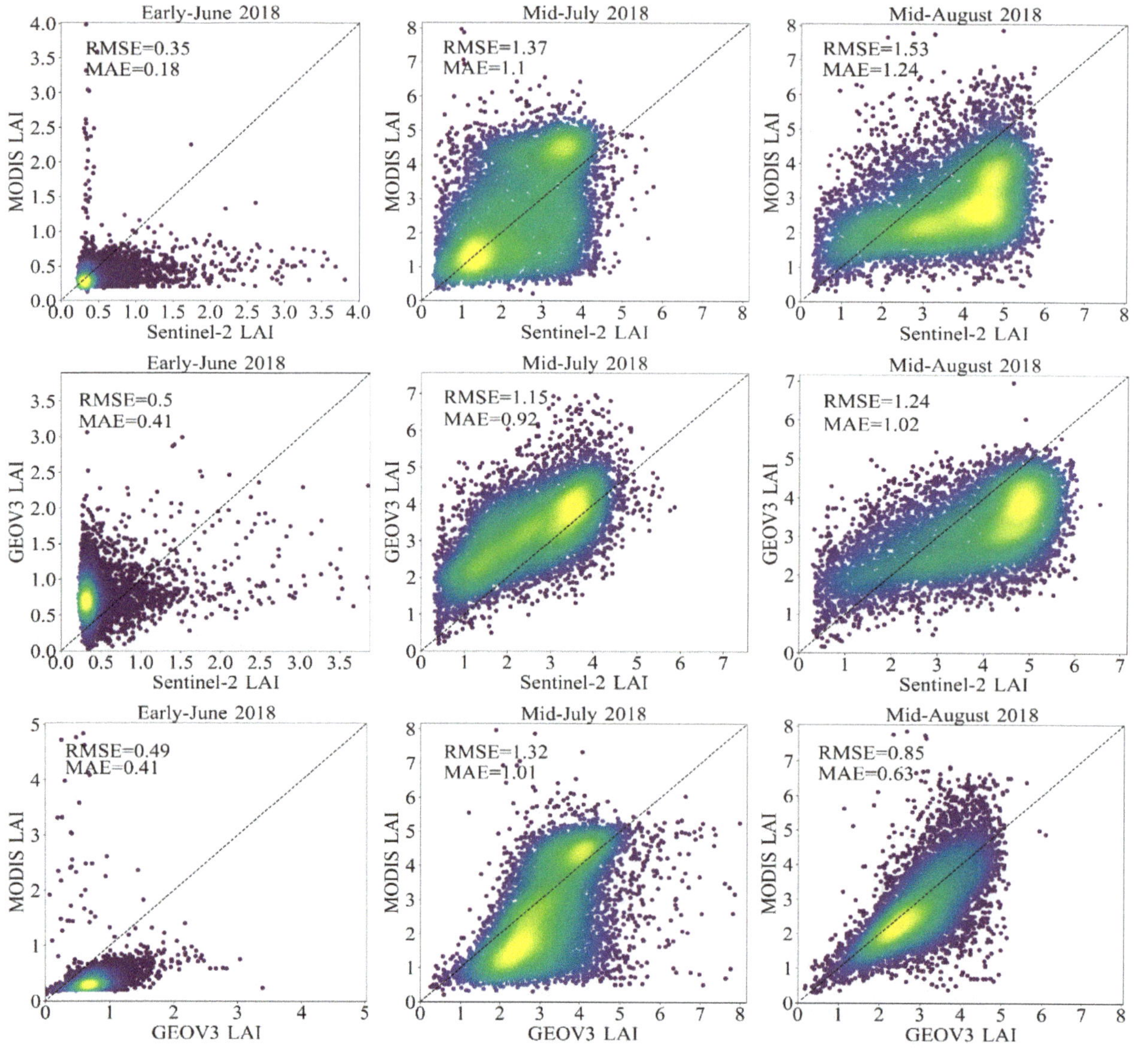

Figure 13. Scatter plots among GEOV3, MODIS LAI products and upscaled Sentinel-2 LAI in the 2018 cotton season.

5. Discussion

Over the last decade, there has been a considerable increase in the introduction of ML algorithms to remote sensing for a wide range of fields. Crop biophysical parameters (e.g., LAI) are key variables for a wide range of applications. With the progress of remote sensing techniques, we are able to acquire high dimensional (spatial, temporal, and spectral) remote sensing data, which demands more efficient, accurate and robust algorithms in a wide variety of applications with remote sensing. In this context, there is great potential for ML algorithms to be used in a wide range of remote sensing applications.

The diversity of available ML algorithms poses great challenges for the selection of MLRAs, as well as the decision of optimal number of training sample size and spectral bands to different MLRAs. Besides, another significant problem that may arise involves hyperparameter tuning in the application of ML algorithms. In general, experience is required to obtain satisfactory results for ML algorithms, and experiments with a large number of datasets may be needed, which are generally not available. Accordingly, it is necessary to perform a comparison of different fashionable ML models to support better remote sensing applications. In this study, we focus on the comparison of five well-known ML algorithms for cotton LAI retrievals with Sentinel-2 imagery, because these algorithms have a wide range of applications in remote sensing. Additionally, the hyperparameter tuning results of five prominent ML models are provided. Our study could provide support for associated remote sensing studies based on ML algorithms.

Furthermore, a comparatively great fluctuation is observed at 7 among all the models except for GPR in Figure 9, the previous removing spectral band is the red-edge band (B7), which indicates that having low dCor values do not necessarily correspond to being less important for LAI retrieval. Notably, ensemble methods (RF and GBRT) and GPR models have a great benefit of delivering feature importance information (not shown in our study), which provides insight into the greatest contributing features of the model. Such information could be used for better model interpretation, and this information is also useful for feature selection processes when applied to models that contain a great number of features.

In related studies, Verrelst, et al. [55] compared four ML algorithms (NN, SVR, KRR and GPR) using simulated Sentinel-2 and 3 data to assess three biophysical parameters (Chlorophyll content, LAI and FVC). GPR outperformed the other regression methods for the majority of Sentinel configurations, whereas in our study, GPR performed worse than the other regression methods. Results that differ from our study may be for the following reasons. First, we used real Sentinel-2 data rather than simulation data, and it may be difficult to represent the performance of real Sentinel-2 data with simulation data. Second, we focus on cotton crops over the entire growth period rather than on various crop types. In addition, there may be different hyperparameter settings between models. Finally, there are differences between the Sentinel-2 bands, and we used 10 spectral bands (SWIR included). Yuan, et al. [83] compared the RF, ANN and SVM regression models for soybean LAI retrieval using unmanned aerial vehicle (UAV) hyperspectral data with different sampling methods. The results showed that RF is suitable for the whole growth period of soybean LAI estimation, while ANN is appropriate for a single growth period. Siegmann and Jarmer [84] compared SVR and RFR for wheat LAI estimation using hyperspectral data, and the results showed that SVR provided the best performance of the entire dataset. Different from previous studies, we considered the GBRT model due to its highly robust performance over a wide range of applications. We further explored the sensitivity of different ML algorithms to training sample size and Sentinel-2 spectral bands. The results show that GBRT achieves the best performance, and GBRT was more robust for the training samples than the other models.

Despite the promising results revealed by five ML regression methods, there are some limitations of our study, and further studies are needed to assess the generalization ability of the five ML regression methods. Firstly, we focus on only one crop type, and the performance of ML regression methods for other crop types has not been validated. It is necessary to evaluate ML algorithms with various crop types, provided there are enough field measurement LAI data of different crop types available.

Secondly, some other crop biophysical parameters (e.g., Chlorophyll content and FVC) demand evaluation as well as ML regression methods, as different results may appear with different ML algorithms and different crop biophysical parameters. Thirdly, there are limitations with Sentinel-2 spectral bands sensitivity analysis, since some spectral bands are highly correlated, and redundancies remain among the spectral bands, so this process would be worth further investigations to explore a better spectral bands sensitivity analysis. Additionally, there are limitations to our quadrats in the field campaigns, as we chose only three sample points to represent a quadrat (20 m $\times$ 20 m). Finally, the performance of five MLRAs has not yet been evaluated for large amounts of data (e.g., tens of thousands of data) due to the limited number of samples in our study, however, it remains to be seen whether a favorable performance can be obtained. In consideration of the ability of ML techniques, it is expected to obtain a better accuracy of the regression models provided there are enough samples, and a turning point should be obtained, as has been demonstrated in previous studies [19]. While before the turning point, the model accuracy increased rapidly with the increase in the training sample size, whereas, after the turning ponits, the model behaves more stable, in other words, the model accuracy remmained unchanged with the increasing of training sample szie. Future studies are required to investigate this problem.

With the advance in ML technology, deep learning, a new subfield of ML, has been applied successfully to many remote sensing domains, especially with a large quantity of data [85]. It is not necessary to construct complex ML models (deep learning-based models) in our study, as traditional ML algorithms have the capability of establishing efficient, accurate and robust estimation models for cotton LAI retrieval. However, as there are greater amounts of data available, deep learning techniques may be promising alternatives for handling such large volume datasets and complex relationships among the input datasets, which deserve further study.

6. Conclusions

With the challenges of selecting ML algorithms for crop LAI estimation, the results of this study have great implications for the selection of appropriate ML models from the diversity of available ML algorithms, and at the same time, these same results provide the optimal number of training sample size and spectral bands of Sentinel-2 for each model required for cotton LAI retrieval. Regarding the comparison of different ML models for cotton LAI retrieval employed in our study, the GBRT model outperforms the other ML models according to our results. Our findings increase the potential for cotton LAI retrieval with Sentinel-2 imagery, and may be transferrable to other associated problems related to agricultural remote sensing applications. On the other hand, the GBRT model is computationally demanding, which may be a significant problem with a large scale of data. However, GBRT can challenge the model accuracy and computational efficiency selection problems. Considering the computational efficiency, SVR exhibits considerable computational superiority over the other ML models.

Regarding the sample size sensitivity of ML models, model accuracy increases with the growth of the training sample size, and the GBRT produces the most robust performance for the training samples with respect to the standard error and the best model accuracy on average.

In terms of spectral band sensitivity analysis, the distance correlation results showed that the SWIR and visible bands have great potential to improve the accuracy of cotton LAI retrievals. By using dCor combined with the backward elimination method, the model accuracy decreases with the reduction in spectral bands. SVR, GPR and RF perform robustly with the spectral bands, and ANN provides the best accuracy on average. The minimum number of bands required for cotton LAI retrieval are recognized as 6 (ANN), 6 (SVR), 5 (RF), 5 (GBRT) and 8 for GPR.

A comprehensive evaluation has been employed to identify the performance of five ML models, considering a combination of model accuracy, computational efficiency, sensitivity to training sample size and sensitivity to spectral bands. The comprehensive performance of the models is identified as GBRT > ANN > SVR > RF > GPR.

Despite the different performances of the five ML regression models, MLRAs are promising ways to retrieve cotton LAI with Sentinel-2 imagery because the MLRAs all achieved encouraging accuracies. With profound applications for a diversity of ML algorithms in remote sensing, MLRAs may provide positive effects for remote sensing applications in terms of classification, regression, and other associated problems.

Author Contributions: Conceptualization, J.M. and H.M.; Methodology, H.M.; Formal Analysis, H.M.; Investigation, H.M., F.J., Q.Z. and H.F.; Data curation, H.M.; Writing—Original Draft Preparation, H.M.; Writing—Review and Editing, J.M.; Visualization, H.M.

Acknowledgments: We are grateful to Agricultural College of Shihezi University for providing help in field campaigns. We would like to acknowledge the authors who contributed to the development of Scikit-learn project. We acknowledge the Copernicus Global Land Service, and the Land Processes Distributed Active Archive Center (LP DAAC) within the NASA Earth Observing System Data and Information System (EOSDIS) for providing the global LAI products. We are thankful to the three anonymous reviewers for their insightful comments and suggestions helped to clarify and improve the manuscript.

References

1. Chen, J.M.; Black, T.A. Defining leaf-area index for non-flat leaves. *Plant Cell Environ.* **1992**, *15*, 421–429. [CrossRef]
2. Garrigues, S.; Lacaze, R.; Baret, F.; Morisette, J.T.; Weiss, M.; Nickeson, J.E.; Fernandes, R.; Plummer, S.; Shabanov, N.V.; Myneni, R.B.; et al. Validation and intercomparison of global Leaf Area Index products derived from remote sensing data. *J. Geophys. Res.-Biogeosci.* **2008**, *113*. [CrossRef]
3. Asner, G.P.; Scurlock, J.M.O.; Hicke, J.A. Global synthesis of leaf area index observations: Implications for ecological and remote sensing studies. *Glob. Ecol. Biogeogr.* **2003**, *12*, 191–205. [CrossRef]
4. Buermann, W.; Dong, J.R.; Zeng, X.B.; Myneni, R.B.; Dickinson, R.E. Evaluation of the utility of satellite-based vegetation leaf area index data for climate simulations. *J. Clim.* **2001**, *14*, 3536–3550. [CrossRef]
5. Yuan, H.; Dai, Y.J.; Xiao, Z.Q.; Ji, D.Y.; Shangguan, W. Reprocessing the MODIS Leaf Area Index products for land surface and climate modelling. *Remote Sens. Environ.* **2011**, *115*, 1171–1187. [CrossRef]
6. Van den Hurk, B.; Viterbo, P.; Los, S.O. Impact of leaf area index seasonality on the annual land surface evaporation in a global circulation model. *J. Geophys. Res.-Atmos.* **2003**, *108*, 4191. [CrossRef]
7. Cheng, Z.Q.; Meng, J.H.; Wang, Y.M. Improving spring maize yield estimation at field scale by assimilating time-series HJ-1 CCD data into the WOFOST model using a new method with fast algorithms. *Remote Sens.* **2016**, *8*, 303. [CrossRef
8. Systematic Observation Requirements for Satellite-Based Products for Climate 2011 Update: Supplemental Details to the Satellite-Based Component of the "Implementation Plan for the Global Observing System for Climate in Support of the UNFCCC (2010 Update)". Available online: https://library.wmo.int/index.php?lvl=notice_display&id=12907 (accessed on 13 December 2018).
9. Dong, Y.Y.; Zhao, C.J.; Yang, G.J.; Chen, L.P.; Wang, J.H.; Feng, H.K. Integrating a very fast simulated annealing optimization algorithm for crop leaf area index variational assimilation. *Math. Comput. Model.* **2013**, *58*, 871–879. [CrossRef]
10. Jego, G.; Pattey, E.; Liu, J.G. Using Leaf Area Index, retrieved from optical imagery, in the STICS crop model for predicting yield and biomass of field crops. *Field Crop. Res.* **2012**, *131*, 63–74. [CrossRef]
11. Baez-Gonzalez, A.D.; Kiniry, J.R.; Maas, S.J.; Tiscareno, M.; Macias, J.; Mendoza, J.L.; Richardson, C.W.; Salinas, J.; Manjarrez, J.R. Large-area maize yield forecasting using leaf area index based yield model. *Agron. J.* **2005**, *97*, 418–425. [CrossRef]
12. Liang, S.L. Recent developments in estimating land surface biogeophysical variables from optical remote sensing. *Prog. Phys. Geogr.* **2007**, *31*, 501–516. [CrossRef

13. Verrelst, J.; Camps-Valls, G.; Munoz-Mari, J.; Rivera, J.P.; Veroustraete, F.; Clevers, J.; Moreno, J. Optical remote sensing and the retrieval of terrestrial vegetation bio-geophysical properties—A review. *ISPRS-J. Photogramm. Remote Sens.* **2015**, *108*, 273–290. [CrossRef]
14. Verrelst, J.; Malenovský, Z.; Van der Tol, C.; Camps-Valls, G.; Gastellu-Etchegorry, J.-P.; Lewis, P.; North, P.; Moreno, J. Quantifying vegetation biophysical variables from imaging spectroscopy data: A review on retrieval methods. *Surv. Geophys.* **2018**, 1–41. [CrossRef]
15. Campos-Taberner, M.; Garcia-Haro, F.J.; Busetto, L.; Ranghetti, L.; Martinez, B.; Gilabert, M.A.; Camps-Valls, G.; Camacho, F.; Boschetti, M. A critical comparison of remote sensing Leaf Area Index estimates over rice-cultivated areas: From Sentinel-2 and Landsat-7/8 to MODIS, GEOV1 and EUMETSAT polar system. *Remote Sens.* **2018**, *10*, 763. [CrossRef]
16. Baret, F.; Buis, S. Estimating canopy characteristics from remote sensing observations: Review of methods and associated problems. In *Advances in Land Remote Sensing: System, Modeling, Inversion and Application*; Liang, S., Ed.; Springer: Dordrecht, The Netherlands, 2008; pp. 173–201. ISBN 978-1-4020-6449-4.
17. Maxwell, A.E.; Warner, T.A.; Fang, F. Implementation of machine-learning classification in remote sensing: An applied review. *Int. J. Remote Sens.* **2018**, *39*, 2784–2817. [CrossRef]
18. Lary, D.J.; Alavi, A.H.; Gandomi, A.H.; Walker, A.L. Machine learning in geosciences and remote sensing. *Geosci. Front.* **2016**, *7*, 3–10. [CrossRef]
19. Wang, T.T.; Xiao, Z.Q.; Liu, Z.G. Performance evaluation of machine learning methods for Leaf Area Index retrieval from time-series MODIS reflectance data. *Sensors* **2017**, *17*, 81. [CrossRef]
20. Noi, P.T.; Kappas, M. Comparison of random forest, k-nearest neighbor, and support vector machine classifiers for land cover classification using Sentinel-2 imagery. *Sensors* **2018**, *18*, 18. [CrossRef]
21. Chen, G.B.; Li, S.S.; Knibbs, L.D.; Hamm, N.A.S.; Cao, W.; Li, T.T.; Guo, J.P.; Ren, H.Y.; Abramson, M.J.; Guo, Y.M. A machine learning method to estimate PM2.5 concentrations across China with remote sensing, meteorological and land use information. *Sci. Total Environ.* **2018**, *636*, 52–60. [CrossRef]
22. Lary, D.J.; Zewdie, G.K.; Liu, X.; Wu, D.; Levetin, E.; Allee, R.J.; Malakar, N.; Walker, A.; Mussa, H.; Mannino, A. Machine learning applications for earth observation. In *Earth Observation Open Science and Innovation*; Mathieu, P.-P., Aubrecht, C., Eds.; Springer: Cham, Switzerland, 2018; pp. 165–218. ISBN 978-3-319-65633-5.
23. Kwon, S.K.; Jung, H.S.; Baek, W.K.; Kim, D. Classification of forest vertical structure in South Korea from aerial orthophoto and lidar data using an artificial neural network. *Appl. Sci.* **2017**, *7*, 1046. [CrossRef]
24. Mohri, M.; Talwalkar, A.; Rostamizadeh, A. *Foundations of Machine Learning*; Dietterich, T., Bishop, C., Heckerman, D., Jordan, M., Kearns, M., Eds.; MIT Press: Cambridge, MA, USA, 2012; ISBN 978-0-262-01825-8.
25. Durbha, S.S.; King, R.L.; Younan, N.H. Support vector machines regression for retrieval of leaf area index from multiangle imaging spectroradiometer. *Remote Sens. Environ.* **2007**, *107*, 348–361. [CrossRef]
26. Karimi, S.; Sadraddini, A.A.; Nazemi, A.H.; Xu, T.R.; Fard, A.F. Generalizability of gene expression programming and random forest methodologies in estimating cropland and grassland leaf area index. *Comput. Electron. Agric.* **2018**, *144*, 232–240. [CrossRef]
27. Verrelst, J.; Alonso, L.; Camps-Valls, G.; Delegido, J.; Moreno, J. Retrieval of vegetation biophysical parameters using Gaussian process techniques. *IEEE Trans. Geosci. Remote Sens.* **2012**, *50*, 1832–1843. [CrossRef]
28. Bacour, C.; Baret, F.; Beal, D.; Weiss, M.; Pavageau, K. Neural network estimation of LAI, fAPAR, fCover and LAIxC(ab), from top of canopy MERIS reflectance data: Principles and validation. *Remote Sens. Environ.* **2006**, *105*, 313–325. [CrossRef]
29. Li, X.; Bai, R.B. Freight Vehicle travel time prediction using gradient boosting regression tree. In Proceedings of the 2016 15th IEEE International Conference on Machine Learning and Applications, Anaheim, CA, USA, 18–20 December 2016; IEEE: Piscataway, NJ, USA, 2016; pp. 1010–1015.
30. Guneralp, I.; Filippi, A.M.; Randall, J. Estimation of floodplain aboveground biomass using multispectral remote sensing and nonparametric modeling. *Int. J. Appl. Earth Obs. Geoinf.* **2014**, *33*, 119–126. [CrossRef]
31. Xiao, Z.B.; Wang, Y.; Fu, K.; Wu, F. Identifying different transportation modes from trajectory data using tree-based ensemble classifiers. *ISPRS Int. Geo-Inf.* **2017**, *6*, 57. [CrossRef]
32. Martinez, B.; Garcia-Haro, F.J.; Camacho-de Coca, F. Derivation of high-resolution leaf area index maps in support of validation activities: Application to the cropland Barrax site. *Agric. For. Meteorol.* **2009**, *149*, 130–145. [CrossRef]

33. Traoré, F. Assessing the impact of China net imports on the world cotton price. *Appl. Econ. Lett.* **2014**, *21*, 1031–1035. [CrossRef]
34. Wang, H.D.; Wu, L.F.; Cheng, M.H.; Fan, J.L.; Zhang, F.C.; Zou, Y.F.; Chau, H.W.; Gao, Z.J.; Wang, X.K. Coupling effects of water and fertilizer on yield, water and fertilizer use efficiency of drip-fertigated cotton in northern Xinjiang, China. *Field Crop. Res.* **2018**, *219*, 169–179. [CrossRef]
35. ESA. GMES Sentinel-2 Mission Requirements Document, Technical Report issue 2 revision 1. Available online: http://esamultimedia.esa.int/docs/GMES/Sentinel-2_MRD.pdf (accessed on 30 March 2019).
36. Jaramaz, D.; Perović, V.; Belanović, S.; Saljnikov, E.; Čakmak, D.; Mrvić, V.; Životić, L. The ESA Sentinel-2 Mission Vegetation Variables for Remote Sensing of Plant Monitoring. In Proceedings of the 2nd International Conference on Regional Development, Spatial Planning and Strategic Governance (RESPAG 2013), Belgrade, Serbia, 22–25 May 2013; Vujošević, M., Milijić, S., Eds.; Institute of Architecture and Urban & Spatial Planning of Serbia (IAUS): Belgrade, Serbia, 2013; pp. 950–961.
37. Delegido, J.; Verrelst, J.; Meza, C.M.; Rivera, J.P.; Alonso, L.; Moreno, J. A red-edge spectral index for remote sensing estimation of green LAI over agroecosystems. *Eur. J. Agron.* **2013**, *46*, 42–52. [CrossRef]
38. Gong, P.; Pu, R.L.; Biging, G.S.; Larrieu, M.R. Estimation of forest leaf area index using vegetation indices derived from Hyperion hyperspectral data. *IEEE Trans. Geosci. Remote Sens.* **2003**, *41*, 1355–1362. [CrossRef]
39. Twele, A.; Erasmi, S.; Kappas, M. Spatially explicit estimation of leaf area index using EO-1 hyperion and landsat ETM+ data: Implications of spectral bandwidth and shortwave infrared data on prediction accuracy in a tropical montane environment. *GISci. Remote Sens.* **2008**, *45*, 229–248. [CrossRef]
40. ESA. Copernicus Open Access Hub. Available online: https://scihub.copernicus.eu/dhus/#/home (accessed on 6 April 2019).
41. Louis, J.; Debaecker, V.; Pflug, B.; Main-Knorn, M.; Bieniarz, J.; Müller-Wilm, U.; Cadau, E.; Gascon, F. SENTINEL-2 SEN2COR: L2A processor for users. In Proceedings of the Living Planet Symposium 2016, Prague, Czech Republic, 9–13 May 2016.
42. Müller-Wilm, U.; Louis, J.; Richter, R.; Gascon, F.; Niezette, M. Sentinel-2 Level-2A prototype processor: Architecture, algorithms and first results. In Proceedings of the ESA Living Planet Symposium 2013, Edinburgh, UK, 9–13 September 2013.
43. Fang, H.L.; Jiang, C.Y.; Li, W.J.; Wei, S.S.; Baret, F.; Chen, J.M.; Garcia-Haro, J.; Liang, S.L.; Liu, R.G.; Myneni, R.B.; et al. Characterization and intercomparison of global moderate resolution leaf area index (LAI) products: Analysis of climatologies and theoretical uncertainties. *J. Geophys. Res.-Biogeosci.* **2013**, *118*, 529–548. [CrossRef]
44. NASA. LAADS DAAC. Available online: https://ladsweb.modaps.eosdis.nasa.gov/ (accessed on 6 April 2019).
45. ESA. Copernicus Global Land Service. Available online: https://land.copernicus.eu/global/ (accessed on 6 April 2019).
46. Myneni, R.; Knyazikhin, Y.; Park, T. MCD15A3H MODIS/Terra+Aqua Leaf Area Index/FPAR 4-day L4 Global 500m SIN Grid V006. NASA EOSDIS Land Processes DAAC. 2015. Available online: http://doi.org/10.5067/MODIS/MCD15A3H.006 (accessed on 6 April 2019).
47. Baret, F.; Weiss, M.; Lacaze, R.; Camacho, F.; Makhmara, H.; Pacholcyzk, P.; Smets, B. GEOV1: LAI and FAPAR essential climate variables and FCOVER global time series capitalizing over existing products. Part1: Principles of development and production. *Remote Sens. Environ.* **2013**, *137*, 299–309. [CrossRef]
48. Baret, F.; Weiss, M.; Verger, A.; Smets, B. ATBD FOR LAI, FAPAR AND FCOVER FROM PROBA-V PRODUCTS AT 300M RESOLUTION (GEOV3). Available online: https://land.copernicus.eu/global/sites/cgls.vito.be/files/products/ImagineS_RP2.1_ATBD-LAI300m_I1.73.pdf (accessed on 30 March 2019).
49. Scikit-Learn Developers. Scikit-learn. Available online: https://scikit-learn.org/stable/index.html (accessed on 6 April 2019).
50. Python Software Foundation. Python. Available online: https://www.python.org/ (accessed on 6 April 2019).
51. Pedregosa, F.; Varoquaux, G.; Gramfort, A.; Michel, V.; Thirion, B.; Grisel, O.; Blondel, M.; Prettenhofer, P.; Weiss, R.; Dubourg, V.; et al. Scikit-learn: Machine learning in Python. *J. Mach. Learn. Res.* **2011**, *12*, 2825–2830.
52. Adelabu, S.; Mutanga, O.; Adam, E. Testing the reliability and stability of the internal accuracy assessment of random forest for classifying tree defoliation levels using different validation methods. *Geocarto Int.* **2015**, *30*, 810–821. [CrossRef]

53. Omer, G.; Mutanga, O.; Abdel-Rahman, E.M.; Adam, E. Performance of support vector machines and artificial neural network for mapping endangered tree species using WorldView-2 data in Dukuduku Forest, South Africa. *IEEE J. Sel. Top. Appl. Earth Observ. Remote Sens.* **2015**, *8*, 4825–4840. [CrossRef]
54. Dube, T.; Mutanga, O.; Adam, E.; Ismail, R. Intra-and-inter species biomass prediction in a plantation forest: Testing the utility of high spatial resolution spaceborne multispectral rapideye sensor and advanced machine learning algorithms. *Sensors* **2014**, *14*, 15348–15370. [CrossRef]
55. Verrelst, J.; Munoz, J.; Alonso, L.; Delegido, J.; Rivera, J.P.; Camps-Valls, G.; Moreno, J. Machine learning regression algorithms for biophysical parameter retrieval: Opportunities for Sentinel-2 and -3. *Remote Sens. Environ.* **2012**, *118*, 127–139. [CrossRef]
56. van Gerven, M. Computational foundations of natural intelligence. *Front. Comput. Neurosci.* **2017**, *11*, 7–30. [CrossRef]
57. Waske, B.; Fauvel, M.; Benediktsson, J.A.; Chanussot, J. Machine learning techniques in remote sensing data analysis. In *Kernel Methods for Remote Sensing Data Analysis*; Camps-Valls, G., Bruzzone, L., Eds.; John Wiley & Sons: Hoboken, NJ, USA, 2009; pp. 3–24. ISBN 978-470-72211-4.
58. Rumelhart, D.E.; Hinton, G.E.; Williams, R.J. Learning representations by back-propagating errors. *Nature* **1986**, *323*, 533–536. [CrossRef]
59. Hornik, K.; Stinchcombe, M.; White, H. Multilayer feedforward networks are universal approximators. *Neural Netw.* **1989**, *2*, 359–366. [CrossRef]
60. Sonmez, H.; Gokceoglu, C.; Nefeslioglu, H.A.; Kayabasi, A. Estimation of rock modulus: For intact rocks with an artificial neural network and for rock masses with a new empirical equation. *Int. J. Rock Mech. Min. Sci.* **2006**, *43*, 224–235. [CrossRef]
61. Madhiarasan, M.; Deepa, S.N. Comparative analysis on hidden neurons estimation in multi layer perceptron neural networks for wind speed forecasting. *Artif. Intell. Rev.* **2017**, *48*, 449–471. [CrossRef]
62. Huang, G.B. Learning capability and storage capacity of two-hidden-layer feedforward networks. *IEEE Trans. Neural Netw.* **2003**, *14*, 274–281. [CrossRef]
63. Cortes, C.; Vapnik, V. Support-vector networks. *Mach. Learn.* **1995**, *20*, 273–297. [CrossRef]
64. Vapnik, V.; Golowich, S.E.; Smola, A. Support vector method for function approximation, regression estimation, and signal processing. In Proceedings of the 9th International Conference on Neural Information Processing Systems, Denver, CO, USA, 3–5 December 1996; Mozer, M.C., Jordan, M.I., Petsche, T., Eds.; MIT Press: Cambridge, MA, USA, 1996; pp. 281–287.
65. Mountrakis, G.; Im, J.; Ogole, C. Support vector machines in remote sensing: A review. *ISPRS-J. Photogramm. Remote Sens.* **2011**, *66*, 247–259. [CrossRef]
66. Basak, D.; Pal, S.; Patranabis, D.C. Support vector regression. *Neural Inf. Process. Lett. Rev.* **2007**, *11*, 203–224.
67. Awad, M.; Khanna, R. *Efficient Learning Machines: Theories, Concepts, and Applications for Engineers and System Designers*; Pepper, J., Weiss, S., Hauke, P., Eds.; Apress: Berkeley, CA, USA, 2015; ISBN 978-1-4302-5990-9.
68. Ramedani, Z.; Omid, M.; Keyhani, A.; Shamshirband, S.; Khoshnevisan, B. Potential of radial basis function based support vector regression for global solar radiation prediction. *Renew. Sust. Energ. Rev.* **2014**, *39*, 1005–1011. [CrossRef]
69. Li, M.; Liu, Y.H. Learning interaction force model for endodontic shaping with support vector regression. In Proceedings of the 2006 IEEE International Conference on Robotics and Automation, Orlando, FL, USA, 15–19 May 2006; IEEE: Piscataway, NJ, USA, 2006; pp. 3642–3647.
70. Rasmussen, C.E.; Williams, C.K. *Gaussian Process for Machine Learning*; Dietterich, T., Bishop, C., Heckerman, D., Jordan, M., Kearns, M., Eds.; MIT Press: Cambridge, MA, USA, 2006; ISBN 0-262-18253-X.
71. Scornet, E. Random forests and Kernel methods. *IEEE Trans. Inf. Theory* **2016**, *62*, 1485–1500. [CrossRef]
72. Breiman, L. Random forests. *Mach. Learn.* **2001**, *45*, 5–32. [CrossRef]
73. Friedman, J.H. Greedy function approximation: A gradient boosting machine. *Ann. Stat.* **2001**, *29*, 1189–1232. [CrossRef]
74. Natekin, A.; Knoll, A. Gradient boosting machines, a tutorial. *Front. Neurorobot.* **2013**, *7*, 21. [CrossRef] [PubMed]
75. Fischer, P.; Etienne, C.; Tian, J.J.; Krauss, T. Prediction of wind speeds based on digital elevation MODELS using boosted regression trees. In Proceedings of the International Conference on Sensors & Models in Remote Sensing & Photogrammetry, Kish Island, Iran, 23–25 November 2015; Arefi, H., Motagh, M., Eds.; Copernicus Gesellschaft Mbh: Göttingen, Germany, 2015; pp. 197–202.

76. Kanungo, T.; Orr, D. Predicting the readability of short web summaries. In Proceedings of the Second ACM International Conference on Web Search and Data Mining, Barcelona, Spain, 9–12 February 2009; Baeza-Yates, R., Boldi, P., Ribeiro-Neto, B., Cambazoglu, B.B., Eds.; ACM: New York, NY, USA, 2009; pp. 202–211.
77. Szekely, G.J.; Rizzo, M.L.; Bakirov, N.K. Measuring and testing dependence by correlation of distances. *Ann. Stat.* **2007**, *35*, 2769–2794. [CrossRef]
78. Li, R.Z.; Zhong, W.; Zhu, L.P. Feature screening via distance correlation learning. *J. Am. Stat. Assoc.* **2012**, *107*, 1129–1139. [CrossRef]
79. Zhong, W.; Zhu, L.P. An iterative approach to distance correlation-based sure independence screening. *J. Stat. Comput. Simul.* **2015**, *85*, 2331–2345. [CrossRef]
80. Kundu, P.P.; Mitra, S. Feature selection through message passing. *IEEE Trans. Cybern.* **2017**, *47*, 4356–4366. [CrossRef] [PubMed]
81. Guyon, I.; Elisseeff, A. An introduction to variable and feature selection. *J. Mach. Learn. Res.* **2003**, *3*, 1157–1182. [CrossRef]
82. Hong, X.; Mitchell, R.J. Backward elimination model construction for regression and classification using leave-one-out criteria. *Int. J. Syst. Sci.* **2007**, *38*, 101–113. [CrossRef]
83. Yuan, H.H.; Yang, G.J.; Li, C.C.; Wang, Y.J.; Liu, J.G.; Yu, H.Y.; Feng, H.K.; Xu, B.; Zhao, X.Q.; Yang, X.D. Retrieving soybean Leaf Area Index from unmanned aerial vehicle hyperspectral remote sensing: Analysis of RF, ANN, and SVM regression models. *Remote Sens.* **2017**, *9*, 309. [CrossRef]
84. Siegmann, B.; Jarmer, T. Comparison of different regression models and validation techniques for the assessment of wheat leaf area index from hyperspectral data. *Int. J. Remote Sens.* **2015**, *36*, 4519–4534. [CrossRef]
85. LeCun, Y.; Bengio, Y.; Hinton, G. Deep learning. *Nature* **2015**, *521*, 436–444. [CrossRef] [PubMed]

7

Learning-Based Colorization of Grayscale Aerial Images using Random Forest Regression

Dae Kyo Seo [1], Yong Hyun Kim [2], Yang Dam Eo [3,*] and Wan Yong Park [4]

1 Department of Advanced Technology Fusion, Konkuk University, Seoul 05029, Korea; tjeory@konkuk.ac.kr
2 Department of Civil and Environmental Engineering, Seoul National University, Seoul 08826, Korea; yhkeen@gmail.com
3 Department of Technology Fusion Engineering, Konkuk University, Seoul 05029, Korea
4 Agency for Defense Development, Daejeon 34060, Korea; wypark@add.re.kr
* Correspondence: eoandrew@konkuk.ac.kr

Abstract: Image colorization assigns colors to a grayscale image, which is an important yet difficult image-processing task encountered in various applications. In particular, grayscale aerial image colorization is a poorly posed problem that is affected by the sun elevation angle, seasons, sensor parameters, etc. Furthermore, since different colors may have the same intensity, it is difficult to solve this problem using traditional methods. This study proposes a novel method for the colorization of grayscale aerial images using random forest (RF) regression. The algorithm uses one grayscale image for input and one-color image for reference, both of which have similar seasonal features at the same location. The reference color image is then converted from the Red-Green-Blue (RGB) color space to the CIE L*a*b (Lab) color space in which the luminance is used to extract training pixels; this is done by performing change detection with the input grayscale image, and color information is used to establish color relationships. The proposed method directly establishes color relationships between features of the input grayscale image and color information of the reference color image based on the corresponding training pixels. The experimental results show that the proposed method outperforms several state-of-the-art algorithms in terms of both visual inspection and quantitative evaluation.

Keywords: colorization; random forest regression; grayscale aerial image; change detection

---

1. Introduction

Image colorization can be described as the process of assigning colors to the pixels of a grayscale image in order to increase the image's visual appeal [1]. This application is often utilized in the image processing community to colorize old grayscale images or movies [2]. Particularly in aerial and satellite imagery, the problem with colorization occurs in a multitude of scenarios that seek to replace topographic maps with vivid, photorealistic renderings of terrain models [3]. There are three main reasons for colorizing aerial and satellite images: (1) grayscale satellite images are available at higher spatial resolutions than their color counterparts are; (2) there are many old grayscale aerial and satellite images that should be represented by color images, typically for monitoring purposes; and (3) grayscale aerial–satellite images can be obtained for approximately one-tenth the cost of color images of the same resolution [3].

In the case of satellite images, modern systems acquire a panchromatic (grayscale) image, with high spatial and low spectral resolutions, and a multispectral (color) image that has complementary properties [4]. In other words, grayscale and color images with different resolutions over the same time period are provided. In order to perform colorization through the color information of the multispectral image while maintaining the high resolution of the panchromatic image, the two components are fused, which is called pansharpening [5]. The fused images provide increased interpretation capabilities and more reliable results [6]. However, this colorization method is confined to satellite images that provide panchromatic and multispectral images of the same time periods, making this type of colorization of aerial images impossible.

On the other hand, grayscale aerial image colorization is a poorly posed problem with more than one solution [7]. As mentioned above, the satellite images fuse the grayscale and color images from the same time period to perform colorization, whereas aerial images do not usually have the two types of imagery available. Furthermore, in contrast to the natural images, this colorization solution depends on the sun elevation angle, season, sensor parameters, etc. It is also problematic that the same intensity may represent different colors, so there is no exact solution [8]. In general, existing colorization methods can be divided into three main categories, all of which have limitations: user-scribbled methods, example-based methods, and those that employ a large number of training images [9]. User-scribbled techniques [10–13] require a user to manually add colored marks to a grayscale image [13]. The colors from these marks are then smoothly propagated across the entire image, based on an optimization framework. A key weakness is that such methods require users to provide a considerable number of scribbles on the grayscale image, which is time-consuming and requires expertise [14]. Moreover, it is almost impossible to add such markings to large volumes (gigabytes) of aerial imagery. In the case of the example-based method [1,8,9,14–16], it typically transfers the color information from a similar reference image to the input grayscale image rather than obtaining chromatic values from the user, thereby reducing the burden on users. However, as feature matching is critical to the quality of the results, satisfactory results cannot be obtained if feature matching is not performed correctly [15]. Moreover, the procedure is very sensitive to image brightness and contrast, whereas real aerial images always include large areas of shadow and low contrast, due to relief, vignetting, and so on. An alternative approach is to employ a large number of training images [17–19], which is a recent example of deep learning. These methods use multiple color images to automatically transfer the color information to the grayscale image, and deep neural networks are used to solve the colorization problem. A large database of color images comprising all kinds of objects is used for training the neural networks. The trained model can then be used to efficiently colorize grayscale images. However, this approach is computationally expensive, and the training is significantly slow [15].

In order to overcome these limitations, this study presents a new, fast learning-based technique for the colorization of grayscale aerial images that colorizes them without any user intervention. The algorithm uses one grayscale image as the input and one-color image for reference, both of which have similar seasonal features at the same location. Then, the reference color image is converted from the Red-Green-Blue (RGB) color space to a CIE L*a*b (Lab) color space in which luminance and two-dimensional (2D) color information are stored. Change detection between the input grayscale image and the luminance of the reference color image is performed, and the unchanged region is selected as training pixels, which allows for the extraction of meaningful training data. For colorization, the relationships are established through learning between features of the input grayscale image and the 2D color information of the reference color image based on training pixels. In other words, for the corresponding unchanged region, the color relationships between the two images are directly established, and the colors for the changed region are predicted. At this time, the study's main technical framework is random forest (RF) regression. Random forest is a data-mining method that has some advantages over most statistical modeling methods [20], including: the ability to model highly nonlinear dimensional relationships; resistance to overfitting; relative robustness with respect to the presence of noise in the data; and the capacity to determine the relevance of the variables used.

The main contributions of this paper can be summarized as follows: (1) to the best of our knowledge, this is the first work that exploits RF regression for aerial imagery colorization, although it has been used for natural image colorization [21–23]; (2) this paper develops a novel algorithm that establishes color relationships based on unchanged regions, which predict the color values of the changed regions; (3) this paper establishes color relationships by directly mapping the features of the input grayscale image with the color information of the reference color image; and (4) this paper performs visual and quantitative analyses that show that our method outperforms the current state-of-the-art methods. The rest of this paper is organized as follows. Section 2 describes the materials used in detail, the background of RF regression, and the proposed algorithm. Section 3 presents the colorization results and a detailed comparison with other state-of-the-art colorization algorithms. Section 4 presents the conclusions of the study.

2. Materials and Methods

2.1. Study Site and Data

The study sites are located in Gwangjin-gu, in the central–western part of South Korea. The input grayscale images were acquired on 10 June 2013, and the reference color images were acquired on 2 June 2016; both are aerial images at 1 m resolution. Coordinate definition and geometric correction were performed on the images using Environment for Visualizing Images (ENVI) geospatial analytical software (version 4.7, HARRIS Geospatial Solutions, Broomfield, CO, USA). The coordinate system of each image was projected as World Geodetic System (WGS) 84 Universal Transverse Mercator Coordinate System (UTM) 52N, and 30 Ground Control Points (GCPs) were selected for image registration. The 30 GCPs returned a total root mean square error of 0.4970, satisfying values within 0.5 m. Then, based on these GCPs, image registration was performed using the "Warp from GCPs: Image to Image" tool in ENVI. Finally, to achieve reasonable computational time, only a portion of the image was extracted prior to conducting the experiments, which was selected to be 1200 × 1200 pixels. A total of four sites were extracted and are shown in Figures 1–4.

(a) (b)

Figure 1. Experimental area of site 1: (**a**) input grayscale image acquired on 10 June 2013, (**b**) reference color image acquired on 2 June 2016.

(a) (b)

Figure 2. Experimental area of site 2: (**a**) input grayscale image acquired on 10 June 2013, (**b**) reference color image acquired on 2 June 2016.

(a) (b)

Figure 3. Experimental area of site 3: (**a**) input grayscale image acquired on 10 June 2013, (**b**) reference color image acquired on 2 June 2016.

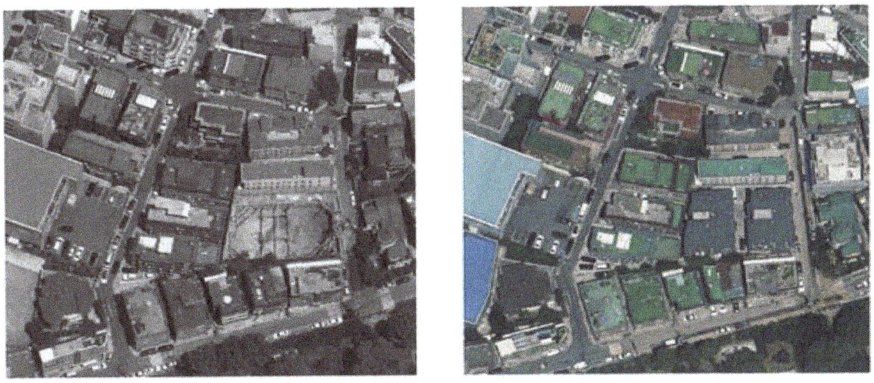

(a) (b)

Figure 4. Experimental area of site 4: (**a**) input grayscale image acquired on 10 June 2013, (**b**) reference color image acquired on 2 June 2016.

2.2. Random Forest

Random forest is a highly versatile ensemble of decision trees that performs well for linear and non-linear prediction by finding a balance between bias and variance [20]. This ensemble learning method constructs and subsequently averages a large number of decision trees for classification or regression purposes [24,25]. At this time, to avoid correlation among the trees, RF increases the diversity of the trees by forcing them to grow from different training data created through a procedure called bootstrap aggregating (bagging) [26]. Bagging refers to aggregating base learners trained through bootstrapping, which creates training data subsets by randomly resampling a given original dataset [27]. That is, as a process of de-correlating trees to train different datasets, it increases stability and makes it more robust when facing slight variations in the input data [28]. Furthermore, approximately one-third of the samples are excluded for the tree training in the bagging process, which is known as the "out-of-bags" (OOB) samples. These OOB samples can be used to evaluate performance, which allows the RF to compute an unbiased estimation of the generalization error without using an external data subset [29]. From the predictions of the OOB samples for every tree in the forest, the mean square error (MSE) is calculated, and the overall MSE is obtained by aggregation, as shown in Equation (1):

$$MSE_{OOB} = \frac{1}{n}\sum_{i=1}^{n}\left(y_i - \overline{\hat{y}_{i_{OOB}}}\right)^2 \tag{1}$$

where n is the number of trees, y_i denotes the prediction for the ith observation, and $\overline{\hat{y}_{i_{OOB}}}$ denotes the average of the OOB predictions for the ith observation.

Furthermore, the use of RF requires the specification of two standard parameters: the number of variables to be selected and tested for the best split at each node (mtry), and the number of trees to be grown (ntree). At each node per tree, the number of mtry variables from the total variables in the model is selected at random, the variable that best splits the input space and the corresponding split are computed, and the input space is split at this point [30]. In a regression problem, the standard value for mtry is one-third of the total number of variables due to computational benefits [31,32]. In the case of ntree, the majority of the studies set the ntree value to 500 since the errors are stabilized before this number of regression trees is achieved [33]. However, recent studies have found that the number of trees does not significantly affect performance improvement, which allows the selection of ntree to consider the performance and training time together [34–36].

Random forest can also be used to assess the importance of each variable during modeling. To determine the importance of input variables, a variable is randomly permuted, and regression trees are grown on the modified dataset. The measure of each variable's importance is then calculated as the difference in the MSE between the original OOB dataset and the modified dataset [37,38]. A key advantage of RF variable importance is that it not only deals with the impact of each variable individually but also looks at multivariate interactions with other variables [39].

2.3. Method

The proposed colorization framework can be decomposed into four steps: (1) pre-processing, (2) feature extraction, (3) colorization, and (4) post-processing, all of which are shown in Figure 5. The first step is to convert the color space and to select the pixels to be used in training for colorization. The second step is to extract feature descriptors of the input grayscale image to be used in learning for color prediction. In the third step, color relationships are established through the proposed method, and colorization is performed on the input grayscale image. The fourth step improves the colorization result by adjusting the histogram. Each of these steps is described below.

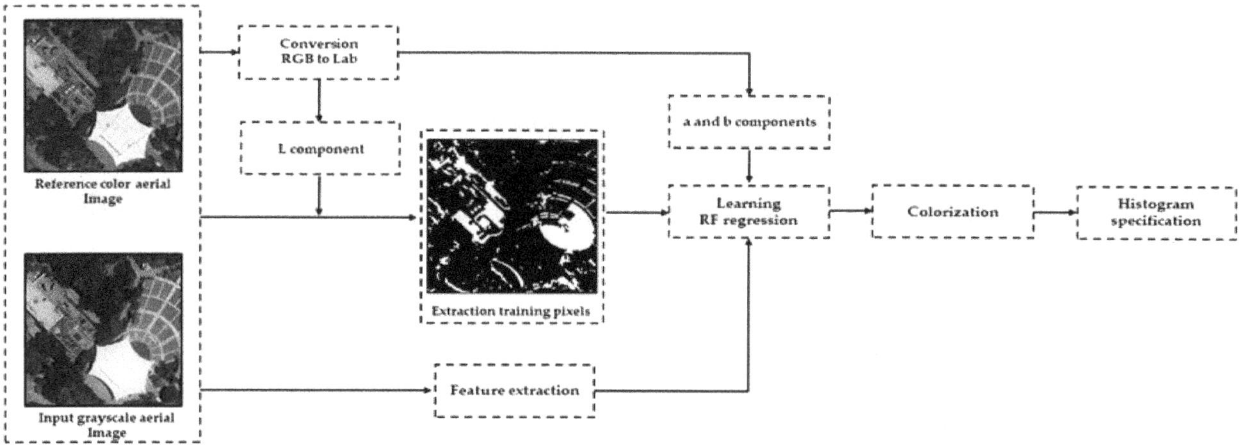

Figure 5. The flowchart of the proposed method. RGB: Red-Green-Blue color space, Lab: CIE L*a*b color space, L component: grayscale axis, a and b components: two-color axes, RF: random forest.

2.3.1. Preprocessing

As mentioned above, the proposed preprocessing step is divided into color-space conversion and extraction of training pixels. First, in this study, a Lab color space is selected since its underlying metric has been designed to express color coherency. Furthermore, Lab has three coordinates: L is the luminance, or lightness, which consequently represents the grayscale axis, whereas a and b represent the two-color axes [16]. In other words, the L component can be known in advance through the input grayscale image, and only the remaining 2D color information, a and b, can be predicted [23]. Thus, the color space of the reference color image is converted from RGB to Lab, and, in the colorization step, only the color relationships for a and b are established through regression. Then, in order to extract a meaningful set of training data, change detection between the input grayscale image and the L component of the reference color image is performed. The change detection method used here comprises two steps for accurate extraction. The first step is a pixel-based method that uses principal component analysis (PCA) [40], and the threshold for distinguishing between changed and unchanged pixels is selected using Otsu's method. However, this process will result in fragmentation and incomplete expression of the change [41]. Therefore, the second step applies the object-based method, which consists of four sub-steps: (1) the morphological closing operation, (2) gap filling, (3) the morphological opening operation, and (4) elimination of small patches.

The morphological closing operation is the combination of dilation followed by erosion, which is used to remove holes in the image [42]. Thus, the closing operation is applied to the image to fill the spaces. Then, gaps within the changed regions that are not filled by the closing operation are additionally filled, which makes the changed information more complete [41]. The morphological opening is then applied, in which erosion is conducted on the image, and it is followed by a dilation operation. The aim of the opening is to remove unnecessary portions. For the structure elements used in the morphological operation, the closing and the opening are set to 3×3 and 5×5, respectively, as selected in Xiao et al. [41]. Small, insignificant patches persist following the opening processing, which can be removed by applying an area threshold. The area threshold is set based on the minimum object size and is acquired through a zero-parameter version of simple linear iterative clustering (SLICO). The SLICO is a spatially localized version of the k-means [43]. To initialize it, the k-cluster centers, which are located on a regular grid and spaced S pixels apart, are sampled [44]. Then, an iterative procedure assigns each pixel to a cluster center using the distance measure D, as defined in Equation (2), which combines the distance of color proximity (Equation (3)) and the distance of spatial proximity (Equation (4)) [45]:

$$\mathrm{D} = \sqrt{\left(\frac{d_c}{m}\right)^2 + \left(\frac{d_s}{S}\right)^2} \tag{2}$$

$$d_c = \sqrt{\sum_{s_i \in S} \left(I(x_1, y_1, s_i) - I(x_2, y_2, s_i)\right)^2} \tag{3}$$

$$d_s = \sqrt{(x_1 - x_2)^2 + (y_1 - y_2)^2} \tag{4}$$

where d_c and d_s represent the color and spatial distance between pixels $I(x_1, y_1, s_i)$ and $I(x_2, y_2, s_i)$ in the spectral band $\mathrm{S_i}$, respectively; and m controls the compactness of the superpixels, which is adaptively chosen for each superpixel. In this study, the number of iterations is set to ten, which is sufficient for error convergence [46], and the initial size of the superpixels is set to 10×10, which represents optimal accuracy and computational time [45]. Finally, the unchanged regions are used for training, which consists of establishing color relationships in the unchanged regions from which those of the changed regions can be predicted.

2.3.2. Feature Extraction

The gray level of one pixel is not informative for color prediction, so additional information such as texture and local context is necessary [15]. In order to extract the maximum information for color prediction, features that describe the textural information of local neighborhoods of pixels are considered. Previous automatic colorization methods used Speed-Up Robust Features, Gabor features, or a histogram of oriented gradients as base tools for textural analysis [14,15]. These descriptors are known to be discriminative but also computationally and memory intensive due to their high number of features. Moreover, recent approaches have separated texture from structure using relative total variation, but their descriptors are not sufficiently accurate to discriminate textures among themselves. Consequently, statistical features are utilized here, as this approach is simple, easy to implement, and has strong adaptability and robustness, among which the gray-level co-occurrence matrix (GLCM) is used [47,48]. The GLCM is used extensively in texture description, and the co-occurrence matrices provide better results than do other forms of texture discrimination [49,50]. For remote sensing images, four types of statistics—angular second moment, contrast, correlation, and entropy—are better suited to texture feature extraction, so they have been selected for statistics in this study [51]. Also, in order to calculate the GLCM values, the window size should be set. The present study has ultimately selected a 5×5 window size, which better reflects coarse and fine textures [48]. Furthermore, the intensity value and the mean and standard deviation of the intensity within a 5×5 neighborhood are included as supplementary components.

2.3.3. Colorization

Like other learning-based approaches, this step consists of two components: (1) a training component and (2) a prediction component, which are described below.

In the training component, image colorization is formulated as a regression problem and is solved using RF. The training data employ seven features of the input grayscale image corresponding to the unchanged regions extracted from the preprocessing step. At this time, these features are trained with the a and b components of the reference color image at the same pixel location. In other words, rather than establishing the color relationships between the L component and the a and b components of the reference color image like other colorization methods do [15,18,19,23], this study establishes color relationships directly between the input grayscale image and the a and b components of the reference color image. Then, the RF regression parameters—mtry and ntree values—are defined. The mtry value is generally set to one-third of the number of features in the regression problem due to computational advantages, but, in this study, the total number of features is utilized, as in the original classification and regression trees procedure, since the number of features is not large enough to affect the computational time [32]. Furthermore, the ntree value is set to 32, which takes into account the performance and training time of the RF regression [35].

The prediction portion of this step uses the RF regression obtained from the training component to predict the colors for the a and b components of the input grayscale image. Then, the input grayscale image is used as the L component, and it is combined with the predicted a and b components. Finally, the Lab image is converted to the RGB color space.

2.3.4. Post-Processing

Finally, the colorized image acquired above is adjusted to reflect global properties based on histogram specification. Histogram specification is a useful technique for modifying histograms via image enhancement without losing the original histogram characteristics [52,53]. Essentially, the input histogram is transformed into a histogram of the specified image to highlight specific ranges. The histogram specification procedure is defined below.

First, the cumulative density functions (CDFs) of the input and specified images are acquired, as shown in Equations (5) and (6):

$$s_k = C_r(r_k) = \sum_{i=0}^{k} P(r_i) = \sum_{i=0}^{k} \frac{n_i}{n} \tag{5}$$

$$v_k = C_z(z_k) = \sum_{i=0}^{k} P(z_i) \tag{6}$$

where s_k and v_k are the respective histograms of the input and specified images, $C_r(r_k)$ and $C_z(z_k)$ are the CDFs of the respective input and specified images, $P(r_i)$ and $P(z_i)$ are the probability density functions of the respective input and specified images, $k = 0, 1, 2, \ldots, L-1$, L is the total number of gray levels, and n_i is the total number of gray levels r_i [53]. Then, the value of z_k, which satisfies Equation (7), is identified:

$$\{ (C_z(z_k) - s_k) = (v_k - s_k) \} \to 0 \tag{7}$$

In other words, the smallest integer between v_k and s_k should be determined. Finally, the mapping table of z_k will be the output of Equation (8):

$$z_k = C_z^{-1}(s_k) = C_z^{-1}(C_r(r_k)) \tag{8}$$

In this study, the colorization image that is converted to the RGB color space is selected as the input image, and the reference color image is selected as the specified image; histogram specification is carried out for the red, green, and blue bands.

3. Results and Discussion

3.1. Implementation and Performance

This section presents the colorization results of the proposed algorithm and compares these with the results of other state-of-the-art colorization algorithms. To ensure a fair comparison, colorization algorithms that use only a single-color image as the reference (exemplar-based method) and that are based on RF regression are compared, so that the methods of Welsh et al. [1], Bugeau et al. [14], Gupta et al. [15], Gupta et al. [21], and Deshpande et al. [22] are selected. When performing the colorization using these other methods, we used the same parameter settings suggested by their respective authors. In addition, two of the latest deep learning-based methods [18,19] are included for visual comparison, using the codes provided by the authors. The results of the various algorithms are compared by visual inspection (see Figures 6–9) and quantitative evaluation with ground-truth images, which are the actual color images of the input grayscale images.

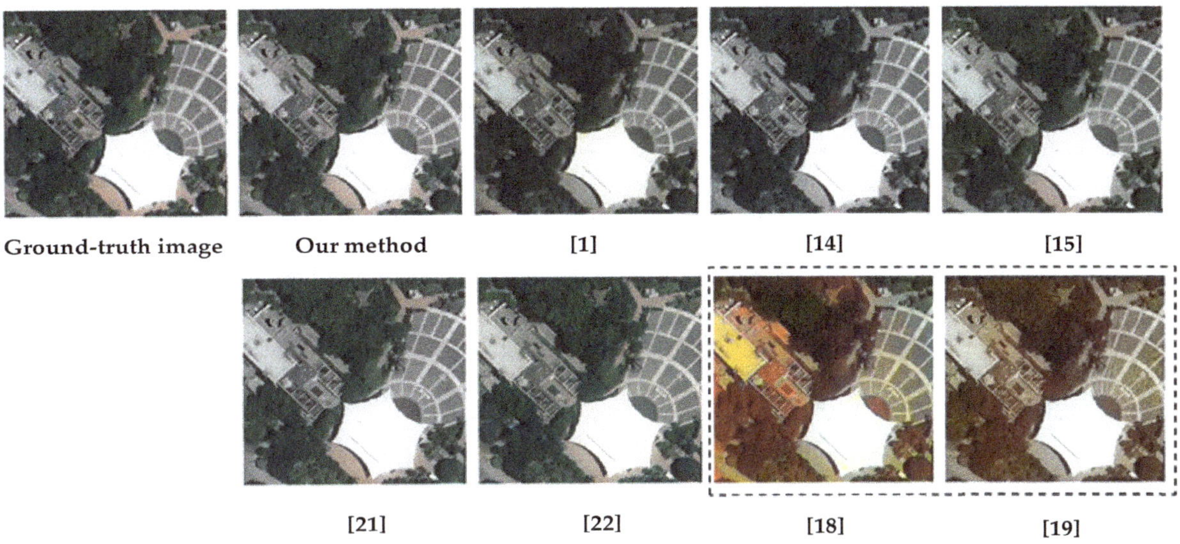

Figure 6. Comparison with existing state-of-the-art colorization methods at site 1.

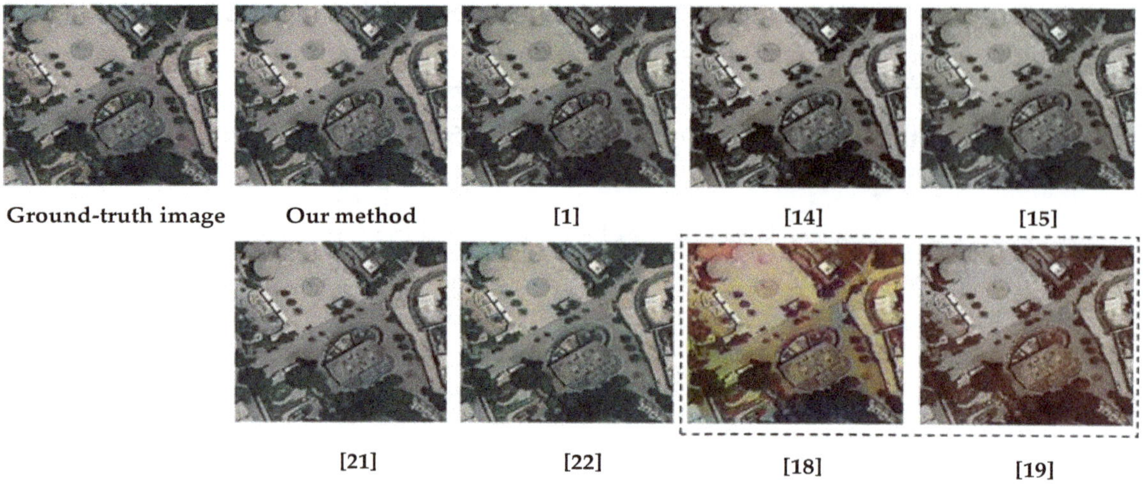

Figure 7. Comparison with existing state-of-the-art colorization methods at site 2.

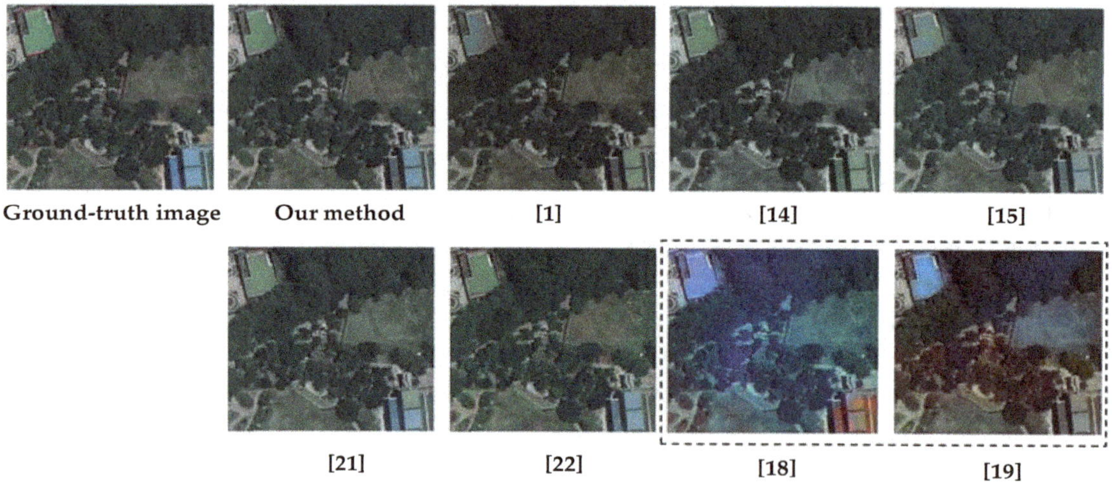

Figure 8. Comparison with existing state-of-the-art colorization methods at site 3.

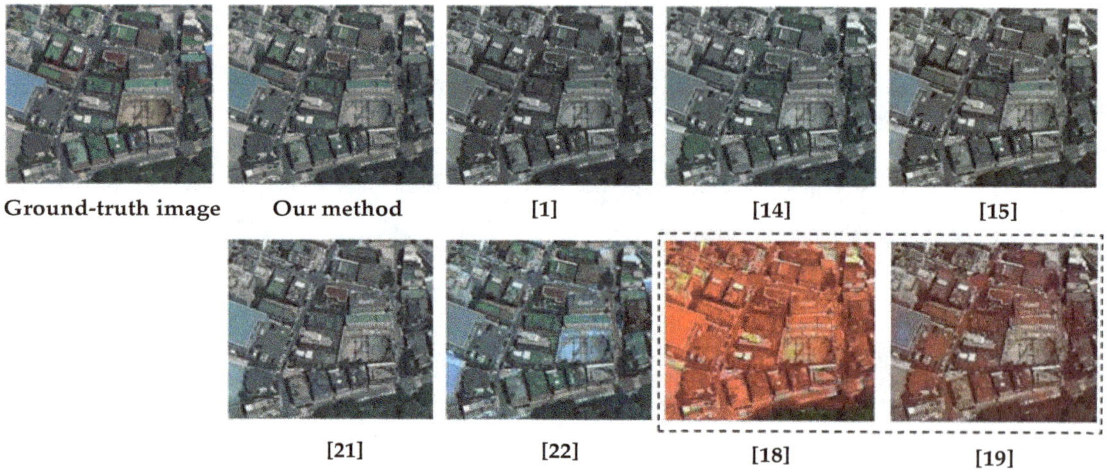

Figure 9. Comparison with existing state-of-the-art colorization methods at site 4.

From the overall visual inspection, the proposed algorithms appear to show better results than do the five existing methods for most cases. The Welsh et al. method [1] is based on transferring color from one initial color image considered as an example. Color prediction is performed by minimizing the distance in the simple statistics of a luminance image. However, as the results show, only a few colors are dealt with, and the results include many artifacts due to the lack of any spatial coherency criteria. In the case of the Bugeau et al. method [14], image colorization is solved by computing the colors using different features and associated metrics. Then, the best colors are automatically selected via a variational framework. Overall, the results fail to select the best colors and have a desaturating effect, confirming the limitations of the variational framework. The Gupta et al. method [15] performs image colorization by automatically exploiting multiple image features. This method transfers color information by performing feature-matching between a reference color image and an input grayscale image—a process that is critical to the quality of the results. This achieves better colorization than the other exemplar-based methods but still only deals with a few colors, resulting in incorrect matches in challenging scenarios.

The other Gupta et al. method [21] is a learning-based method in which learning is performed in superpixel units to enforce higher color consistency. Superpixels are extracted, features are computed for each superpixel, and learning is performed based on an RF regression. The results of this method contain more color information than do the results of the other exemplar-based methods, but the approach still does not retrieve certain colors such as blue or red (Figure 7), and it sometimes predicts completely different colors (as in Figures 8 and 9). The Deshpande et al. method [22] is also based on RF regression, which is performed within the LEArning to seaRCH (LEARCH) framework. Furthermore, histogram correction is performed on the colorization image to improve the visual appeal of the results. In the state-of-the-art-methods used for comparison, this method predicts sufficient color information. However, halo effects exist in the object boundaries, especially in Figure 7. Moreover, as shown in Figure 9, the more complex the structure, the more halo effects are added, leading to many artifacts.

In addition to exemplar-based and RF regression-based methods, deep learning-based methods [18,19] are used for comparison. These colorization algorithms use millions of images for training neural networks, which are based on ImageNet and convolutional neutral networks (CNN). Both results contain color information that is completely different from the ground-truth or reference color images, as shown in Figures 6–9. For example, although the colors of the tree are somewhat predicted, the colors of the buildings or the roads are not predicted at all, which suggests that artifacts are more obvious when the structure is complex. In other words, it is impossible to colorize aerial images through a model that is trained with natural images.

As can be seen in Figures 6–9, our approach more accurately predicts colors than do the other methods, producing results with fewer artifacts. In Figure 6, the color determined by our method is much clearer, especially in the red portion of the ground that is correctly recovered without artifacts. Figure 7 is a site with many human-made objects, and our method demonstrates remarkably high performance in color prediction, while other methods completely fail to correctly predict colors or contain halo effects. Figure 8 also has many human-made objects that contain multiple colors in the area of vegetation, which makes it more difficult to predict the color values at this site. Except for our method and that of Deshpande et al. [22], the colors of human-made objects are not correctly predicted, which indicates that our method is robust for color prediction. Figure 9, the urban area, contains the most complex structure and the greatest variety of colors for prediction. Although the results of the proposed method also contain slightly turbid colors, our method produces significantly better results than do the other methods. In other words, our method retrieves more color values with fewer artifacts from the reference image; these details can be confirmed in Figures 10–13.

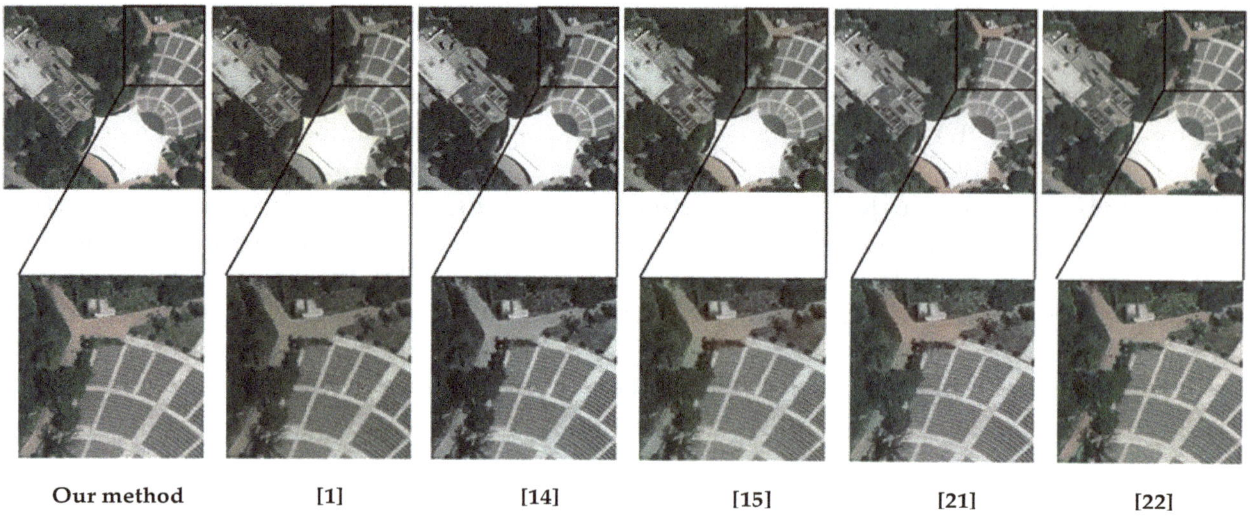

Figure 10. Enlargement of site 1: our method retrieves color values well in this region, compared with other methods.

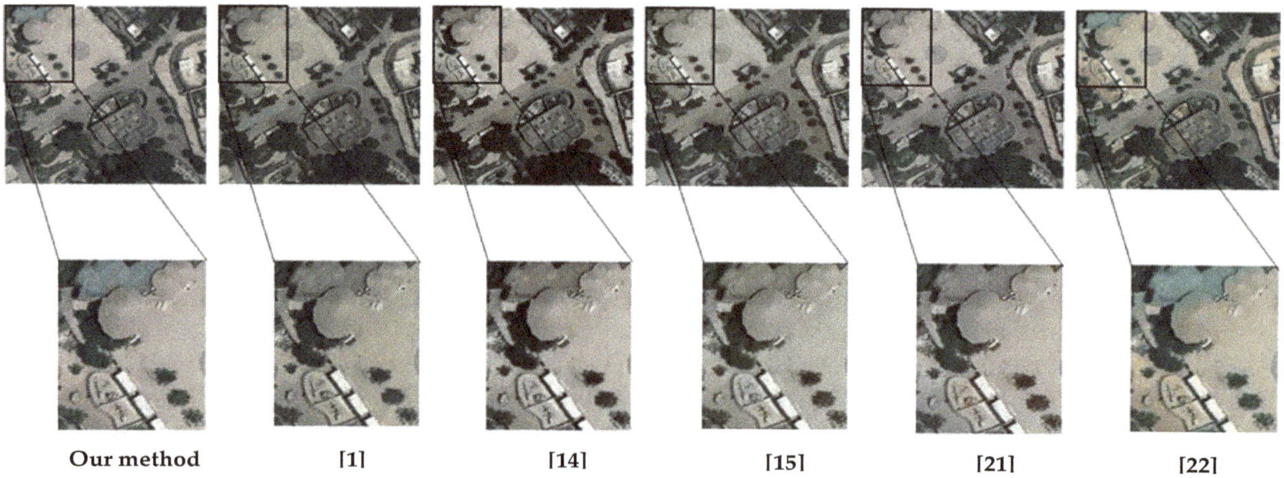

Figure 11. Enlargement of site 2: our method retrieves color values well in this region, compared with other methods.

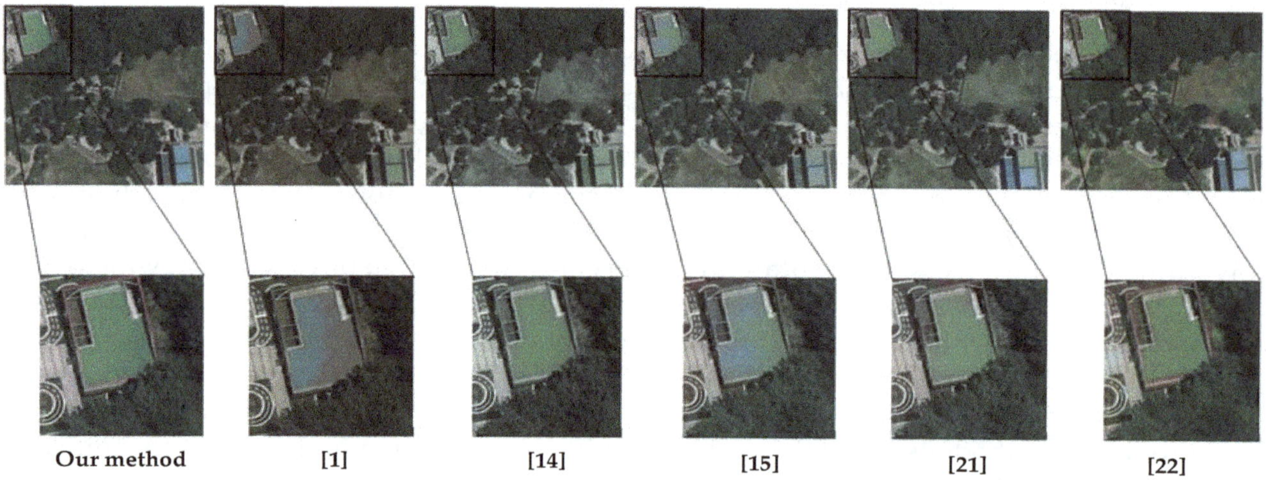

Figure 12. Enlargement of site 3: our method retrieves color values well in this region, compared with other methods.

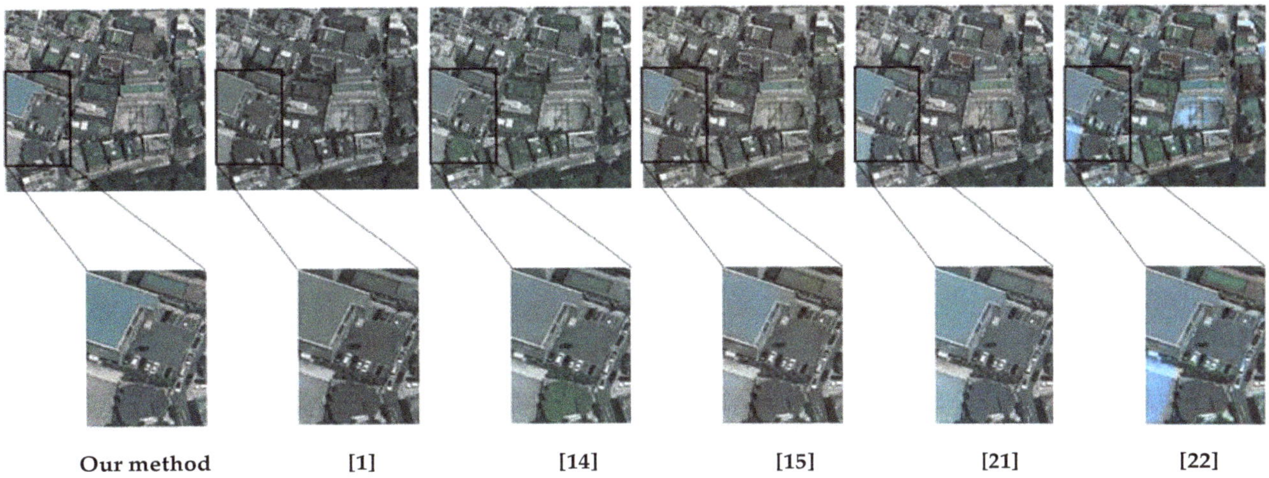

Figure 13. Enlargement of site 4: our method retrieves color values well in this region, compared with other methods.

Although visual inspection is a simple and direct way of appreciating the quality of the colorization results, it is highly subjective and cannot accurately evaluate the results of the various colorization methods. Therefore, for quantitative evaluation of the results, we employ the standard peak signal-to-noise ratio (PSNR) and normalized color difference (NCD). The PSNR, which is expressed in terms of the decibel (dB), is an estimate of the quality of the reconstructed (colorization) image compared with the ground-truth color image [54]. Given an $m \times n$ ground-truth color image u_0 and a colorization result u, PSNR is defined as:

$$\mathrm{PSNR} = 10 \cdot \log\left(\frac{3mn\,(MAX)^2}{\sum_{RGB}\sum_{i=0}^{m-1}\sum_{j=0}^{n-1}\,[u(i,j) - u_0(i,j)]^2}\right) \tag{9}$$

where MAX_I is the maximum possible pixel value of the image (i.e., 255 with standard 8-bit samples), and $\sum_{\mathrm{RGB}}()$ denotes summation over the red, green, and blue bands. The higher the PSNR value, the better the reconstruction process. Normalized color difference is used to measure the color quality degradation in color images [55]. For the NCD calculation, Lab space is used and is defined as:

$$\mathrm{NCD} = \frac{\sum_{i=0}^{m-1}\sum_{i=0}^{n-1}[\,(\Delta L)^2 + (\Delta a)^2 + (\Delta b)^2]^{1/2}}{\sum_{i=0}^{m-1}\sum_{i=0}^{n-1}[\,(L_{u0})^2 + (a_{u0})^2 + (b_{u0})^2]^{1/2}}, \tag{10}$$

where ΔL, Δa and Δb are the differences between the components of the ground-truth color image and the colorization result, and L_{u0}, a_{u0}, and b_{u0} are each component values of the ground-truth color image [56]. The lower the NCD value, the better the color quality.

The PSNRs and NCDs of the various algorithms are shown in Tables 1 and 2. In the case of PSNR, the proposed method (Max: 35.0906, Min: 29.6542, Average (Avg): 32.8773) significantly outperforms those of Welsh et al. [1] (Max: 27.1564, Min: 23.5535, Avg: 25.9514), Bugeau et al. [14] (Max: 29.1172, Min: 26.2594, Avg: 27.5596), Gupta et al. [15] (Max: 30.6962, Min: 27.1006, Avg: 29.3068), Gupta et al. [21] (Max: 32.6280, Min: 28.0476, Avg: 30.6694), and Deshpande et al. [22] (Max: 31.5879, Min: 24.2211, Avg: 29.6766). The performance difference from the state-of-the-art methods ranges from 2.4626 to 7.9342 for the maximum PSNR, 1.6066–6.1007 for the minimum PSNR, and 2.1779–6.9259 for the average PSNR, which indicates high performance for all results, regardless of site. That is, it is possible to colorize images stably regardless of the object included in the image or the complexity of the included structure.

Table 1. Quantitative evaluation of algorithm performance using standard peak signal-to-noise ratio (PSNR). dB: decibel.

	PSNR (dB)					
Method	**Our Method**	**[1]**	**[14]**	**[15]**	**[21]**	**[22]**
Site 1	35.0906	26.9364	29.1172	30.6962	32.628	31.5879
Site 2	34.8936	27.1564	26.2594	29.9741	31.6331	31.8681
Site 3	31.8709	26.1595	28.0316	29.4565	30.369	31.0295
Site 4	29.6542	23.5535	26.8302	27.1006	28.0476	24.2211

Table 2. Quantitative evaluation of algorithm performance using normalized color difference (NCD).

	NCD					
Method	**Our Method**	**[1]**	**[14]**	**[15]**	**[21]**	**[22]**
Site 1	0.0707	0.1472	0.127	0.0963	0.0858	0.1042
Site 2	0.0716	0.1408	0.1573	0.1334	0.0929	0.1069
Site 3	0.1098	0.2094	0.1722	0.1472	0.1386	0.1373
Site 4	0.1244	0.2304	0.1677	0.1679	0.1422	0.1801

The NCD of the proposed method (Max: 0.0707, Min: 0.1244, Avg: 0.0941) also show better performance than do Welsh et al. [1] (Max: 0.1408, Min: 0.2304, Avg: 0.1819), Bugeau et al. [14] (Max: 0.1270, Min: 0.1722, Avg: 0.1561), Gupta et al. [15] (Max: 0.0963, Min: 0.1679, Avg: 0.1362), Gupta et al. [21] (Max: 0.0858, Min: 0.1422, Avg: 0.1148), and Deshpande et al. [22] (Max: 0.1042, Min: 0.1801, Avg: 0.1231), in which the range of the improved performance difference is 0.0151–0.0701 for the maximum NCD, 0.0178–0.1244 for the minimum NCD, and 0.0207–0.0878 in the average NCD. This means that the degradation in color quality is lowest when performing colorization through the proposed method. In other words, both visual and quantitative evaluations confirm the superiority of the method proposed herein.

3.2. Limitations

The results show that the proposed algorithm can realize better results than can the existing methods; however, there remain several limitations. Firstly, if there are errors in orthorectification or image registration, incorrect extraction can be performed during selection of the training pixels in the preprocessing step. Although RF regression is robust to training data, some color relationships can be established incorrectly. Secondly, our method retrieves more color values than do the other methods, but, if the structure is complex, it contains somewhat turbid colors. The histogram specification for the reference color image is performed by post-processing, but there are still limitations. Thirdly, in this study, aerial images three years apart are used. However, further verification is needed to determine the extent of the period in which the colorization can properly be performed. Finally, our method is established by directly correlating color relationships between the input grayscale image and the reference color image, making it dependent on the availability of reference color aerial imagery of the same input area with matching seasonal characteristics. Consequently, when suitable color aerial images are unavailable, colorization may fail.

4. Conclusions

This paper presents a colorization algorithm for aerial imagery. The proposed method uses a reference color image with similar seasonal features at the same location as an input grayscale image. The color space of the reference color image is converted to Lab, and unchanged regions are selected by applying change detection to the input grayscale image and the L component of the reference color image, which serves as meaningful training data. Moreover, color relationships are established in direct correspondence between the feature descriptors of the input grayscale image and the a and b components of the reference color image based on the RF regression. Finally, histogram specification is applied to the colorization image to improve the results and is compared with state-of-the-art methods. Experimental results for multiple sites show that our method achieves visually appealing colorizations with significantly improved quantitative performance. In other words, the proposed algorithm performs well and outperforms existing colorization approaches for aerial images.

Future work will include other complex descriptors or features in order to retrieve more color values for complex structures. In particular, we intend to find the combination of features that best describes the characteristics of the aerial images for colorization. We will also extend our application of the technique by applying satellite images obtained from various sensors other than aerial images. Furthermore, to overcome the limitations that may prevent colorization from being performed when reference color images are unavailable, a method of using reference color images that are obtained with different sensors or contain different seasons or resolutions will be sought out. Finally, we plan to colorize past grayscale aerial images using a time-series approach, possibly by incorporating monitoring frameworks.

Author Contributions: All authors contributed to the writing of the manuscript. D.K.S. carried out the experiments and analyzed the results; Y.H.K. designed the experiments and presented the direction of this study; Y.D.E. supervised this study; W.Y.P. provided scientific counsel.

References

1. Welsh, T.; Ashikhmi, M.; Mueller, K. Transferring Color to Grayscale Images. *ACM Trans. Graph.* **2002**, *21*, 277–280. [CrossRef]
2. Bugeau, A.; Ta, V. Patch-Based Image Colorization. In Proceedings of the 21st International Conference on Pattern Recognition, Tsukuba, Japan, 11–15 November 2012; pp. 3058–3061.
3. Lipowezky, U. Grayscale Aerial and Space Image Colorization Using Texture Classification. *Pattern Recogn. Lett.* **2005**, *27*, 275–286. [CrossRef]
4. Yang, Y.; Wan, W.; Huang, S.; Lin, P.; Que, Y. A Novel Pan-Sharpening Framework Based on Matting Model and Multiscale Transform. *Remote Sens.* **2017**, *9*, 391. [CrossRef]
5. Li, S.; Kang, X.; Fang, L.; Hu, J.; Yin, H. Pixel-Level Image Fusion: A Survey of the State of the Art. *Inf Fusion.* **2017**, *33*, 100–112. [CrossRef]
6. Ghassemian, H. A Review of Remote Sensing Image Fusion Methods. *Inf Fusion.* **2016**, *32*, 75–89. [CrossRef]
7. Horiuchi, T. Estimation of Color for Gray-Level Image by Probabilistic Relaxation. In Proceedings of the Object Recognition Supported by User Interaction for Service Robots, Quebec, Canada, 11–15 August 2002; pp. 867–870.
8. Arbelot, B.; Vergne, R.; Hurtut, T.; Thollot, J. Automatic Texture Guided Color Transfer and Colorization. In Proceedings of the Joint Symposium on Computational Aesthetics and Sketch Based Interfaces and Modeling and Non-Photorealistic Animation and Rendering, Lisbon, Portugal, 7–9 May 2016; pp. 21–32.
9. Li, B.; Lai, Y.K.; Rosin, P.L. Example-Based Image Colorization via Automatic Feature Selection and Fusion. *Neurocomputing* **2017**, *266*, 687–698. [CrossRef]
10. Levin, A.; Lischinski, D.; Weiss, Y. Colorization Using Optimization. *ACM Trans. Graph.* **2004**, *23*, 689–694. [CrossRef]
11. Huang, Y.C.; Tung, Y.S.; Chen, J.C.; Wang, S.W.; Wu, J.L. An Adaptive Edge Detection Based Colorization Algorithm and Its Applications. In Proceedings of the 13th ACM International Conference on Multimedia, Hilton, Singapore, 6–11 November 2005; pp. 351–354.
12. Irony, R.; Cohen-Or, D.; Lischinski, D. Colorization by Example. In Proceedings of the Sixteen Eurographics Conference on Rendering Techniques, Konstanz, Germany, 29 June–1 July 2005; pp. 201–210.
13. Yatziv, L.; Sapiro, G. Fast Image and Video Colorization Using Chrominance Blending. *IEEE Trans. Image Process.* **2006**, *15*, 1120–1129. [CrossRef] [PubMed]

14. Bugeau, A.; Ta, V.T.; Papadakis, N. Variational Exemplar-Based Image Colorization. *IEEE Trans. Image Process.* **2014**, *23*, 298–307. [CrossRef] [PubMed]
15. Gupta, R.L.; Chia, A.Y.S.; Rajan, D.; Ng, E.S.; Zhiyoung, H. Image Colorization Using Similar Images. In Proceedings of the 20th ACM International Conference on Multimedia, Nara, Japan, 29 October–2 November 2012; pp. 369–378.
16. Charpiat, G.; Hofmann, M.; Scholkopf, B. Automatic Image Colorization via Multimodal Predictions. In Proceedings of the 10th European Conference on Computer Vision, Marseille, France, 12–18 October 2008; pp. 126–139.
17. Cheng, Z.; Yang, Q.; Sheng, B. Deep Colorization. In Proceedings of the IEEE International Conference on Computer Vision, Santiago, Chile, 7–13 December 2015; pp. 415–423.
18. Zhang, R.; Isola, P.; Efros, A.A. Colorful Image Colorization. In Proceedings of the Computer Vision—ECCV 2016—14th European Conference, Amsterdam, The Netherlands, 11–14 October 2016; pp. 649–666.
19. Larrson, G.; Maire, M.; Shakhnarovich, G. Learning Representations for Automatic Colorization. In Proceedings of the Computer Vision—ECCV 2016—14th European Conference, Amsterdam, The Netherlands, 11–14 October 2016; pp. 577–593.
20. Brieman, L. Random Forests. *Mach. Learn.* **2001**, *45*, 5–32. [CrossRef]
21. Gupta, R.K.; Chia, A.Y.; Rajan, D.; Zhiyong, H. A Learning-based Approach for Automatic Image and Video Colorization. In Proceedings of the Computer Graphics International, Bournemouth, UK, 12–15 June 2012; pp. 1–10.
22. Deshpande, A.; Rock, J.; Forsyth, D. Learning Large-Scale Automatic Image Colorization. In Proceedings of the 2015 IEEE International Conference on Computer Vision, Santiago, Chile, 7–13 December 2015; pp. 567–575.
23. Mohn, H.; Caebelein, M.; Hansch, R.; Hellwich, O. Towards Image Colorization with Random Forests. In Proceedings of the 13th International Joint Conference on Computer Vision, Funchal, Madeira, Portugal, 27–29 January 2018; pp. 270–278.
24. Culter, D.R.; Edwards, T.C.; Beard, K.H.; Culter, A.; Hess, K.T.; Gibson, J.C. Random Forests for Classification in Ecological Society of America. *Ecology* **2007**, *88*, 2783–2792. [CrossRef]
25. Yang, Y.; Cao, C.; Pan, X.; Li, X.; Zhu, X. Downscaling Land Surface Temperature in an Arid Area by Using Multiple Remote Sensing Indices with Random Forest Regression. *Remote Sens.* **2017**, *9*, 789. [CrossRef]
26. Shataee, S.; Kalbi, S.; Fallah, A.; Pelz, D. Forest Attribute Imputation Using Machine Learning Methods and ASTER Data: Comparison of K-NN, SVR, Random Forest Regression Algorithms. *Int. J. Remote Sens.* **2012**, *33*, 6254–6280. [CrossRef]
27. Peters, J.; Baets, B.D.; Verhoest, N.E.C.; Samson, R.; Degroeve, S.; Becker, P.D.; Huybrechts, W. Random Forests as a Tool for Ecohydrological Distribution Modeling. *Ecol. Model.* **2007**, *207*, 304–318. [CrossRef]
28. Rodriguez-Galiano, V.; Sanchez-Castillo, M.; Olmo-Chica, M.; Chica-Rivas, M. Machine Learning Predictive Models for Mineral Prospectivity: An Evaluation of Neural Networks, Random Forest, Regression Trees and Support Vector Machines. *Ore Geol. Rev.* **2015**, *71*, 804–818. [CrossRef]
29. Prasad, A.M.; Iverson, L.R.; Liaw, A. Newer Classification and Regression Tree Techniques: Bagging and Random Forests for Ecological Prediction. *Ecosystems* **2006**, *9*, 181–199. [CrossRef]
30. Hutengs, C.; Vohland, M. Downscaling Land Surface Temperature at Regional Scales with Random Forest Regression. *Remote Sens. Environ.* **2016**, *178*, 127–141. [CrossRef]
31. Changas, C.S.; Junior, W.C.; Behring, S.B.; Filho, B.G. Spatial Prediction of Soil Surface Texture in a Semiarid Region Using Random Forest and Multiple Linear Regressions. *Catena* **2016**, *139*, 232–240. [CrossRef]
32. Scornet, E. Tuning Parameters in Random Forests. In Proceedings of the ESAIM: Proceedings and Surveys, Grenoble, France, 29–31 August 2017; pp. 144–162.
33. Lawrence, R.L.; Wood, S.D.; Sheley, R.L. Mapping Invasive Plants Using Hyperspectral Imagery and Breiman Cutler Classification (RandomForest). *Remote Sens. Environ.* **2016**, *100*, 356–362. [CrossRef]
34. Belgiu, M.; Dragut, L. Random Forest in Remote Sensing: A Review of Applications and Future Directions. *ISPRS J. Photogramm. Remote Sens.* **2015**, *114*, 24–31. [CrossRef]
35. Seo, D.K.; Kim, Y.H.; Eo, Y.D.; Park, W.Y.; Park, H.C. Generation of Radiometric, Phenological Normalized Image Based on Random Forest Regression for Change Detection. *Remote Sens.* **2017**, *9*, 1163. [CrossRef]
36. Sug, H. Applying Randomness Effectively Based on Random Forest for Classification Task of Datasets of Insufficient Information. *J. Appl. Math.* **2012**, *2012*, 1–13. [CrossRef]

37. Palmer, D.S.; O'Boyle, N.M.; Glen, R.C.; Mitchell, J.B.O. Random Forest Models to Predict Aqueous Solubility. *J. Chem. Inf. Model.* **2007**, *47*, 150–158. [CrossRef] [PubMed]
38. Dye, M.; Mutanga, O.; Ismail, R. Combining Spectral and Textural Remote Sensing Variables Using Random Forests: Predicting the Age of Pinus Forests in KwaZulu-Natal, South Africa. *J. Spat Sci.* **2012**, *57*, 193–211. [CrossRef]
39. Quintana, D.; Saez, Y.; Isasi, P. Random Forest Prediction of IPO Underpricing. *Appl. Sci.* **2017**, *7*, 636. [CrossRef]
40. Hussain, M.; Chen, D.; Cheng, A.; Wei, H.; Stanley, D. Change Detection from Remotely Sensed Images: From Pixel-Based to Object-Based Approaches. *ISPRS J. Photogramm. Remote Sens.* **2013**, *80*, 91–106. [CrossRef]
41. Xiao, P.; Zhang, X.; Wang, D.; Yuan, M.; Feng, X.; Kelly, M. Change Detection of Built-up Land: A Framework of Combining Pixel-Based Detection and Object-Based Recognition. *ISPRS J. Photogramm. Remote Sens.* **2016**, *119*, 402–414. [CrossRef]
42. Wang, J.; Qin, Q.; Gao, Z.; Zhao, J.; Ye, X. A New Approach to Urban Road Extraction Using High-Resolution Aerial Image. *Int. J. Geo-Inf.* **2016**, *5*, 114. [CrossRef]
43. Crommelinck, S.; Bennett, R.; Gerke, M.; Koeva, M.N.; Yang, M.Y.; Vosselman, G. SLIC Superpixels for Object Delineation from UAV Data. In Proceedings of the International Conference on Unmanned Aerial Vehicles in Geomatics, Bonn, Germany, 4–7 September 2017; pp. 9–16.
44. Mei, T.; An, L.; Li, Q. Supervised Segmentation of Remote Sensing Image Using Reference Descriptor. *IEEE Geosci. Remote Sens. Lett.* **2015**, *12*, 938–942. [CrossRef]
45. Csillik, O. Fast Segmentation and Classification of Very High Resolution Remote Sensing Data Using SLIC Superpixels. *Remote Sens.* **2017**, *9*, 243. [CrossRef]
46. Achanta, R.; Shaji, A.; Smith, K.; Luchi, A.; Fua, P.; Susstrunk, S. SLIC Superpixels Compared to State-of-the-Art Superpixels Methods. *IEEE Trans. Pattern Anal. Mach. Intell.* **2012**, *34*, 2274–2281. [CrossRef] [PubMed]
47. Zainal, Z.; Ramli, R.; Mustafa, M.M. Grey-Level Cooccurence Matrix Performance Evaluation for Heading Angle Estimation of Movable Vision System in Static Environment. *J. Sens.* **2013**, *2013*, 1–15. [CrossRef]
48. Zhang, X.; Cui, J.; Wang, W.; Lin, C. A Study for Texture Feature Extraction of High-Resolution Satellite Images Based on a Direction Measure and Gray Level Co-Occurrence Matrix Fusion Algorithm. *Sensors* **2017**, *17*, 1474. [CrossRef] [PubMed]
49. Huang, X.; Liu, X.; Zhang, L. A Multichannel Gray Level Co-Occurrence Matrix for Multi/Hyperspectral Image Texture Representation. *Remote Sens.* **2014**, *6*, 8424–8445. [CrossRef]
50. Jia, B.; Wang, W.; Yoon, S.C.; Zhuang, H.; Li, Y.F. Using a Combination of Spectral and Texture Data to Measure Water-Holding Capacity in Fresh Chicken Breas Fillets. *Appl. Sci.* **2018**, *8*, 343. [CrossRef]
51. Zheng, S.; Zheng, J.; Shi, M.; Li, X. Classification of Cultivated Chinese Medicinal Plants Based on Fractal Theory and Gray Level Co-Occurrence Matrix Textures. *J. Remote Sens.* **2014**, *18*, 868–886. [CrossRef]
52. Sun, C.C.; Ruan, S.J.; Shie, M.C.; Pai, T.W. Dynamic Contrast Enhancement Based on Histogram Specification. *IEEE Trans. Consum. Electron.* **2005**, *51*, 1300–1305. [CrossRef]
53. Xie, L.; Wang, G.; Zhang, X.; Xiao, B.; Zhou, B.; Zhang, F. Remote Sensing Image Enhancement Based on Wavelet Analysis and Histogram Specification. In Proceedings of the 2014 IEEE 3rd International Conference on Cloud Computing and Intelligence Systems, Shenzhen, China, 27–29 November 2014; pp. 55–59.
54. Al-Najjar, Y.; Chen, S.D. Comparison of Image Quality Assessment: PSNR, HVS, SSIM, UIQI. *Int. J. Sci. Eng. Res.* **2012**, *3*, 1–5.
55. Senthilkumaran, N.; Saromary, J. Detailed Performance Evaluation of Bilateral Filters for De-noising Chromosome Image. *Int. J. Inf. Technol.* **2017**, *3*, 64–70.
56. Szczepanski, M.; Smolka, B.; Plataniots, K.N.; Ventesanopoulos, A.N. On the Distance Function Approach to Color Image Enhancement. *Discret.Appl. Math.* **2004**, *139*, 283–305. [CrossRef]

8

Land Subsidence Susceptibility Mapping using Bayesian, Functional and Meta-Ensemble Machine Learning Models

Hyun-Joo Oh [1], Mutiara Syifa [2], Chang-Wook Lee [2,*] and Saro Lee [3,4,*]

1 Geo-Environmental Hazard Research Center, Korea Institute of Geoscience and Mineral Resources (KIGAM), 124, Gwahak-ro Yuseong-gu, Deajeon 34132, Korea; ohj@kigam.re.kr
2 Division of Science Education, Kangwon National University, Chuncheon Campus, 1 Gangwondaehakgil, Chuncheon-si, Gangwon-do 24341, Korea; mutiarasyifa@kangwon.ac.kr
3 Geoscience Platform Research Division, Korea Institute of Geoscience and Mineral Resources (KIGAM), 124, Gwahak-ro Yuseong-gu, Daejeon 34132, Korea
4 Department of Geophysical Exploration, Korea University of Science and Technology, 217 Gajeong-ro Yuseong-gu, Daejeon 34113, Korea
* Correspondence: cwlee@kangwon.ac.kr (C.-W.L.); leesaro@kigam.re.kr (S.L.)

Abstract: To effectively prevent land subsidence over abandoned coal mines, it is necessary to quantitatively identify vulnerable areas. In this study, we evaluated the performance of predictive Bayesian, functional, and meta-ensemble machine learning models in generating land subsidence susceptibility (LSS) maps. All models were trained using half of a land subsidence inventory, and validated using the other half of the dataset. The model performance was evaluated by comparing the area under the receiver operating characteristic (ROC) curve of the resulting LSS map for each model. Among all models tested, the logit boost, which is a meta-ensemble machine leaning model, generated LSS maps with the highest accuracy (91.44%), i.e., higher than that of the other Bayesian and functional machine learning models, including the Bayes net (86.42%), naïve Bayes (85.39%), logistic (88.92%), and multilayer perceptron models (86.76%). The LSS maps produced in this study can be used to mitigate subsidence risk for people and important facilities within the study area, and as a foundation for further studies in other regions.

Keywords: land subsidence; Bayes net; naïve Bayes; logistic; multilayer perceptron; logit boost

1. Introduction

Coal mining was once the driving force of the national industry and economic development in Korea, but this situation changed as demand for coal decreased. Gangwon Province was once Korea's largest coal mining area but most of its mines were closed in the early 1990s. Among the environmental problems that follow mine closures, land subsidence events can threaten human life and damage property and infrastructure, including buildings, houses, railroads, and roads [1–4]. Recovery of surface structures following land subsidence is difficult and costly; therefore, it is necessary to predict land subsidence susceptibility (LSS) zones before subsidence occurs, and to implement management strategies in these zones [3].

Generally, prediction of subsidence susceptibility zones requires the input of several environmental factors and the application of perdition models [5]. Several previous studies have developed quantitative and qualitative models that have been successfully applied in various hazard susceptibility zones worldwide [3–11]. These include logistic regression (LR) [3], frequency ratio (FR) [3,6], weight of evidence (WOE) [3], evidential belief function (EBF) [4], artificial neural network

(ANN) [3,5,7,8], support vector machine (SVM) [9], random forest (RF) [10], and fuzzy logic (FL) [8,11] models. Single LSS mapping models can be combined to form ensemble models, which provide more precise and meaningful results [9]. Ensemble models based on machine learning have recently improved the prediction accuracy and performance of single classifiers [12]. The main advantages of this approach are the ability to represent complex relationships between influential factors, and to incorporate spatial data of various scales [13].

Based on existing studies, probability and statistical models using geographical information systems (GIS) have been applied extensively to predict the susceptibly of geohazards, such as landslides, floods, subsidence, and rockfalls [3,14–16]. Recently, data mining and machine learning models for addressing nonlinear problems have been developed, which have been applied frequently and had their performances compared in landslide susceptibility mapping [17–20]. In ground subsidence hazard mapping, ground subsidence hazard maps around abandoned underground coal mines (AUCMs) have been constructed by integrating the adaptive neuro-fuzzy inference system and GIS [21]. In addition, a fuzzy operator, decision tree with the CHAID and QUEST algorithms, and the frequency ratio have been applied to construct subsidence susceptibility maps at AUCMs in Korea [2,11]. In this study, we investigated the performance of some models that have never been applied to land subsidence prediction. Therefore, in this study, we generated LSS maps for a South Korean district containing abandoned subsurface coal mines using machine learning methods, including a logit boost meta-ensemble model, two Bayesian models (Bayes net and NB models) and two functional models (logistic and multilayer perceptron models). The reliability and accuracy of all models were assessed by comparing their area under the receiver operating characteristic (ROC) curves. Data processing was performed using WEKA 3.9.2 and ArcGIS 10.5 software to produce five machine learning algorithms.

2. Land Subsidence in the Study Area

The study area, Hwajeon, is located in the city of Taebaek, South Korea (Figure 1), at 37°11′07″–37°11′07″ N, 128°56′40″–128°57′43″ E. Underground coal mining activities were carried out in Taebaek for nearly 20 years. The coal seams in this area were irregularly disturbed and inclined with various widths by reverse and thrust faults [22]. Therefore, the slant-chute block caving method was mainly used. About 10 million tons of coal were mined from the study area between 1953–1991 [22], and coal was transported to other areas by railroad beginning in 1973 (Figure 1). Since 1990, most of the coal mines have been closed due to reduced coal demand. However, the abandoned underground coal mines are currently causing land subsidence in the study area [11,21–23]. Additionally, infrastructure has been damaged by the land subsidence, as shown in photographs in a previous report [11].

Subsidence is caused by a variety of contributing factors, including geological discontinuities, presence of water, mining depth, and weak overburden [24,25]. The two forms of subsidence caused by underground coal mining are trough and sinkhole subsidence [25]. In the study area, a very irregular sinkhole occurred due to many complex underground coal mine pits excavated via slant–chute block caving in combination with the aforementioned factors [22]. After a mine cavity is excavated, roof stability becomes unstable over time due to changes in the strength and stress of the roof strata. Under such conditions, additional contributing factors can lead to the occurrence of sinkholes [25]. The Coal Industry Promotion Board [11,26] has reported 24 land subsidence events within the study area. Figure 1 shows a representative land subsidence from location S1 to location S6 of a subsidence event reported in 1999. Table 1 provides a description of the land subsidence. Locations S1 to S5 of this land subsidence mainly occurred along railways and at elevations above 800 m. Location S6 occurred in residential areas and at a lower elevation than S1–S5. Also, the depth of subsidence of S6 is the deepest (508 mm). Some photographs providing evidence of the land subsidence have been published [11,23].

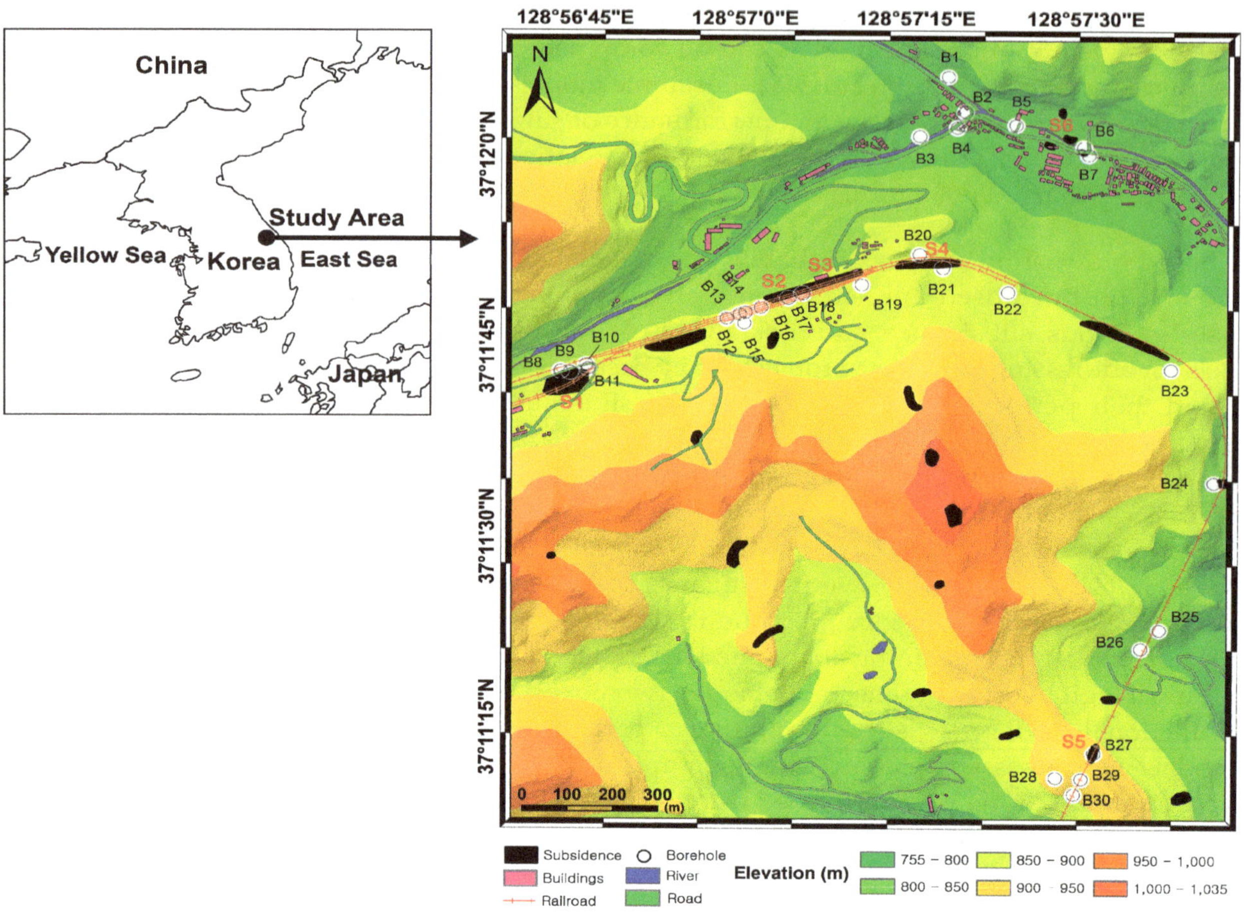

Figure 1. The study area in Taebaek, South Korea.

Table 1. Description of representative land subsidence in the study area [22].

Location	Structure	Elevation (m)	Mining Depth (m)	Thickness (m) and Slope of Coal Seam	Subsidence Depth (mm)	Other
S1	Railway	885	20–30	1–2 40–50°	90	-The coal seam is oblique to the railroad. -Shallow depth of mine -Sinkhole-type subsidence
S2	Railway	885	0	-	72	-Progression of cavity by mining -Subsidence by limestone cavity
S3	Railway	885	30–50	1–2 20°	329	-Subsidence along railway
S4	Railway	885	40–65	2 20°	223	-Shallow depth of mine -Coal bonanza
S5	Tunnel Railway	810	30–260	105 50–70°	65	-The tunnel is located above the mine cavity. -Vertical cracks and leakage in tunnel
S6	Road	765	60–98	3 20°	508	-Residential area and elementary school -Differential subsidence

3. Construction of Spatial Database

It is necessary to determine the factors affecting the land subsidence of a coal mine area. The lithology of the overburden rocks, geological discontinuities, ground slope, scope of the mined cavity, extent and depth of mining, mechanical characteristics of the rock mass rating (RMR), and flow of groundwater are considered the main factor [11,25,27,28]. Spatial data for all of these factors may be difficult to collect and may not be available. The available spatial databases used in this study were constructed using ArcGIS 10.5.

The surface geology with cross section lines was constructed using a digital geological map with 1:50,000 scale [29] published by the Korea Institute of Geoscience and Mineral Resources (KIGAM). The geological formations include the Manhan, Jangseong, Hambaegsan, Dosagog, and Alluvium horizons (Figure 2a, Figure 3). Most of the coal was mined from the Jangseong Formation with a thickness of 80–15 m [22,30]. This formation includes four to five cyclothems consisting of dark-gray sandstone, black shale, and coal seam (Table 2). The land use was constructed from a digital land characteristics map with 1:5,000 scale [31] supplied by the National Geographic Information Institute (NGII). Land use for the study area was classified into 10 categories: wood land, railroad, river, field, plot, road, school, hybrid land, brook and unclassified area (Figure 2b). The rate of land subsidence compared to the area of each category was higher in the railroad and school classes [21]. The surface slope was calculated from a digital elevation model (DEM) constructed from a digital elevation contour line with 1:5000 scale [32] published by NGII (Figure 2c). Surface slope was considered an affecting factor because land subsidence can change surface slope, differential horizontal strain, and vertical displacement [33]. Distance from drift was calculated from a digital drift map provided by the Mine Reclamation Corporation (MIRECO) [26] (Figure 2d). The map is important because it identifies the areas of mining activity in this region. Geological discontinuities are considered to be factors affecting land subsidence, but no geological lineaments appear in the study area on the available 1:50,000 geological map. Therefore, geomorphological lineament was visually extracted from an IKONOS satellite image by a field geologist (Figure 2e). If the location is near a lineament, the value of distance from lineament is low.

The borehole data in the study area, provided by the Mine Reclamation Organization (MIRECO) in 1996 [26], were collected from 29 boreholes (Figure 1 and Table 3). The depths of the boreholes ranged from a minimum of 19.5 m to a maximum of 200 m. The data included hydrologic properties and rock mass information [34]. The depth of groundwater, rock mass rating (RMR), and permeability were obtained from 16, 19, and 6 boreholes, respectively (Table 3 and Figure 2f,g,h). The maximum depth of groundwater was 42.5 m. On the railroad, the upper part of the railroad had a deeper groundwater depth and lower elevation than the lower part of the railway. The RMR was classified as classes 1–5, representing very good, good, fair, poor, and very poor, respectively. In this study, the RMR ranged from 2–4.5. The lowest RMRs appeared in the northwest and southeast portions of the railroad. Permeability was classified as classes 1–6, representing very highly (>1 cm/s), highly (1–10^{-2} cm/s), moderately (10^{-2}–10^{-3} cm/s), slightly (10^{-3}–10^{-5} cm/s), and very slightly (10^{-5}–10^{-7} cm/s) permeable and practically impermeable (<10^{-7} cm/s), respectively. In this study, the permeability grade ranged from 4–4.5 (slightly permeable). The groundwater data were collected from a report published in May 1996 by the Coal Industry Promotion Board. Borehole point data should be converted into raster data for spatial analysis, and the accuracy of a raster map depends on the number of data points. However, the available borehole data were limited in this study. Therefore, raster maps from the limited borehole data were constructed using an inverse distance weighting (IDW) interpolation method, which is useful for predicting values at unmeasured locations where data are insufficient [11].

Eight control factors influencing land subsidence were constructed with 2 m × 2 m grid data, resulting in 775 columns and 860 rows, for a total of 666,500 cells within the study area. In total, 24 land subsidence areas as 24 vector-type polygons were converted to 2 m × 2 m grid data for a total of 3863 cells with a value of 1. The 3863 cells of land subsidence were randomly classified into training and validation sets, with a 50% (1931 cells) and 50% (1932 cells) distribution, respectively, to evaluate model performance.

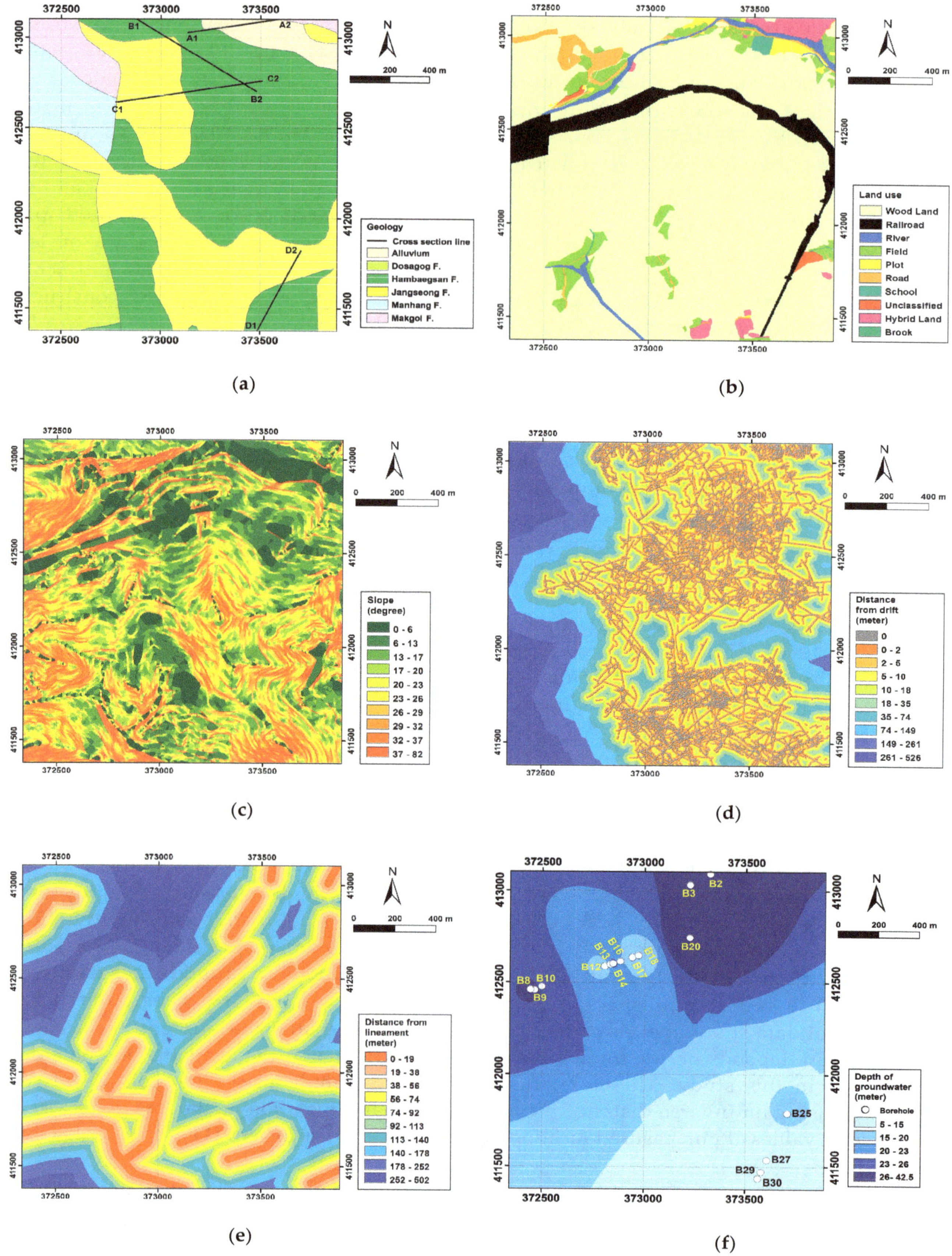

Figure 2. *Cont.*

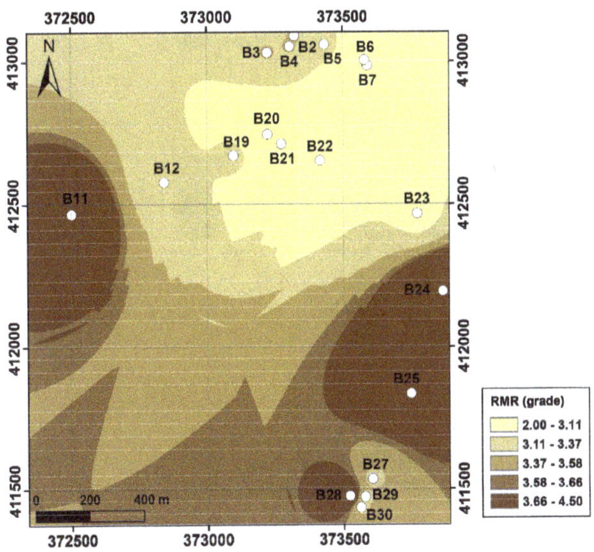

(g)

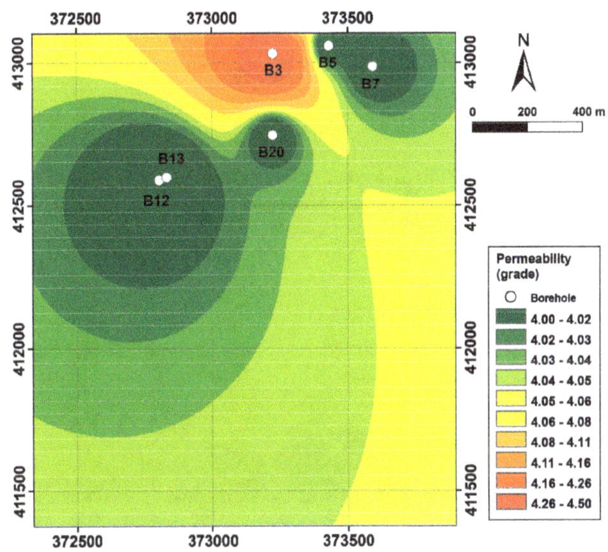

(h)

Figure 2. Eight factors influencing coal mine subsidence were used as input data in this study: (**a**) Geology, (**b**) land use, (**c**) slope, (**d**) distance from drift, (**e**) distance from lineament, (**f**) groundwater depth, (**g**) rock mass rating (RMR), and (**h**) permeability.

Table 2. Description of geological stratigraphy in Taebaek [30].

Geological Age	Formation	Thickness (m)	Description
Quaternary	Alluvium (Qa)	~20	- Gravel, sand, and clay
	Dosagog (Pd)	250–350	- Mainly milky white–light green coarse–very coarse sandstone with greenish-gray–gray shale interbeds. Intercalations of pinkish sandstone, purple shale, and grayish-green sandy shale in the upper part. The sandstone is less compact than that of the Hambaegsan Formation.
Permian	Hambaegsan (Ph)	70–250	- Mainly milky white–light gray coarse sandstone with some interbeds of black shale with thickness of 2–3 m. Some pebbly sandstones occur at the base.
	Jangseong (Pj)	80–150	- Four–five cyclothems consisting of dark-gray sandstone, black shale, and coal seam. Abundant plant fossils occur in the shale above the coal seam, the most valuable anthracite bed, of the 3rd–4th cyclothem from the bottom.
	Geumcheon (Cg)	50–100	- Mainly dark-gray–black shale and dark-gray fine sandstone intercalated with dark-gray limestone lenses and two to three thin coal seams
Carboniferous	Manhang (Cm)	250–300	- Mainly purple, greenish-gray, or light-green shale and light-green–green or light-gray medium–very coarse sandstone intercalated with three–four limestone lenses. Conglomerates with a thickness of a few meters occur at the base in some places.
Ordovician	Makgol (Om)		- In the upper part, gray–dark gray limestone intercalated with dolomite

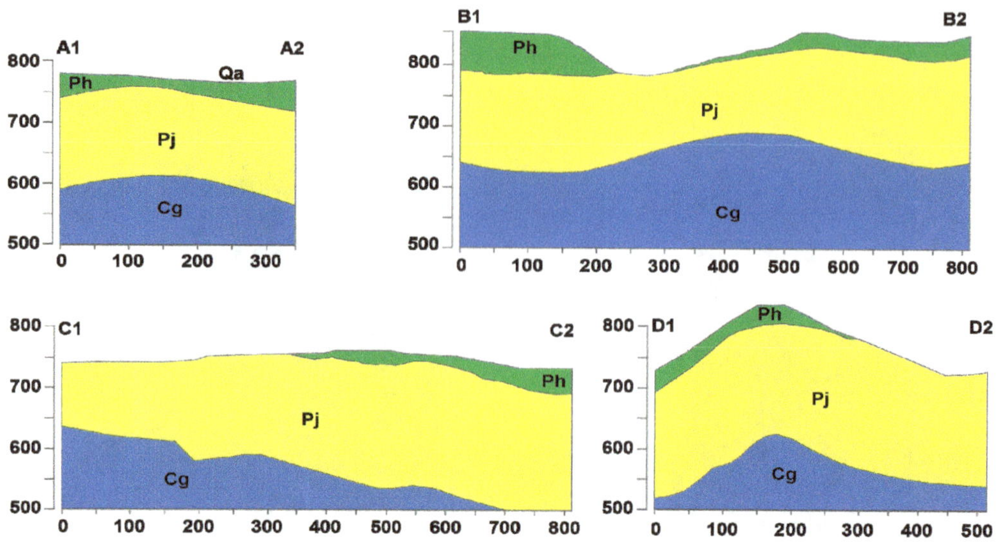

Figure 3. Geological cross sections in the study area.

Table 3. Borehole data in the study area.

ID	Depth of Borehole	Depth of Groundwater (m)	RMR (grade)	Permeability (grade)	Geology
B1	50.0	32.0	3.4	-	Alluvium-Hambaegsan
B2	50.0	27.2	3.4	4.5	Alluvium-Hambaegsan
B3	30.0	-	3.4	-	Alluvium-Hambaegsan
B4	60.2	-	3.4	4	Alluvium-Hambaegsan
B5	86.3	-	2.0	-	Alluvium-Hambaegsan
B6	80.0	-	2.0	4	Alluvium-Hambaegsan
B7	33.0	27.5	-	-	Jangseong
B8	20.5	27.7	-	-	Jangseong
B9	40.0	26.1	-	-	Jangseong
B10	35.5	-	4.4	-	Jangseong
B11	30.0	15.7	-	4	Jangseong
B12	40.5	21.6	-	4	Jangseong
B13	41.1	29.4	-	-	Jangseong
B14	22.0	-	3.2	-	Jangseong
B15	35.7	20.0	-	-	Jangseong
B16	40.8	20.0	-	-	Jangseong
B17	50.5	14.7	-	-	Jangseong
B18	58.0	-	3.2	-	Jangseong
B19	54.0	42.5	2.5	4	Hambaegsan-Jangseong
B20	60.0	-	3.0	-	Hambaegsan-Jangseong
B21	115.0	-	3.0	-	Hambaegsan-Jangseong
B22	80.0	-	3.0	-	Hambaegsan-Jangseong
B23	80.0	-	4.5	-	Hambaegsan-Jangseong
B24	84.0	-	4.3	-	Jangseong
B25	80.4	18.0	-	-	Jangseong
B26	19.5	5.0	3.3	-	Hambaegsan
B27	200.0	-	4.3	-	Hambaegsan-Jangseong
B28	40.0	5.0	3.3	-	Hambaegsan-Jangseong
B29	35.0	5.5	3.3	-	Hambaegsan-Jangseong

4. Methods

As shown in Figure 4, the mapping process consisted of five steps: (a) Spatial database construction, (b) random categorization of land subsidence locations into training and validation datasets at a ratio of 1:1, (c) selection of land subsidence conditioning factors, (d) application of machine learning methods to map LSS, and (d) validation and comparison of the five models.

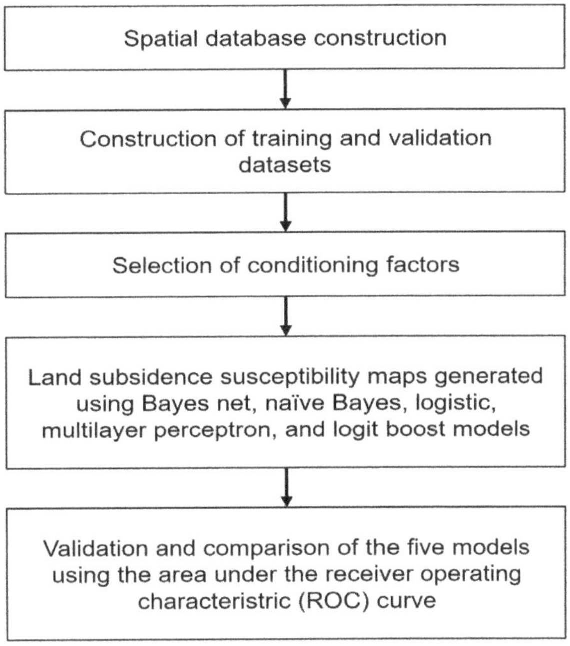

Figure 4. Flowchart for the generation of land subsidence susceptibility (LSS) maps using various machine learning models including Bayes net, naïve Bayes (NB), logistic, multilayer perceptron, and logit boost models.

4.1. Models

4.1.1. Bayes Net (BN)

The BN algorithm applies Bayes' theorem to produce graphical representations of the probability distribution [35]. BN is commonly used to model complex systems [36]. BN has not yet been used to model land subsidence; however, Pham et al. (2016) [37] applied this algorithm to evaluate landslide risk. The distinct universal probability of a subsidence event for a set of input factors can be estimated as follows:

$$P_B(X_1, X_2, \ldots, X_n) = \Pi_{i=1}^{n} = P_B\left(X_1 |\Pi_{x_i}\right) = \Pi_{i=1}^{n}\theta_{x_i|\Pi_{x_i}} \quad (1)$$

where $X = (X_1, X_2, \ldots, X_n)$ represents the subsidence input factors, $P_B\left(X_1 |\Pi_{x_i}\right) = \theta_{x_i|\Pi_{x_i}}$ is a common probability distribution for input factors X_i, and n is the number of subsidence input factors [37].

4.1.2. Naïve Bayes (NB)

The NB algorithm is a classification system that applies Bayes' theorem under the assumption of conditional independence for all attributes [10,38]. The NB classifier is easy to build, without any need for complicated iterative parameter-estimation schemes [38]. The NB algorithm estimates the probability $P(y_j/x_i)$ for all possible output classes as shown in Equation (2). The class with the largest posterior probability is predicted as follows:

$$y = \text{argmax}\, P\left(y_j\right) \prod_{i=1}^{n} P(x_i/y_j) \quad (2)$$

{subsidence, no subsidence}

where x_i is the input factor, y_j is the output class, $P(y_j)$ is the prior probability, and $P(y_j/x_i)$ is the conditional probability.

The conditional probability is calculated as

$$P\left(\frac{x_i}{y_j}\right) = \frac{1}{\sqrt{2\pi\sigma}} e^{-(x_i-\mu)^2/2\sigma^2} \tag{3}$$

where μ is the mean and σ is the standard deviation of x_i.

4.1.3. Logistic Regression (LR)

LR is a statistical technique that allows the predictor to analyze several types of variables [39–41]. LR does not require the normality assumption, which is an advantage over linear and log-linear regression. The inclusion of multiple parameters offers the user the ability to select the best predictors for use in the model [39]. The LR model is formulated as follows [42]:

$$f(x) = \text{logit}\,(P) = \ln\left[\frac{P}{1-P}\right] == c_0 + c_1x_1 + \cdots + c_nx_n \tag{4}$$

$$P = \frac{1}{1+e^{-f(x)}} = \frac{1}{1+e^{-(c_0+c_1x_1+\cdots+c_nx_n)}} \tag{5}$$

where $x_1, x_2, \ldots, x_n$ are the input factors, c_0 is the model intercept, and $c_1, \ldots, c_n$ are the regression coefficients to be approximated. In this study, P is the probability of subsidence occurrence and $1 - P$ is probability that subsidence will not occur. The function f(x) is represented as logit (P).

4.1.4. Multilayer Perceptron (MLP)

MLP is an artificial neural network classifier that is widely used in various fields [12,43]. MLP neural nets consist of three structures: Input, hidden, and output layers. In this study, the input layers represent factors that affect land subsidence, and the inputs are processed to become outputs within the hidden layers. The classification results, dividing land subsidence and non-subsidence, are shown in the output layers [12,44]. Two processes are required to train data from MLP neural nets: 1) Forward propagation of the inputs through the hidden layers to obtain output and compare output values to initial values, and 2) adjustment of the connection weights using differences between subsequent values to generate the best results [44,45]. In this study, $t = t_i$, $i = 1, 2, \ldots, 8$ is a vector containing eight land-subsidence conditioning factors, and $\phi = \phi_j$, $j = 1,2$ represents the land subsidence and non-subsidence classes. The MLP neural net function is then determined as follows:

$$\phi = f(t) \tag{6}$$

where $f(t)$ is an unknown function that is improved during the training process by adjustable network weights for a given network architecture.

An advantage of MLP is that the user is not required to decide the relative importance of the various input measurements; most inputs can be selected during the training process, based on weight adjustment [46]. Additionally, MLP does not require assumptions about the distribution of the training dataset.

4.1.5. Logit Boost (LB)

LB is a famous machine-learning algorithm introduced by Friedman et al., 2000 [47] that effectively reduces bias and variance; it is a slight modification of the most popular boosting method (AdaBoost) for handling noisy data [48], which reduces training errors and improves classification accuracy [49]. LB has been widely applied in binary classification problems [50], medical science [51], and computer science [52]; however, it has not yet been applied to land-subsidence problems [53].

In the current study, we create a vector $x_i = x_1, x_2, \ldots, x_n$, where n is the number of input factors; y = [1, 0] represents two output classes (subsidence or non-subsidence). The LB algorithm is trained in the following steps [47]:

1. Assign weights $\omega_i = \frac{1}{n}$, $i = 1, 2, \ldots, n$, $f(x) = 0$ and probability estimates $p_e(x_i) = 1/2$.
2. For m = 1, 2, ..., m, repeat the following steps:

    a. Compute the working response and weights:

$$r_i = \frac{[y_i^* - p_e(x_i)]}{[p_e(x_i)(1 - p(x_i)]}$$

$$\omega_i = p_e(x_i)(1 - p(x_i)$$

    b. Fit the function by weighted least-squares regression of r_i to x_i using weights ω_i.
    c. Update the function as:

$$f(x) \leftarrow f(x) + \frac{1}{2} f_m(x)$$

$$p(x) \leftarrow \frac{e^{f(x)}}{e^{f(x)} + e^{-f(x)}}$$

3. Output the classifier.

$$\text{sign}[f(x)] = \text{sign}\left[\sum_{m=1}^{M} f_m(x)\right]$$

$$= \begin{cases} 1 \text{ (subsidence) if } f(x) < 0 \\ -1 \text{ (non subsidence)if } f(x) \geq 0 \end{cases}$$

4.2. Model Evaluation and Comparison

During the modeling and validation phases, model efficiency should be evaluated and compared [44]. We quantitatively evaluated and compared the efficiency of the models according to the area under the ROC curve (%). This technique has been applied to assess risk models of various hazards including subsidence [9], landslides [54], and sinkholes [55]; it is a standard method to quantitatively evaluate the quality of probabilistic and statistical models [56]. The x and y axes of the curve are sensitivity and specificity, respectively [56], and the area under the curve ranges from 0.5–1, with higher values indicating higher model accuracy and prediction capability.

5. Results

5.1. LSS Mapping

Figure 5 shows the LSS maps produced by the five algorithms: Bayes net (Figure 5a), NB (Figure 5b), logistic (Figure 5c), multilayer perceptron (Figure 5d), and logit boost (Figure 5e). To generate the LSS maps, we used the LSS index (LSSI) to classify susceptibility events into four classes: Very high (5% of total area), high (5%), moderate (5%), and low (85%). The probability of land subsidence was predicted for each class, and subsidence hazard was predicted for residential areas. The susceptibility indexes from the five algorithms were similar. The region with very high susceptibility appeared from the western part of the region to the eastern part as railroad area, which is marked by the red color. In the Bayes net result, the very high susceptibility area did not appear as often as in the other models. In the middle of the region, the Bayes net result has a low index, whereas the rest of the models have a very high or high index. Some very high indexes also appear in the northeastern part of the region, as elementary school area, but most of the region has a low susceptibility index rank for subsidence.

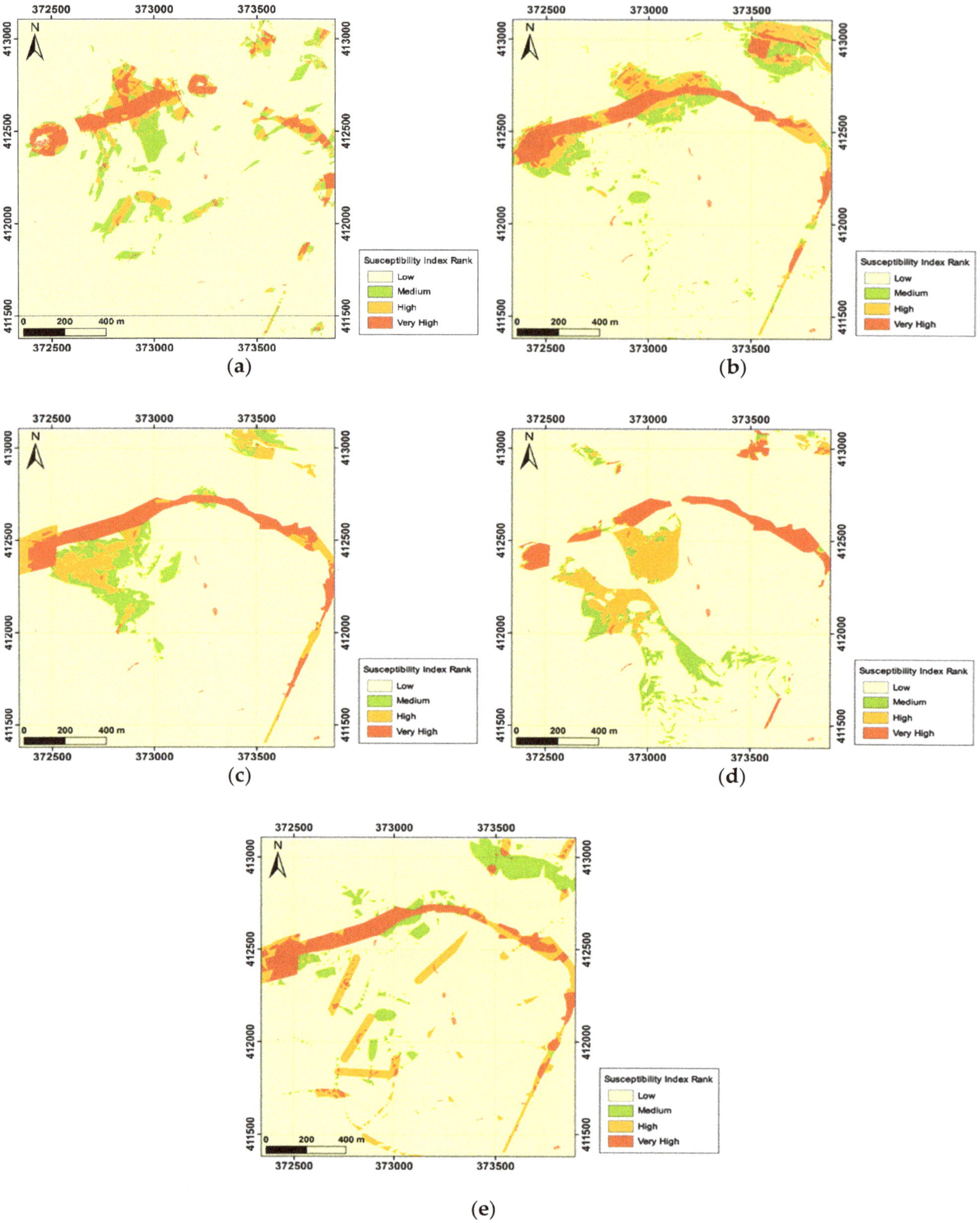

Figure 5. LSS maps generated using the five algorithms: (**a**) Bayes net, (**b**) NB, (**c**) logistic, (**d**) multilayer perceptron, and (**e**) logit boost.

However, there are some differences for the medium-susceptibility index rank, marked by the green color. The area with medium susceptibility of land subsidence is spreading and has a different pattern in each model result. For example, the NB and logit boost results show the northern part of the region is mostly covered by the medium susceptibility index. In contrast, the multilayer perceptron shows the medium index in the southern part of the region. Meanwhile, in the Bayes net and logistic models, the medium index is diffusely distributed from the northern to the middle part of the study area.

5.2. Validation

The land subsidence susceptibility (LSS) analysis results were validated by comparison with 1932 land subsidence cells (i.e., 50% of the total subsidence data) that had not been used in the analysis. A quantitative comparison among all models of the receiver operating characteristic (ROC) curves for model performance is shown in Figure 6. The land subsidence susceptibility index (LSSI) values of all cells were sorted in descending order, divided into 100 classes [57], and associated with the cumulative number of subsidence events for each class (Figure 6). The model with the highest area under the ROC curve was considered to be the model with the best predictive performance. The area under the curve values for the Bayes net, naïve Bayes (NB), logistic, multilayer perceptron, and logit boost models were 0.8640, 0.8539, 0.8892, 0.8676, and 0.9144, respectively; thus, the respective LSS mapping accuracy rates were 86.42, 85.39, 88.92, 86.76, and 91.44%. Although all models had sufficient performance, the different applied models had different prediction performances using same training data. In particular, the logit boost model had a higher predictive accuracy (by about 2.52, 4.68, 5.02, and 6.05%, respectively) than the logistic, multilayer perceptron, Bayes net, and NB. Therefore, model reliability followed the order logit boost > logistic > multilayer perceptron > Bayes net > NB. The percentage differences of the validation result are discussed in Section 6.

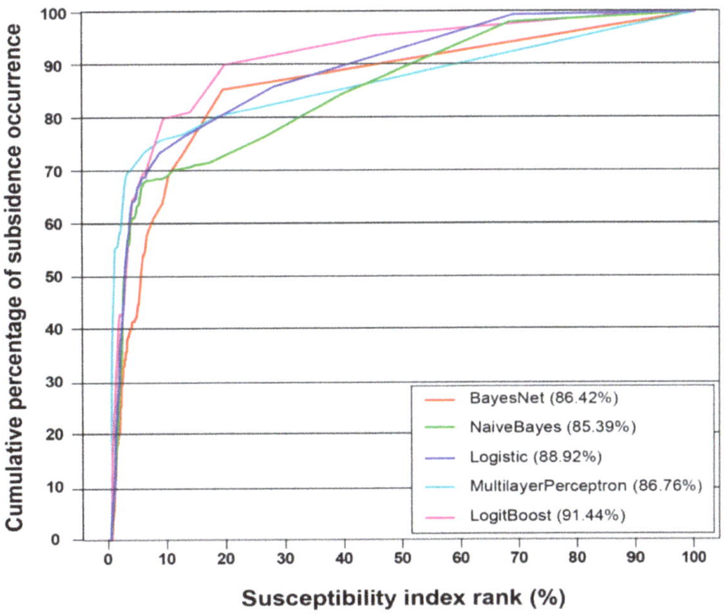

Figure 6. Susceptibility index rank (x-axis) and subsidence occurrence (y-axis) of the five algorithms.

6. Discussion

Recently, there has been great interest within the hazard prediction community toward improving the performance of hazard susceptibility models. In various fields, machine learning techniques have been shown to be effective in terms of performance [58–62]. In particular, ensemble learning has improved machine learning results by combining several models [17,63,64]. The results of different applied models under the same conditions (i.e., study area, input data, ratio of training, and validation datasets) can be compared to the quantitative accuracy values of the area under the ROC to present the predictive power of the model. Models with similar (different) accuracy values can be said to have similar (differing) performances. Therefore, the reliabilities of the models can be ordered according to the accuracies of the models.

In this study, the logit boost model, based on ensemble machine learning, had a 91.44% accuracy and a predictive accuracy that was higher (by 2.52–6.05%) than those of the logistic, multilayer perceptron, Bayes net, and NB based on machine learning. Similarly, a previous study [2] found that a decision tree model (the CHAID algorithm) produced LSS maps with higher accuracy (94.01%) than the QUEST decision tree (90.37%) and frequency ratio (86.70%). The other algorithms examined in the current study also exhibited high accuracy. Thus, the Bayes net, NB, logistic, and multilayer perceptron models can also be used as alternative models for mapping land subsidence hazard risk. Even though the logit boost model, as an ensemble model, had not been used to predict land subsidence in previous research, the results of the current study indicate that it can achieve high accuracy.

However, some limitations of the models might be a consideration for future studies. For example, the Bayes net model assumes no missing values, and this model also needs to be updated, especially for estimating the conditional probabilities [65]. The benefits and drawbacks of the machine learning models are influenced by several factors, such as the availability of datasets, characteristics of the study area, and condition of the region [18]. The use of Bayesian algorithms, such as the Bayes net and Naïve Bayes, has not been fully verified in natural hazard assessments [18]. According to Mezaal [66], the multilayer perceptron algorithm also has limitations, such as overlearning and high computational complexity.

It has been reported that the sinkhole subsidence attributable to underground mining is caused by shallow depth, weak overburden, geological discontinuities, solution of rocks, rainfall, groundwater, and earthquakes [25]. However, this study used a spatial database obtained from previous studies due to the limitation of available data. No further surveys or new surveys on land subsidence have been conducted in the study area for 14 years. If real-time monitoring data and additional data are obtained in the study area, a 4D underground subsidence model [67] with 3D geological modeling could be constructed to predict land subsidence hazard areas accurately. Thus, continuous monitoring and detailed new surveying for causative factors are essential in the study area. The maps produced in this study can be used as basic data for policymakers and further research. Future studies should develop alternative models and methods to determine the relative influence of factors affecting LSS, so that these methods can be applied in other regions.

7. Conclusions

Land subsidence is a hazardous effect of coal mine abandonment, including that in Korea. To prevent damage and loss of life in the Taebaek region, it is necessary to predict areas with high subsidence risk effectively. In this study, we used Bayesian (i.e., Bayes net and NB), functional (i.e., logistic, multilayer perceptron), and meta-ensemble (i.e., logit boost) machine learning models to perform LSS assessments. Although all models had sufficient performance, the logit boost meta-ensemble machine learning model had the highest accuracy (91.44%) among the five models. The logit boost model also had higher predictive accuracy (by 2.52%, 4.68%, 5.02%, and 6.05%, respectively) than the logistic, multilayer perceptron, Bayes net, and NB models. According to previous studies [11,57] in the same study area, the fuzzy operator with 84.40–88.98% accuracy, frequency ratio with 86.70% accuracy, CHAID decision tree with 94.01% accuracy, and QUEST decision tree with 90.37% accuracy have been applied to the subsidence hazard assessment, but the five models used in this study had been rarely applied. Based on these case studies, the land subsidence hazard rating can be applied to future policy decisions using additional data.

Author Contributions: Conceptualization, S.L. and C.-W.L.; methodology, S.L. and H.-J.O.; software, H.-J.O. and M.S.; validation, H.-J.O. and M.S.; formal analysis, H.-J.O. and M.S.; investigation, S.L. and H.-J.O.; resources, S.L. and H.-J.O.; writing—original draft preparation, H.-J.O. and M.S.; writing—review and editing, S.L. and C.-W.L.; visualization, H.-J.O.; supervision, S.L. and C.-W.L.; project administration, S.L. and C.-W.L.; funding acquisition, S.L. and C.-W.L.

References

1. Ghorbanzadeh, O.; Rostamzadeh, H.; Blaschke, T.; Gholaminia, K.; Aryal, J. A New Gis-Based Data Mining Technique Using an Adaptive Neuro-Fuzzy Inference System (Anfis) and K-Fold Cross-Validation Approach for Land Subsidence Susceptibility Mapping. *Nat. Hazards* **2018**, *94*, 497–517. [CrossRef]
2. Lee, S.; Park, I. Application of decision tree model for the ground subsidence hazard mapping near abandoned underground coal mines. *J. Environ. Manag.* **2013**, *127*, 166–176. [CrossRef] [PubMed]
3. Oh, H.-J.; Lee, S. Integration of ground subsidence hazard maps of abandoned coal mines in Samcheok, Korea. *Int. J. Coal Geol.* **2011**, *86*, 58–72. [CrossRef]
4. Pradan, B.; Abokharima, M.H.; Jebur, M.N.; Tehrany, M.S. Land Subsidence Susceptibility Mapping at Kinta Valley (Malaysia) Using the Evidential Belief Function Model in Gis. *Nat. Hazards* **2014**, *73*, 1019–1042. [CrossRef]
5. Lee, S.; Park, I.; Choi, J.K. Spatial prediction of ground subsidence susceptibility using an artificial neural network. *Environ. Manag.* **2012**, *49*, 347–358. [CrossRef] [PubMed]
6. Oh, H.-J.; Ahn, S.-C.; Choi, J.-K.; Lee, S. Sensitivity analysis for the GIS-based mapping of the ground subsidence hazard near abandoned underground coal mines. *Environ. Earth Sci.* **2011**, *64*, 347–358. [CrossRef]
7. Pishro, M.; Khosravi, S.; Tehrani, S.M.; Mousavi, S.R. Modeling and zoning of land subsidence in the southwest of Tehran using artificial neural networks. *Int. J. Hum. Cap. Urban Manag.* **2016**, *1*, 159–168.
8. Rafie, M.; Samimi Namin, F. Prediction of subsidence risk by FMEA using artificial neural network and fuzzy inference system. *Int. J. Min. Sci. Technol.* **2015**, *25*, 655–663. [CrossRef]
9. Tien Bui, D.; Shahabi, H.; Shirzadi, A.; Chapi, K.; Pradhan, B.; Chen, W.; Khosravi, K.; Panahi, M.; Bin Ahmad, B.; Saro, L. Land Subsidence Susceptibility Mapping in South Korea Using Machine Learning Algorithms. *Sensors* **2018**, *18*, 2464. [CrossRef]
10. Ilia, I.; Loupasakis, C.; Tsangaratos, P. Land subsidence phenomena investigated by spatiotemporal analysis of groundwater resources, remote sensing techniques, and random forest method: The case of Western Thessaly, Greece. *Environ. Monit. Assess.* **2018**, *190*, 623. [CrossRef]
11. Choi, J.-K.; Kim, K.-D.; Lee, S.; Won, J.-S. Application of a fuzzy operator to susceptibility estimations of coal mine subsidence in Taebaek City, Korea. *Environ. Earth Sci.* **2010**, *59*, 1009–1022. [CrossRef]
12. Pham, B.T.; Tien Bui, D.; Prakash, I.; Dholakia, M.B. Hybrid integration of Multilayer Perceptron Neural Networks and machine learning ensembles for landslide susceptibility assessment at Himalayan area (India) using GIS. *CATENA* **2017**, *149*, 52–63. [CrossRef]
13. Kanevski, M.; Parkin, R.; Pozdnukhov, A.; Timonin, V.; Maignan, M.; Demyanov, V.; Canu, S. Environmental data mining and modeling based on machine learning algorithms and geostatistics. *Environ. Model. Softw.* **2004**, *19*, 845–855. [CrossRef]
14. Baillifard, F.; Jaboyedoff, M.; Sartori, M. Rockfall hazard mapping along a mountainous road in Switzerland using a GIS-based parameter rating approach. *Nat. Hazards Earth Syst. Sci.* **2003**, *3*, 435–442. [CrossRef]
15. Samanta, S.; Pal, D.K.; Palsamanta, B. Flood susceptibility analysis through remote sensing, GIS and frequency ratio model. *Appl. Water Sci.* **2018**, *8*, 66. [CrossRef]
16. Lee, S.; Pradhan, B. Landslide hazard mapping at Selangor, Malaysia using frequency ratio and logistic regression models. *Landslides* **2007**, *4*, 33–41. [CrossRef]
17. Kadavi, P.R.; Lee, C.-W.; Lee, S. Application of Ensemble-Based Machine Learning Models to Landslide Susceptibility Mapping. *Remote Sens.* **2018**, *10*, 1252. [CrossRef]
18. Pham, B.T.; Prakash, I.; Khosravi, K.; Chapi, K.; Trinh, P.T.; Ngo, T.Q.; Hosseini, S.V.; Bui, D.T. A comparison of Support Vector Machines and Bayesian algorithms for landslide susceptibility modelling. *Geocarto Int.* **2018**, 1–23. [CrossRef]

19. Darabi, H.; Choubin, B.; Rahmati, O.; Torabi Haghighi, A.; Pradhan, B.; Kløve, B. Urban flood risk mapping using the GARP and QUEST models: A comparative study of machine learning techniques. *J. Hydrol.* **2019**, *569*, 142–154. [CrossRef]
20. Truong, X.L.; Mitamura, M.; Kono, Y.; Raghavan, V.; Yonezawa, G.; Truong, X.Q.; Do, T.H.; Tien Bui, D.; Lee, S. Enhancing Prediction Performance of Landslide Susceptibility Model Using Hybrid Machine Learning Approach of Bagging Ensemble and Logistic Model Tree. *Appl. Sci.* **2018**, *8*, 1046. [CrossRef]
21. Park, I.; Choi, J.; Jin Lee, M.; Lee, S. Application of an adaptive neuro-fuzzy inference system to ground subsidence hazard mapping. *Comput. Geosci.* **2012**, *48*, 228–238. [CrossRef]
22. Bang, G.M.; Choi, S.S.; Oh, S.H.; Sin, J.S.; Jeon, M.G.; Woo, S.U.; Heo, J.E.; Cheon, M.N.; Jo, N.S.; Kim, B.C.; et al. *A Detailed Survey of Monitoring in Whajeon Region*; Coal Industry Promotion Board: Seoul, Korea, 1999; p. 681.
23. Kim, J.N. *A Case Study on Stability Analysis of Ground Subsidence in Abandoned Mine Area—Focused on the Vicinity of the Chunjeon Station*; Seoul National University of Science and Technology: Seoul, Korea, 2011.
24. Canbulat, I.; Zhang, C.; Black, K.; Johnston, J.; McDonald, S. Assessment of Sinkhole Risk in Shallow Coal Mining. In Proceedings of the 10th Triennial Conference of Mine Subsidence: Adaptive Innovation for Managing Challenges, Pokolbin, Australia, 5–7 November 2017.
25. Singh, K.B.; Dhar, B.B. Sinkhole subsidence due to mining. *Geotech. Geol. Eng.* **1997**, *15*, 327–341. [CrossRef]
26. Board, C.I.P. *Ground Stability Investigation for Hwajeon (Korean Edn)*; Coal Industry Promotion Board: Seoul, Korea, 1996; pp. 9–84.
27. Sahu, P.; Lokhande, R.D. An Investigation of Sinkhole Subsidence and its Preventive Measures in Underground Coal Mining. *Procedia Earth Planet. Sci.* **2015**, *11*, 63–75. [CrossRef]
28. Lee, F.T.; Abel, J.F. *Subsidence from Underground Mining; Environmental Analysis and Planning Considerations*; USGS: Reston, VA, USA, 1983. [CrossRef]
29. The Geological Society of Korea. *Geologic Altas of Taebaegsan Region: Homyeong Geological Sheet*; The Geological Society of Korea: Seoul, Korea, 1962.
30. Seo, H.K.; Kim, D.S.; Park, S.H.; Park, J.S.; Bae, D.J.; Yu, Y.S.; Lee, D.Y.; Im, S.B.; Jang, Y.H.; Jo, M.J. *Geologic Atlas of the Samcheog Coalfield*; Korea Institute of Geoscience and Mineral Resources: Daejeon, Korea, 1979; p. 22.
31. National Geographic Information Institute. *Digital Land Characteristics Map with 1:5000 Scale: No. of index: 37816018, 37816019, 37816028, 37816029*; National Geographic Information Institute: Suwon, Korea, 1996.
32. National Geographic Information Institute. *Digital Topographical Map with 1:5000 Scale: No. of index: 37816018, 37816019, 37816028, 37816029*; National Geographic Information Institute: Suwon, Korea, 1996.
33. Tripathi, N.; Singh, R.S. Underground Coal Mine Subsidence Impacts Forest Ecosystem. 2010. Available online: https://www.researchgate.net/publication/275207695 (accessed on 25 January 2019).
34. Öge, İ.F.; Çırak, M. Relating rock mass properties with Lugeon value using multiple regression and nonlinear tools in an underground mine site. *Bull. Eng. Geol. Environ.* **2017**. [CrossRef]
35. Marcot, B.; Steventon, J.; Sutherland, G.; McCann, R. Guidelines for Developing and Updating Bayesian Belief Networks Applied to Ecological Modeling and Conservation. *Can. J. Forest Res.* **2006**, *36*, 3063–3074. [CrossRef]
36. Song, Y.; Gong, J.; Gao, S.; Wang, D.; Cui, T.; Li, Y.; Wei, B. Susceptibility assessment of earthquake-induced landslides using Bayesian network: A case study in Beichuan, China. *Comput. Geosci.* **2012**, *42*, 189–199. [CrossRef]
37. Pham, B.T.; Pradhan, B.; Tien Bui, D.; Prakash, I.; Dholakia, M.B. A comparative study of different machine learning methods for landslide susceptibility assessment: A case study of Uttarakhand area (India). *Environ. Model. Softw.* **2016**, *84*, 240–250. [CrossRef]
38. Tien Bui, D.; Pradhan, B.; Lofman, O.; Revhaug, I. Landslide susceptibility assessment in vietnam using support vector machines, decision tree, and nave bayes models. *Math. Probl. Eng.* **2012**, *2012*. [CrossRef]
39. Erener, A.; Mutlu, A.; Sebnem Düzgün, H. A comparative study for landslide susceptibility mapping using GIS-based multi-criteria decision analysis (MCDA), logistic regression (LR) and association rule mining (ARM). *Eng. Geol.* **2016**, *203*, 45–55. [CrossRef]
40. Ozdemir, A.; Altural, T. A comparative study of frequency ratio, weights of evidence and logistic regression methods for landslide susceptibility mapping: Sultan Mountains, SW Turkey. *J. Asian Earth Sci.* **2013**, *64*, 180–197. [CrossRef]

41. Mertler, C.A.; Reinhart, R.V. *Advanced and Multivariate Statistical Methods: Practical Application and Interpretation: Sixth Edition*; Routledge: New York, NY, USA, 2016; pp. 1–374. [CrossRef]
42. Hosmer, D.W., Jr.; Lemeshow, S.; Sturdivant, R.X. *Applied Logistic Regression*; John Wiley & Sons: Hoboken, NJ, USA, 2013; Volume 398.
43. Haykin, S.S. *Neural Networks and Learning Machines/Simon Haykin*; Prentice Hall: New York, NY, USA, 2009.
44. Pham, B.T.; Tien Bui, D.; Pourghasemi, H.R.; Indra, P.; Dholakia, M.B. Landslide susceptibility assesssment in the Uttarakhand area (India) using GIS: A comparison study of prediction capability of naïve bayes, multilayer perceptron neural networks, and functional trees methods. *Theor. Appl. Climatol.* **2017**, *128*, 255–273. [CrossRef]
45. Bui, D.T.; Pradhan, B.; Revhaug, I.; Nguyen, D.B.; Pham, H.V.; Bui, Q.N. A novel hybrid evidential belief function-based fuzzy logic model in spatial prediction of rainfall-induced shallow landslides in the Lang Son city area (Vietnam). *Geomat. Nat. Hazards Risk* **2015**, *6*, 243–271. [CrossRef]
46. Gardner, M.W.; Dorling, S.R. Artificial neural networks (the multilayer perceptron)—A review of applications in the atmospheric sciences. *Atmos. Environ.* **1998**, *32*, 2627–2636. [CrossRef]
47. Friedman, J.; Tibshirani, R.; Hastie, T. Additive Logistic Regression: A Statistical View of Boosting (With Discussion and a Rejoinder by the Authors). *Ann. Stat.* **2000**, *28*, 337–407. [CrossRef]
48. Zhang, G.; Fang, B. LogitBoost classifier for discriminating thermophilic and mesophilic proteins. *J. Biotechnol.* **2007**, *127*, 417–424. [CrossRef] [PubMed]
49. Song, J.; Lu, X.; Liu, M.; Wu, X. Stratified Normalization LogitBoost for Two-Class Unbalanced Data Classification. *Commun. Stat. Simul. Comput.* **2011**, *40*, 1587–1593. [CrossRef]
50. Fraz, M.M.; Remagnino, P.; Hoppe, A.; Uyyanonvara, B.; Rudnicka, A.R.; Owen, C.G.; Barman, S.A. An ensemble classification-based approach applied to retinal blood vessel segmentation. *IEEE Trans. Biomed. Eng.* **2012**, *59*, 2538–2548. [CrossRef]
51. Cai, Y.-D.; Feng, K.-Y.; Lu, W.-C.; Chou, K.-C. Using LogitBoost classifier to predict protein structural classes. *J. Theor. Biol.* **2006**, *238*, 172–176. [CrossRef]
52. Lutz, R.W. Logitboost with trees applied to the wcci 2006 performance prediction challenge datasets. In Proceedings of the 2006 IEEE International Joint Conference on Neural Network, Vancouver, BC, Canada, 16–21 July 2006; pp. 1657–1660.
53. Pham, B.T.; Tien Bui, D.; Dholakia, M.B.; Prakash, I.; Pham, H.V. A Comparative Study of Least Square Support Vector Machines and Multiclass Alternating Decision Trees for Spatial Prediction of Rainfall-Induced Landslides in a Tropical Cyclones Area. *Geotech. Geol. Eng.* **2016**, *34*, 1807–1824. [CrossRef]
54. Conforti, M.; Pascale, S.; Robustelli, G.; Sdao, F. Evaluation of prediction capability of the artificial neural networks for mapping landslide susceptibility in the Turbolo River catchment (Northern Calabria, Italy). *CATENA* **2014**, *113*, 236–250. [CrossRef]
55. Ozdemir, A. Sinkhole susceptibility mapping using logistic regression in Karapınar (Konya, Turkey). *Bull. Eng. Geol. Environ.* **2016**, *75*, 681–707. [CrossRef]
56. Zweig, M.H.; Campbell, G. Receiver-operating characteristic (ROC) plots: A fundamental evaluation tool in clinical medicine. *Clin. Chem.* **1993**, *39*, 561–577.
57. Park, I.; Lee, S. Spatial prediction of landslide susceptibility using a decision tree approach: A case study of the Pyeongchang area, Korea. *Int. J. Remote Sens.* **2014**, *35*, 6089–6112. [CrossRef]
58. Korup, O.; Stolle, A. Landslide Prediction from Machine Learning. *Geol. Today* **2014**, *30*, 26–33. [CrossRef]
59. McGaughey, W.J.; Laflèche, V.; Howlett, C.; Sydor, J.L.; Campos, D.; Purchase, J.; Huynh, S. Automated, Real-Time Geohazard Assessment in Deep Underground Mines. In *Proceedings of the Eighth International Conference on Deep and High Stress Mining*; Wesseloo, J., Ed.; Australian Centre for Geomechanics: Perth, Australia, 2017; pp. 521–528.
60. Tayfur, G.; Singh, V.P.; Moramarco, T.; Barbetta, S. Flood Hydrograph Prediction Using Machine Learning Methods. *Water* **2018**, *10*, 968. [CrossRef]
61. Karpatne, A.; Ebert-Uphoff, I.; Ravela, S.; Babaie, H.A.; Kumar, V. Machine Learning for the Geosciences: Challenges and Opportunities. *IEEE Trans. Knowl. Data Eng.* **2018**. [CrossRef]
62. Canli, E.; Mergili, M.; Thiebes, B.; Glade, T. Probabilistic Landslide Ensemble Prediction Systems: Lessons to Be Learned from Hydrology. *Nat. Hazards Earth Syst. Sci.* **2018**, *18*, 2183–2202. [CrossRef]

63. Mojaddadi, H.; Pradhan, B.; Nampak, H.; Ahmad, N.; Ghazali, A.H.B. Ensemble Machine-Learning-Based Geospatial Approach for Flood Risk Assessment Using Multi-Sensor Remote-Sensing Data and Gis. *Geomat. Nat. Hazards Risk* **2017**, *8*, 1080–1102. [CrossRef]
64. Chen, W.; Sun, Z.; Han, J. Landslide Susceptibility Modeling Using Integrated Ensemble Weights of Evidence with Logistic Regression and Random Forest Models. *Appl. Sci.* **2019**, *9*, 171. [CrossRef]
65. Bouckaert, R.R. *Bayesian Network Classifiers in Weka for Version 3-5-7*; The University of Waikato: Waikato, New Zealand, 2008; Volume 11, pp. 369–387.
66. Mezaal, M.; Pradhan, B.; Shafri, H.; Md Yusoff, Z.; Al-Zuhairi, M. Optimized Neural Architecture for Automatic Landslide Detection from High-Resolution Airborne Laser Scanning Data. *Appl. Sci.* **2017**, *7*, 730. [CrossRef]
67. Mira Geoscience. *Geohazmap Workflow Earth Modelling Solutions for Mining. Montreal: Mira Geoscience*; Mira Geoscience: Montreal, QC, Canada, 2007.

9

Mapping Areal Precipitation with Fusion Data by ANN Machine Learning in Sparse Gauged Region

Guoyin Xu [1], Zhongjing Wang [1,2,3,*] and Ting Xia [4]

1 Department of Hydraulic Engineering, Tsinghua University, Beijing 100084, China; xgy13@mails.tsinghua.edu.cn
2 State Key Laboratory of Hydro-Science and Engineering, Tsinghua University, Beijing 100084, China
3 State Key Lab of Plateau Ecology and Agriculture, Qinghai University, Xining 810016, China
4 China Renewable Energy Engineering Institute, Beijing 100120, China; yifuling@126.com
* Correspondence: zj.wang@tsinghua.edu.cn

Abstract: Focusing on water resources assessment in ungauged or sparse gauged areas, a comparative evaluation of areal precipitation was conducted by remote sensing data, limited gauged data, and a fusion of gauged data and remote sensing data based on machine learning. The artificial neural network (ANN) model was used to fuse the remote sensing precipitation and ground gauge precipitation. The correlation coefficient, root mean square deviation, relative deviation and consistency principle were used to evaluate the reliability of the remote sensing precipitation. The case study in the Qaidam Basin, northwest of China, shows that the precision of the original remote sensing precipitation product of Tropical Precipitation Measurement Satellite (TRMM)-3B42RT and TRMM-3B43 was 0.61, 72.25 mm, 36.51%, 27% and 0.70, 64.24 mm, 31.63%, 32%, respectively, comparing with gauged precipitation. The precision of corrected TRMM-3B42RT and TRMM-3B43 improved to 0.89, 37.51 mm, −0.08%, 41% and 0.91, 34.22 mm, 0.11%, 42%, respectively, which indicates that the data mining considering elevation, longitude and latitude as the main influencing factors of precipitation is efficient and effective. The evaluation of areal precipitation in the Qaidam Basin shows that the mean annual precipitation is 104.34 mm, 186.01 mm and 174.76 mm based on the gauge data, corrected TRMM-3B42RT and corrected TRMM-3B43. The results show many differences in the areal precipitation based on sparse gauge precipitation data and fusion remote sensing data.

Keywords: Qaidam Basin; remote sensing; TRMM; artificial neural network

1. Introduction

Precipitation is one of the essential links in the water cycle process and varies significantly whether it is spatial or temporal [1,2]. Traditionally, the measurement of precipitation is based on a ground gauge station such as a hydrometric station or meteorological station. The gauge precipitation is identified in terms of both effectiveness and accuracy due to its direct measurement. The spatial distribution of precipitation is mostly interpolated from the gauged data. However, the accuracy of interpolation in the sparse and uneven gauged area is generally not reliable [3]. Therefore, the fusion of remote sensing data and gauged data for evaluation has become a challenging topic [4–8].

There are many high-resolution rainfall products at both the global and regional scales which have been released successively [9,10], such as the Global Precipitation Climate Program (GPCP), Global Satellite Mapping Precipitation Program (GSMaP), Tropical Precipitation Measurement Satellite (TRMM) and Global Precipitation Measurement (GPM) [11–13]. Many remote sensing precipitation products [14–17] are widely used to compensate for the shortage of gauged data areas [18–21].

However, the remote sensing precipitation production is not highly reliable due to its indirect observation which needs adjusting and evaluation [22–25]. There are many achievements published about the evaluation of remote sensing precipitation products. [26–30]. The precision evaluation index of remote sensing precipitation products mainly includes a correlation coefficient, determination coefficient, scatter slope, fuzzy comprehensive score, etc. [31–34]. For those remote sensing precipitation products with low precision, it is necessary to be corrected. The most used method is machine learning, such as the classification and regression tree (CART), random forest (RF), multi-factor data mining set correction, etc. [18,35]. An artificial neural network (ANN) is a powerful machine learning algorithm with a complex network structure formed by the interconnection of a large number of processing units (neurons) [36]. It is an information processing system based on imitating the structure and function of the brain neural network [37]. The theory of an ANN has made significant progress in pattern recognition [38,39], automatic control, signal processing [40], assistant decision-making, artificial intelligence [41], networking and healthcare [42–45]. It has been successfully introduced into the field of hydrology and water resources [46,47].

There are many publications on TRMM remote sensing precipitation products for applicability in specific areas [48–50]. Dominque et al. [51] found that the accuracy of both the total and the monthly precipitation of TRMM in the Amazon basin are high enough. Ji et al. [52] validated the accuracy of TRMM precipitation products and found it has a high accuracy in Sichuan and Chongqing in China on seasonal and monthly scales. Wang et al. [53] analyzed TRMM precipitation products with the observation data of meteorological stations in the Tianshan Mountains and its surrounding areas, and the results showed that the TRMM products had good applicability. A large number of research results showed that the accuracy of TRMM precipitation products was higher on monthly and annual scales [54,55], which could be used for analyzing the dynamic variability of a long-time precipitation sequence [56].

However, Qu et al. [57] evaluated the daily precipitation products of TRMM in the Irrawaddy River basin and found that the remote sensing precipitation and the measured values had a high correlation but a large deviation. Xu et al. [58] evaluated the TRMM precipitation in the southern part of the Qinghai–Tibet Plateau by gauged data from high-density rainfall stations and found that TRMM overestimated the amount of light rains. The altitude, slope, direction, latitude, longitude and other factors impact the accuracy of TRMM precipitation [18]. Therefore, TRMM precipitation products should be corrected before being applied in some areas, especially high mountain areas. Based on the evaluation of TRMM 3B42RT and 3B43 in the Qaidam Basin, northwest of China, this paper fused the gauged data and remote sensing data of precipitation by machine learning and assessed the rainfall resources in the Qaidam Basin. The methodology can be used in other sparse gauged areas.

2. Data and Methods

2.1. Study Area and Data Sources

2.1.1. Study Area

The Qaidam Basin is located in the northeastern edge of the Tibetan Plateau. The geographical coordinates are 34°41′–39°20′ N and 87°48′–99°18′ E, spanning the Gansu Province, Qinghai Province and Xinjiang Uygur Autonomous Region. The vast majority of the Qaidam Basin is in the Qinghai Province with an area of 234.14 thousand km^2. The area in Xinjiang Uygur Autonomous Region is 17.42 thousand km^2 and that in the Gansu Province is 17.89 thousand km^2. The northwest, northeast and south of the Qaidam Basin are surrounded by the Altun Mountains, Qilian Mountains and Kunlun Mountains, respectively, as shown in Figure 1. The Qaidam Basin is the only large plateau inland basin in the world and its elevation ranges from 2653 m to 6748 m. The basin is deep in the mainland and surrounded by mountains. It is hard for the warm and humid airflow from the southwest to reach the basin, forming the typical cold-dry continental climate. Affected by the topography and latitude, the temperature of the basin is high in the central portion, but low all around. The lowest temperature occurs in January with −9.8–−16.1 °C in the basin area and −14.7–−17.2 °C in the mountainous area.

The highest temperature is in July with 13.5–19.2 °C in the basin area and 5.6–10.4 °C in the mountainous area. The annual sunshine duration is generally above 3100 h.

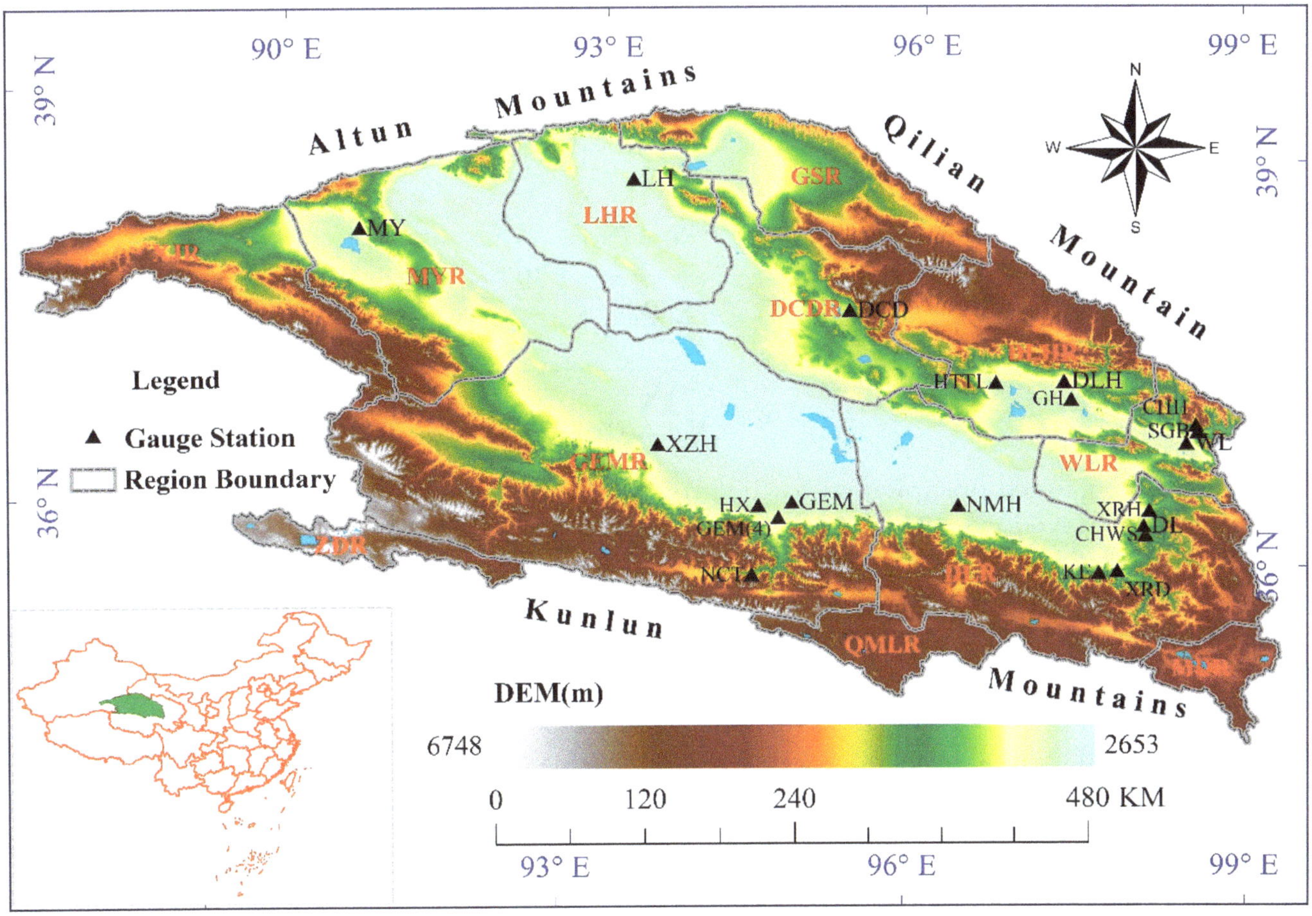

Figure 1. The research area of the Qaidam Basin.

2.1.2. Data Sources

The data used in this study include Digital Elevation Model (DEM), representing ground elevation, TRMM 3B42RT and 3B43, representing remote sensing precipitation, and gauged precipitation data by surface ground meteorological and hydrometric stations. The DEM data is from the Geographical Information Monitoring Cloud Platform (GIM Cloud) [59], and the spatial resolution is 1 km × 1 km. The TRMM 3B42RT and 3B43 are from the NASA website [60] with a spatial resolution of 0.25° × 0.25° and a time resolution of 3 h. The TRMM products were processed by ArcGIS to fit the Qaidam Basin and the data is from 2001 to 2016. In order to consider the effect of elevation, the 1 km × 1 km spatial resolution matched with DEM data was used when resampling the TRMM data. The gauged precipitation data, including 9 meteorological stations and 11 hydrometric stations from 2001 to 2016, was from China Meteorological Science Data Sharing Service Network [61] and Hydrological Red Book of the People's Republic of China [62], respectively.

2.2. Methodology

2.2.1. Evaluation of Remote Sensing Precipitation Precision

(1) The TRMM products' precision in gauged grids:

The correlation coefficient (R^2), relative deviation (Bias) and root mean square deviation (RMSD) were calculated in the grids where a gauge station was located. The data series is from 2001 to 2016 and the temporal resolution is all on an annual scale. The formulas are as follows:

$$R^2 = \frac{\left[\sum_{i=1}^{n}\left(Ps_i - \overline{Ps}\right)\left(Pt_i - \overline{Pt}\right)\right]^2}{\sum_{i=1}^{n}\left(Ps_i - \overline{Ps}\right)^2 \cdot \sum_{i=1}^{n}\left(Pt_i - \overline{Pt}\right)^2}, \tag{1}$$

$$Bias = \frac{\sum_{i=1}^{n}(Pt_i - Ps_i)}{\sum_{i=1}^{n} Ps_i} \times 100\%, \tag{2}$$

$$RMSD = \sqrt{\frac{1}{n}\sum_{i=1}^{n}(Pt_i - Ps_i)^2}, \tag{3}$$

where, Ps is the gauged precipitation by the ground gauge stations (mm); Pt is the remote sensing precipitation retrieved from TRMM products (mm); $\overline{Ps}$ and $\overline{Pt}$ are the average value of Ps and Pt, respectively; n is the number of years (2001–2016). The R reflects the correlation between the gauged precipitation and the TRMM precipitation. The relative deviation (Bias) and root mean square deviation (RMSD) reflect the deviation degree between the gauged precipitation data and the TRMM precipitation.

(2) The TRMM products' precision in ungauged grids:

There are only 20 gauges in the study area which means the gauged grids are equal or less than 20. All the other grids are ungauged. According to Xia's achievements [9], the accuracy of the remote sensing precipitation on the grids without a gauge station can be evaluated by the criteria of the consistency rate (CR). The formulas are as follows:

$$Count_i = \begin{cases} 1, & \text{if } Pt_i \in D \\ 0, & \text{if } Pt_i \notin D \end{cases} \quad i = 1, 2, \cdots, N, \tag{4}$$

$$S = \sum_{i=1}^{N} Count_i, \tag{5}$$

$$CR = \frac{S}{N} \times 100\%, \tag{6}$$

where, D is the precipitation-elevation mask (PEM) derived from the relationship of gauged precipitation and the elevation of the gauge stations; N is the total number of grids without gauge stations; S is the total number of grids in which remote sensing precipitation falls into the PEM. If the remote sensing precipitation value falls into the mask, then the remote sensing precipitation is considered consistent with the gauged precipitation at the same elevation region, that is, the remote sensing precipitation is reliable and vice versa.

2.2.2. Correction of Remote Sensing Precipitation by ANN Model

A three-layer ANN model was set up for remote sensing precipitation correction. The model has five input variables (i.e., gauged precipitation (Ps), elevation (DEM), longitude (X) and latitude (Y) of the gauge stations, and TRMM precipitation). Specifically, the Ps is the target value of the model, and the four others are the variables of the model input layer. The output variable is only the corrected precipitation in the model output layer. The hidden layer nodes are set to 20 by the preferred selection. The structure of the ANN model in this paper is shown in Figure 2.

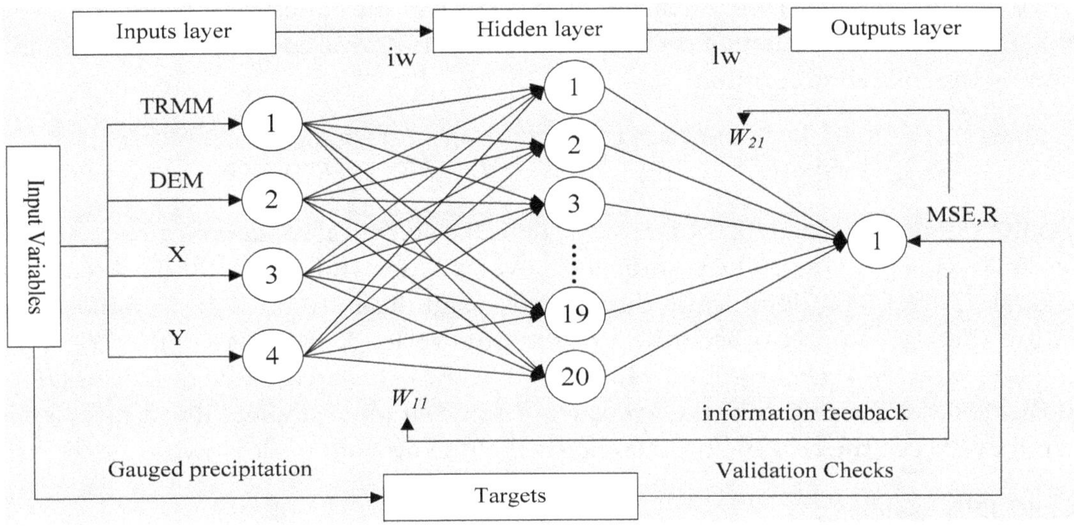

Figure 2. The artificial neural network (ANN) structure diagram.

The model training function is 'trainlm' which updates weight and bias values according to Levenberg–Marquardt optimization. At the same time, we chose the 'tansig' as the model neural transfer function to calculate a layer's output from its net input. The divide function is accessed automatically whenever the network is trained, which is used to divide the data into training, validation and testing subsets. In this study, the net divide function is set to 'dividerand', and the ratios for training, validation and testing are 0.7, 0.0 and 0.3 according to the needs of the training function of 'trainlm'. The prepared sequence data is randomly divided into the training subset (70%) and testing subset (30%) using the division parameters by the divide function. The fusion models for 3B42RT and 3B43 correction completed 103 and 127 times of training, and the convergence error (mean square error) was 0.0036 and 0.0041, respectively.

The process of data fusion mainly includes four steps: (1) the Pt is extracted by combining ArcGIS technology according to the spatial locations of the gauge stations. Moreover, the data of Ps, DEM, X, Y and Pt are stored in a one-to-one correspondence; (2) training and testing the model after setting each function of the artificial neural network. When the parameters such as the R^2 reach certain conditions, the construction of the model is completed; (3) correction remote sensing precipitation based on the fusion model which is established in the previous step; (4) the fused precipitation data are inversed to the research area on a spatial–temporal scale according to the spatial information of the grids.

2.2.3. Assessment of Rainwater Resources

The areal rainwater resources (i.e., precipitation) is evaluated by the mean value. The variation trend of the areal precipitation is predicted by the improved Mann–Kendall method [63–65].

Corresponding to the time series X with n sample sizes, the order column is constructed as follows:

$$S_k = \sum_{i=1}^{k} r_i \quad r_i = \begin{cases} 1 & x_i > x_j \\ 0 & x_i \le x_j \end{cases} j = 1, 2, \cdots, i. \tag{7}$$

It can be seen that the order sequence S is the cumulative number of values at the ith moment greater than that at the jth moment. Under the assumption that the time series is randomly independent, the statistics are defined as follows:

$$UF_k = \frac{S_k - E(S_k)}{\sqrt{Var(S_k)}} k = 1, 2, \cdots, n, \tag{8}$$

where $UF_k = 0$, $E(S_k)$ and $Var(S_k)$ are the mean and variance of the cumulative number S_k, respectively. When $x_1, x_2, \ldots, x_n$ are independent of each other and have the same continuous distribution, they can be calculated by the following formula:

$$E(S_k) = \frac{n(n+1)}{4} Var(S_k) = \frac{n(n-1)(2n+5)}{72}, \tag{9}$$

where UF_i is the standard normal distribution, which is the statistical sequence calculated according to the time series X $(x_1, x_2, \ldots, x_n)$. The significance level α is determined and the normal distribution table is checked. If $|UF_i| > U_\alpha$, it indicates that there is a significant trend change in the sequence.

Repeating the above process according to the time series X in inverse order $x_n, x_{n-1}, \ldots, x_1$, while letting $UB_k = -UF_k$, $k = n, n-1, \ldots, 1$, $UB_1 = 0$. If the calculated value of UB_k or UF_k is greater than 0, it indicates that the sequence has an upward trend; if lower than 0, it indicates a downward trend. When they exceed the critical line, it suggests that the upward or downward trend is significant.

3. Results and Discussion

3.1. Precision of Original TRMM Products

3.1.1. Grids with the Gauge Station

According to the methods above, the precision criteria of original TRMM products on the 20 grids with the gauge station are shown in Table 1. The names of the gauge stations are LengHu (LH), XiaoZaoHuo (XZH), GeErMu (GEM), HeXi (HX), MangYa (MY), NuoMuHong (NMH), GeErMu4 (GEM4), DaChaDan (DCD), HuaiTouTaiLa (HTTL), GaHai (GH), NaChiTai (NCT), WuLan (WL), DeLingHa (DLH), DuLan (DL), ChaHanWuSu (CHWS), XiaRiHa (XRH), XiangRiDe (XRD), KeEr (KE), ShangGaBa (SGB), ChaHanHe (CHH), respectively.

Table 1. The precision of original Tropical Precipitation Measurement Satellite (TRMM) products (2001–2016). LH: LengHu; XZH: XiaoZaoHuo; GEM: GeErMu; HX: HeXi; MY: MangYa; NMH: NuoMuHong; GEM4: GeErMu4; DCD: DaChaDan; HTTL: HuaiTouTaiLa; GH: GaHai; NCT: NaChiTai; WL: WuLan; DLH: DeLingHa; DL: DuLan; CHWS: ChaHanWuSu; XRH: XiaRiHa; XRD: XiangRiDe; KE: KeEr; SGB: ShangGaBa; CHH: ChaHanHe; MAP: Mean Annual Precipitation.

Gauges						Original 3B42RT			Original 3B43		
No.	**Name**	**X (°)**	**Y (°)**	**DEM (m)**	**MAP (mm)**	**MAP (mm)**	**Bias (%)**	**RMSD (mm)**	**MAP (mm)**	**Bias (%)**	**RMSD (mm)**
1	LH	93.33	38.75	2777	19.88	41.10	106.79	23.17	36.19	82.07	18.42
2	XZH	93.68	36.80	2772	30.34	81.60	168.97	57.76	77.49	155.44	52.74
3	GEM	94.90	36.42	2812	47.32	126.01	166.30	81.28	113.31	139.46	68.68
4	HX	94.60	36.38	2822	47.34	121.63	156.91	77.60	111.21	134.89	68.22
5	MY	90.85	38.25	2942	50.16	68.37	34.55	26.85	60.59	19.79	20.07
6	NMH	96.42	36.43	2796	55.54	118.66	113.65	67.30	117.21	111.05	65.48
7	GEM4	94.78	36.30	2957	61.76	126.01	104.05	67.61	113.31	83.48	55.44
8	DCD	95.37	37.85	3190	104.00	122.89	18.17	38.35	119.28	14.69	28.18
9	HTTL	96.73	37.35	2867	107.94	160.30	48.51	65.39	162.39	50.45	63.29
10	GH	97.43	37.23	2877	161.24	157.18	−2.52	32.02	168.12	4.27	26.18
11	NCT	94.57	35.87	3966	182.31	230.40	26.38	70.91	222.21	21.89	65.31
12	WL	98.48	36.92	2959	222.26	157.31	−29.22	83.38	174.37	−21.55	66.26
13	DLH	97.37	37.37	2988	228.68	201.19	−12.02	41.53	203.84	−10.86	35.93
14	DL	98.10	36.30	3190	240.73	183.92	−23.60	73.11	195.62	−18.74	58.43
15	CHWS	98.12	36.23	3273	244.97	211.97	−13.47	54.70	218.89	−10.65	45.53
16	XRH	98.15	36.42	3143	252.91	183.92	−27.28	95.16	195.62	−22.65	81.60
17	XRD	97.87	35.97	3100	285.18	264.48	−7.26	57.26	265.41	−6.93	55.43
18	KE	97.70	35.95	3269	307.43	278.26	−9.49	70.71	277.94	−9.59	70.63
19	SGB	98.58	37.00	3168	332.05	170.04	−48.79	171.22	187.25	−43.61	151.87
20	CHH	98.57	37.05	3351	432.30	257.26	−40.49	189.76	258.21	−40.20	187.08
–	Average	–	–	–	170.72	163.12	36.51	72.25	163.92	31.63	64.24

It is clear in Table 1 that the average mean annual precipitation (MAP) of 3B42RT and 3B43 on the 20 grids are 163.12 mm and 163.92 mm, respectively, which look quite close to that of the gauged precipitation (170.72 mm). However, it can be found that the TRMM precipitation is higher in the low gauged precipitation areas (LGPA), including the stations of LH, XZH, GEM, HX, MY, NMH, GEM4, DCD and HTTL. On the contrary, the TRMM precipitation is lower in the high gauged precipitation area (HGPA), including the stations of WL, DLH, DL, CHWS, XRH, XRD, KE, SGB and CHH. As a result, the average MAP of TRMM is approximately equal to that of the gauged precipitation, but the underestimated HGPA and overestimated LGPA will lead to systematic bias. The systematic error is considered the result of planarization of the original TRMM when calibration was done with too limited gauged data.

The average RMSD of the original 3B42RT and 3B43 in the Qaidam Basin is 72.25 mm and 64.24 mm, respectively. Considering with the bias together, the minimum RMSD (23.17 mm of 3B43RT and 18.42mm of 3B43) relates to the overestimation (bias of 106.79%, 82.07% and value of 21.26 mm, 16.31 mm) of 3B42RT and 3B43 in the LGPA (LH station), while the maximum RMSD (189.76 mm of 3B43RT and 187.08 mm of 3B43) relates to the underestimation (bias of –40.49%, –40.20% and value of –175.04 mm, –174.09 mm) in the HGPA (CHH station). It is obvious that the bias of underestimation is smaller than that of overestimation, but the absolute amount of underestimation is far greater than that of overestimation, as shown in Figure 3. Therfore, it is supposed that the average TRMM precipitation in the Qaidam Basin would be an underestimation overall.

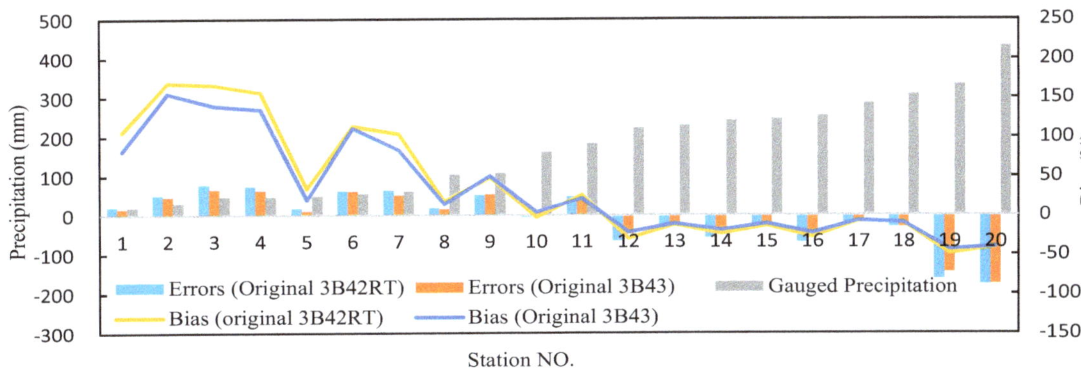

Figure 3. The Original TRMM Precipitation Products Errors.

3.1.2. Grids without a Gauge Station

Most of the grids have no ground gauge so that we could not evaluate the precision by precipitation itself. Some other validation principle is introduced. Here, it is the consistency principle (CR), which is a relationship rule of rainfall and elevation retrieved from the gauged data. The situation of 3B42RT and 3B43 precipitation falling into the PEM is shown in Figure 4.

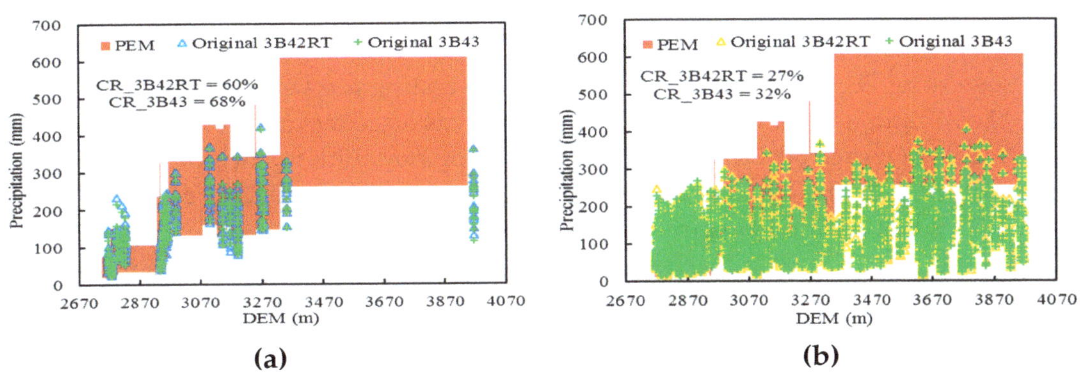

Figure 4. The gauged precipitation – elevation mask (PEM) and original TRMM precipitation filter from 2001 to 2016. (**a**) Grids with gauge station. (**b**) Grids without gauge station.

It can be found that the CR value of original 3B42RT and 3B43 precipitation on the grids with gauge is 60% and 68%, while on the grids without gauge it is only 27% and 32%, respectively. It means the TRMM precipitation on the gauged grids is more reliable than that on the ungauged grids.

3.2. Correction of TRMM Products

3.2.1. Calibration and Validation

The regression machine learning by the ANN model mentioned above was carried out. In the process, we have a total of 20 gauges with 16 years of data sets, of which 224 sets of data are used for the model training, and the remaining 96 sets are used for model testing. The result of a comparison of output and target is shown in Figure 5.

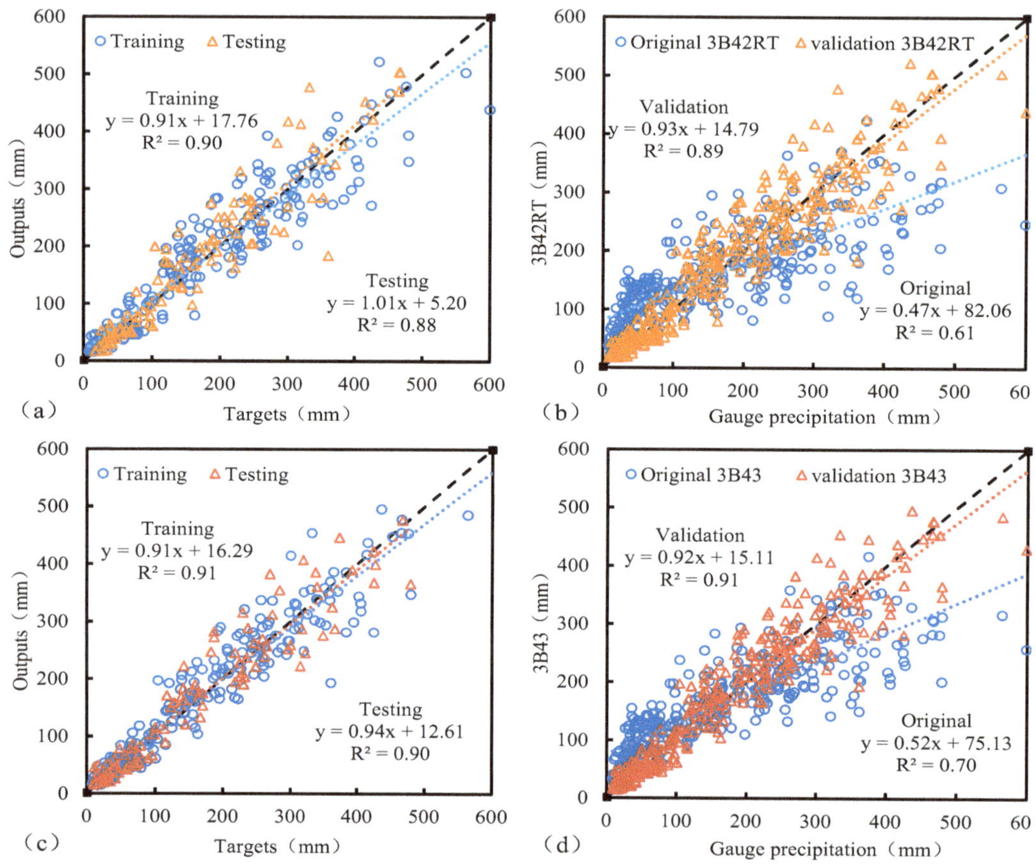

Figure 5. The calibration and validation of TRMM precipitation by artificial neural networks (ANN). (**a**) Training and testing for 3B42RT; (**b**) correction for 3B42RT; (**c**) training and testing for 3B43; (**d**) correction for 3B43.

It can be seen from Figure 5a,c that the training and testing are very good with the determinant coefficients 0.90, 0.91 and 0.88, 0.90, respectively. There was also improvement with the determinant coefficients 0.89, 0.91 of corrected TRMM precipitation compared with the original TRMM determinant coefficients of 0.61, 0.70. It also can be seen that the values of original 3B42RT and 3B43 are all mostly above the 1:1 line when gauged precipitation was less than 100 mm, while a lot of the values of original TRMM precipitation are under the 1:1 line when gauged precipitation was more than 200 mm. This indicates that the original TRMM precipitation in the LGPA was overestimated and the TRMM precipitation in the HGPA was underestimated. On the whole, the dispersion of TRMM precipitation points on both sides of the 1:1 line is large and uneven, the R^2 is only 0.61 and 0.70, respectively. Fortunately, the 3B42RT and 3B43 precipitation improved significantly both in the LGPA and HGPA. They closely dispersed on the both sides of the 1:1 line after the correction by the ANN model. The R^2

increased to 0.89 and 0.91, respectively. It was proven that the fusion model based on an ANN is effective for the correction of TRMM products.

3.2.2. The Precision of Corrected TRMM Products

After the correction, the precision of TRMM precipitation in the Qaidam Basin on the grids with the gauge station is shown in Table 2. It can be found that the average bias is significantly reduced, from 36.51% to –0.08% (3B42RT) and 31.63% to 0.11% (3B43), respectively. It also can be seen from the RMSD that the fusion model has a significant correction effect on TRMM products. The average RMSD value decreased from 72.25 mm and 64.24mm to 37.51 mm and 34.22 mm after correction.

Table 2. The precision evaluation of corrected 3B42RT and 3B43 (2001–2016). - LH: LengHu; XZH: XiaoZaoHuo; GEM: GeErMu; HX: HeXi; MY: MangYa; NMH: NuoMuHong; GEM4: GeErMu4; DCD: DaChaDan; HTTL: HuaiTouTaiLa; GH: GaHai; NCT: NaChiTai; WL: WuLan; DLH: DeLingHa; DL: DuLan; CHWS: ChaHanWuSu; XRH: XiaRiHa; XRD: XiangRiDe; KE: KeEr; SGB: ShangGaBa; CHH: ChaHanHe; MAP: Mean Annual Precipitation.

Gauges					Corrected 3B42RT			Corrected 3B43		
Name	X (°)	Y (°)	DEM (m)	MAP (mm)	MAP (mm)	Bias (%)	RMSD (mm)	MAP (mm)	Bias (%)	RMSD (mm)
LH	93.33	38.75	2777	19.88	19.03	−12.15	8.21	17.80	−10.44	7.21
XZH	93.68	36.80	2772	30.34	31.60	4.15	11.71	32.25	6.32	12.21
GEM	94.90	36.42	2812	47.32	45.53	−3.78	11.06	44.48	−6.00	10.49
HX	94.60	36.38	2822	47.34	45.53	−3.82	14.92	45.09	−4.77	15.71
MY	90.85	38.25	2942	50.16	48.87	−2.46	22.2	52.54	4.52	22.01
NMH	96.42	36.43	2796	55.54	54.30	−2.23	17.24	53.51	−3.66	17.53
GEM4	94.78	36.30	2957	61.76	55.26	−10.52	13.57	56.97	−7.74	12.57
DCD	95.37	37.85	3190	104.00	98.31	−5.47	29.43	105.38	1.33	23.10
HTTL	96.73	37.35	2867	107.94	123.67	14.57	38.85	123.21	14.15	35.36
GH	97.43	37.23	2877	161.24	160.28	−0.60	31.44	158.83	−1.50	26.12
NCT	94.57	35.87	3966	182.31	182.35	0.02	35.08	185.47	1.74	34.19
WL	98.48	36.92	2959	222.26	243.09	9.38	56.33	231.36	4.10	48.45
DLH	97.37	37.37	2988	228.68	226.20	−1.08	31.62	223.36	−2.32	24.75
DL	98.11	36.30	3190	240.73	246.18	2.26	48.60	247.77	2.93	40.58
CHWS	98.12	36.23	3273	244.97	290.87	18.74	63.93	289.24	18.07	58.15
XRH	98.15	36.42	3143	252.91	247.61	−2.11	66.42	247.41	−2.17	59.82
XRD	97.87	35.97	3100	285.18	271.46	−4.81	54.38	275.56	−3.37	52.80
KE	97.7	35.95	3269	307.43	304.7	−0.89	62.78	308.31	0.29	63.06
SGB	98.58	37	3168	332.05	316.19	−4.78	57.49	305.51	−7.99	51.4
CHH	98.57	37.05	3351	432.3	449.37	3.95	74.96	426.29	−1.27	68.8
Average	–	–	–	170.72	173.02	−0.08	37.51	171.52	0.11	34.22

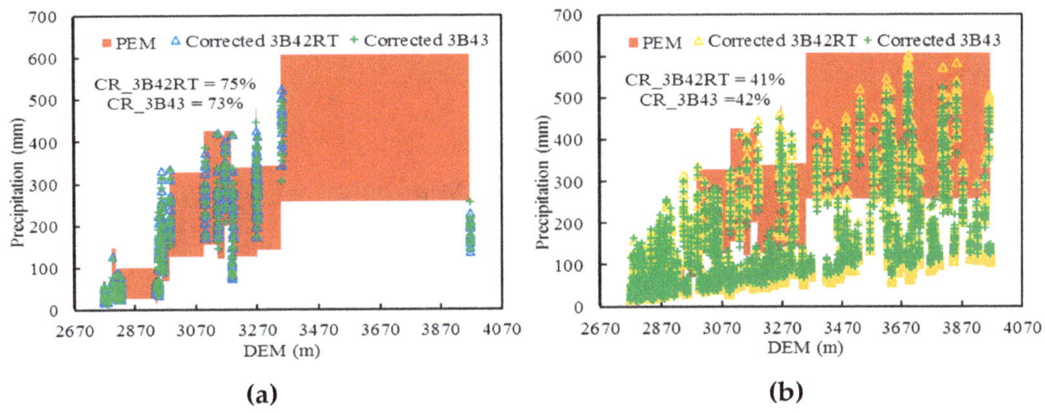

Figure 6. The consistency rate (CR) elevation for corrected TRMM from 2001 to 2016 in the Qaidam Basin. (**a**) Gauged grids. (**b**) Ungauged grids.

For those grids without gauges, the CR values were also evaluated. The CR of 3B42RT and 3B43 on the grids with and without the gauge station improved up to 75%, 73% and 41%, 42% after correction

(Figure 6). This means the precision of TRMM precipitation on grids with and without gauge station are also improved.

3.3. *Assessment of Rainwater Resources of the Qaidam Basin*

3.3.1. The Average Amount of Precipitation

The spatial distribution of the mean annual precipitation from 2001 to 2016 before and after the correction of 3B42RT and 3B43 in the Qaidam Basin is shown in Figure 7. It can be seen that the precipitation on the eastern, southern and southeastern edges of the Qaidam Basin is high, while in the center and northwest it is small. The original distribution of precipitation in the east and west of the Kunlun Mountains is not even, but the trend of decreasing from east to west is enhanced significantly after the correction. The main reason for this is the increase of precipitation in the eastern Kunlun Mountains after the TRMM products corrected. The precipitation in the Qilian Mountains and the southeastern edges of the Qaidam Basin was also relatively high and increased after correction. At the same time, there was a higher consistency between precipitation and elevation in those regions, that is, precipitation increased with elevation.

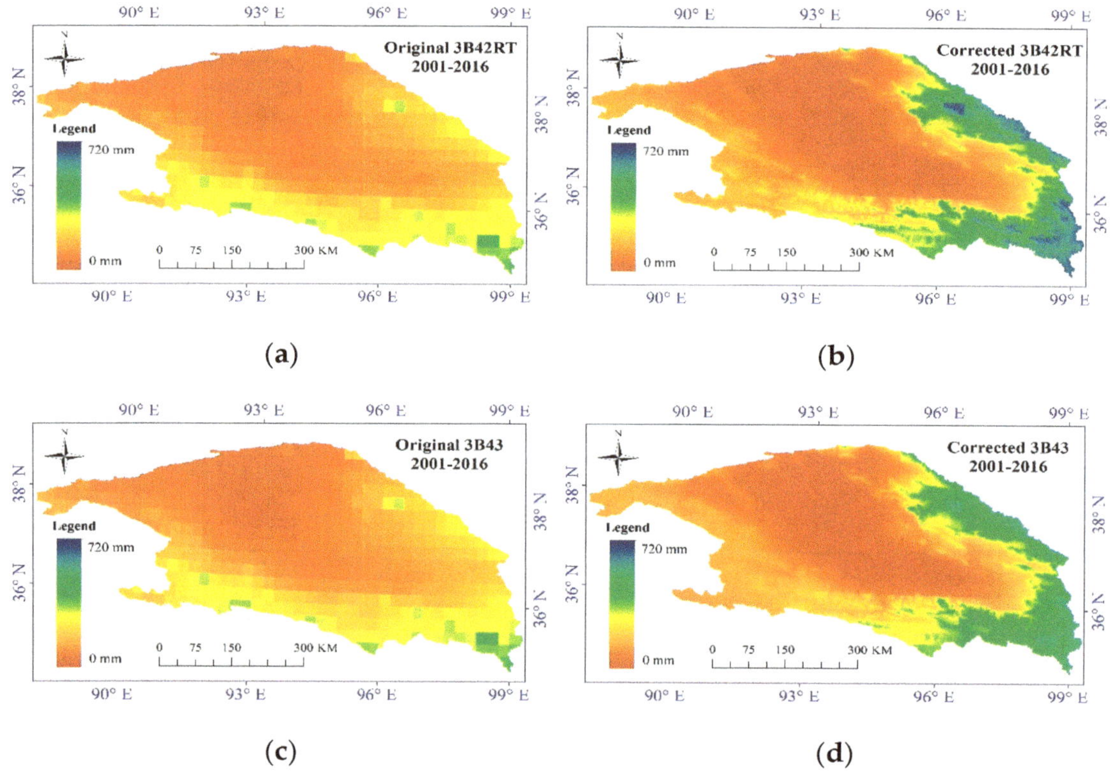

Figure 7. Comparison of spatial distribution of TRMM precipitation before and after correction. (**a**) Original 3B42RT precipitation; (**b**) Corrected 3B42RT precipitation; (**c**) Original 3B43 precipitation; (**d**) Corrected 3B43 precipitation.

For the 12 subregions of the basin, including MA, LE, DCD, DLH, WL, DL, GEM, QML, MD, ZD, XJR and GSR as shown in Figure 1, the areal precipitation in each region is shown in Table 3.

We can see from Table 3 that the average annual precipitation of the Qaidam Basin based on the original 3B42RT and 3B43 are 148.45 mm and 146.26 mm, respectively, while the new assessment value is 186.01 mm and 174.76 mm based on corrected TRMM 3B42 and 3B43. It is obvious that the annual precipitation (104.34 mm) based on the interpolated precipitation by gauges is significantly less than that of original and corrected TRMM products.

The previous studies on precipitation in the Qaidam Basin were mostly based on the gauged precipitation data to get the areal precipitation of the Qaidam Basin according to the traditional interpolation method. The pity is that the gauge distribution is too sparse and uneven, causing an unreliable interpolation result. It is obvious that the precipitation in the area was underestimated in the past. The fusion remote sensing precipitation with local ground information in high mountainous areas is helpful.

Table 3. Assessment of annual precipitation in each region of the Qaidam Basin.

Region Name	Area (10^3 km^2)	Interpolated Precipitation by Gauges	Original 3B42RT	Corrected 3B42RT	Original 3B43	Corrected 3B43
LE	19.3	29.65	43.79	22.42	39.43	25.3
MY	31.3	42.43	67.72	55.27	62.22	62.48
GEM	69.4	65.64	162.73	118.69	157.5	114.27
DCD	21.3	69.55	100.92	128.46	97.61	126.22
XJ	17.4	74.31	114.1	131.95	108.28	133.53
ZD	4.8	83.98	239.7	171.39	238.79	141.03
GS	17.9	86.57	112.33	221.02	107.09	208.24
QML	6.5	157.57	275.74	363.65	273.22	321.82
DLH	22.7	168.45	190.01	369.43	192.47	343.08
DL	43.8	183.01	203.43	302.37	205.56	273.62
WL	10.4	259.57	171.17	299.41	185.38	288.63
MD	4.7	292.9	354.29	602.71	358.21	523.32
The entire basin	269.4	104.34	148.45	186.01	146.26	174.76

3.3.2. The Precipitation Variation Trend

The annual precipitation time series and its variation trend of the gauged precipitation, original and corrected 3B42RT and 3B43 products at the 20 gauge stations are shown in Figure 8 and Table 4. From Figure 8 we can know that the original 3B42RT and 3B43 precipitation at the GEM, LH, NMH, XZH. HX, NCT and GEM4 stations, where the precipitation is low, is obviously higher than the gauged precipitation. On the contrary, the original 3B42RT and 3B43 precipitation at the CHH and SGB stations, where the precipitation is high, is obviously less than the gauged precipitation. This again indicates that the original 3B42RT and 3B43 are significantly overestimated in the LGPA and underestimated in the HGPA. What is nice is that the 3B42RT and 3B43 decreases or increases significantly and tends to the gauged precipitation values after the fusion model was corrected.

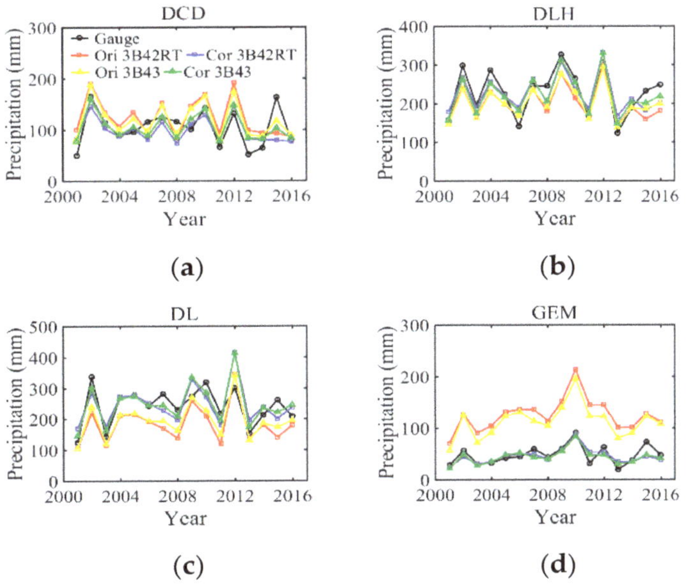

(a) (b)

(c) (d)

Figure 8. *Cont.*

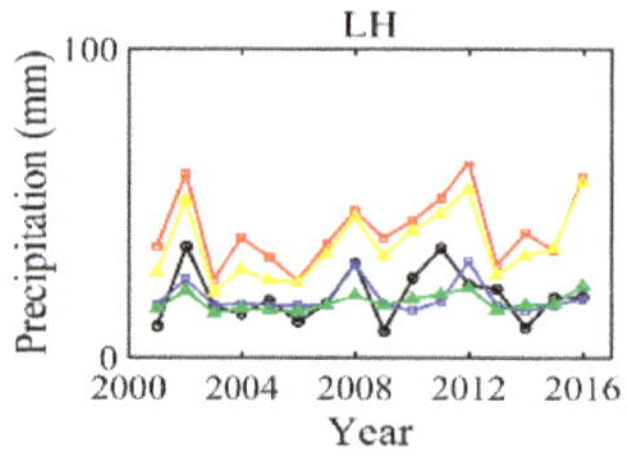

(e)

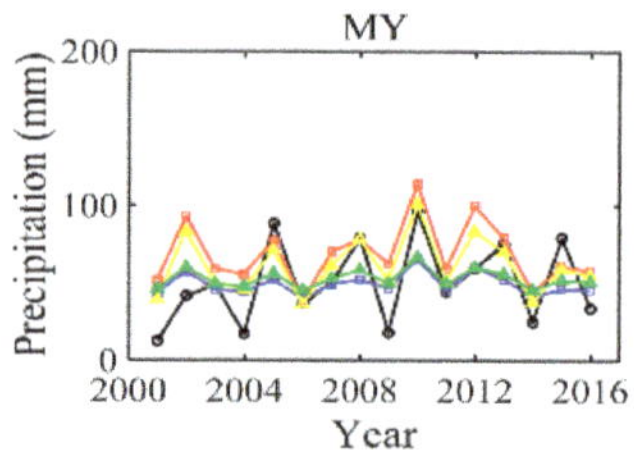

(f)

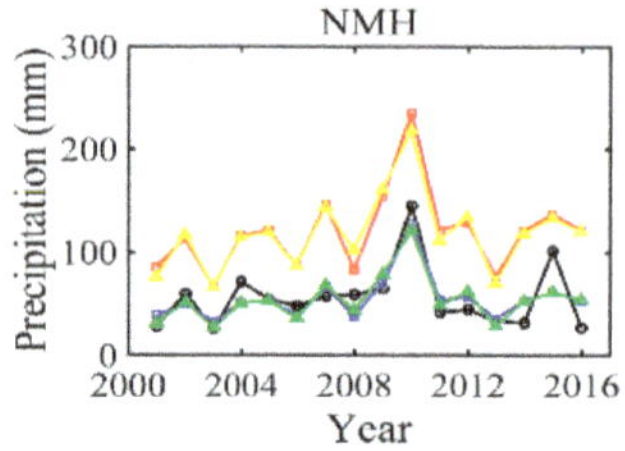

(g)

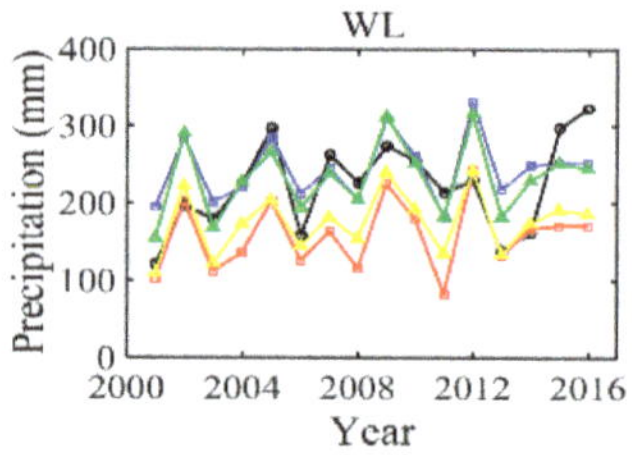

(h)

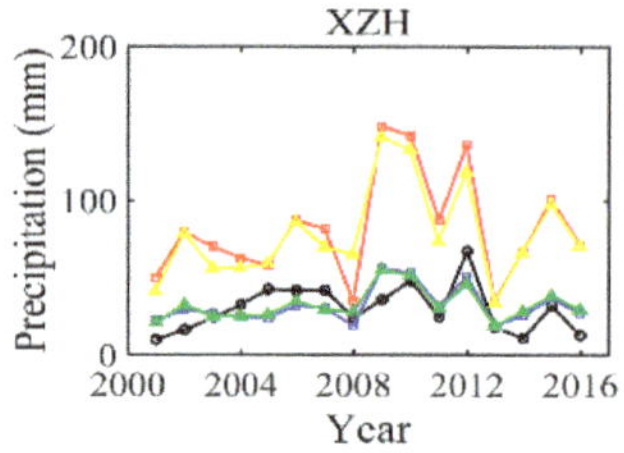

(i)

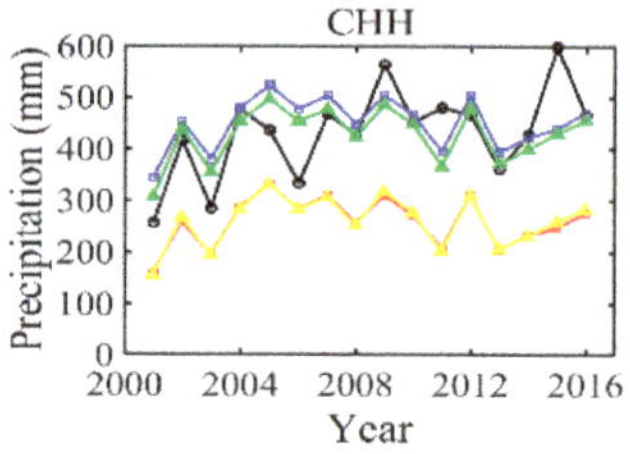

(j)

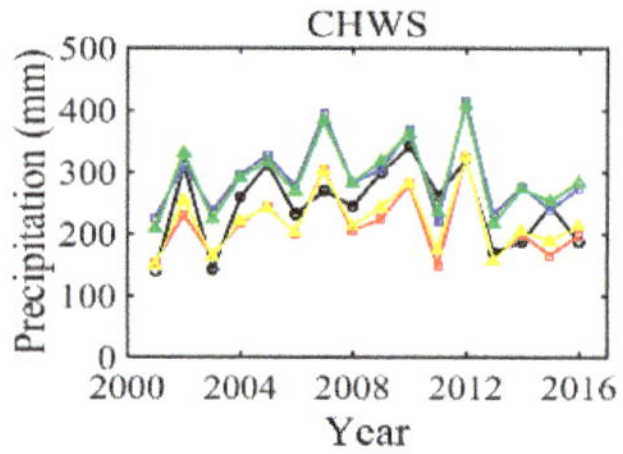

(k)

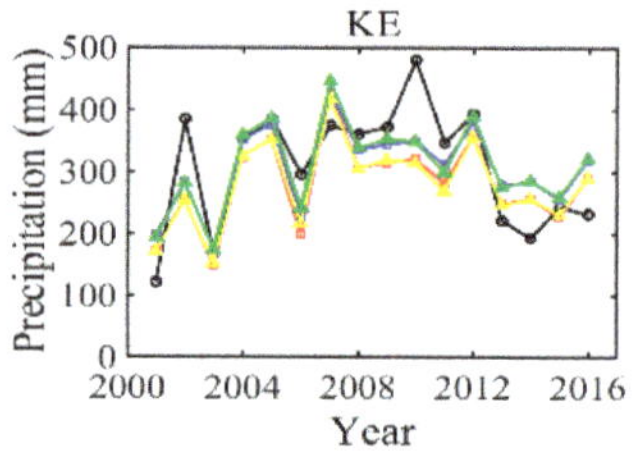

(l)

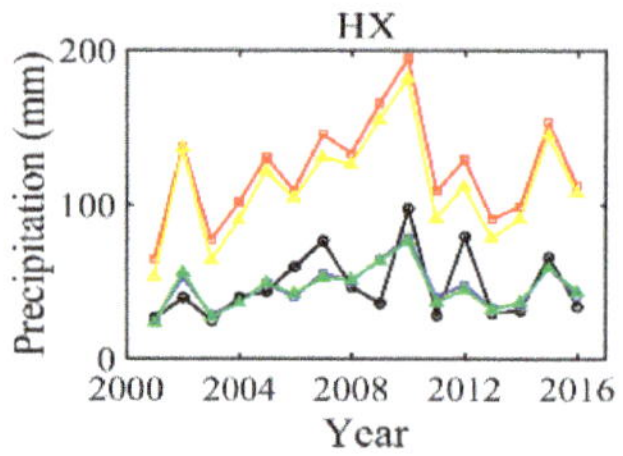

(m)

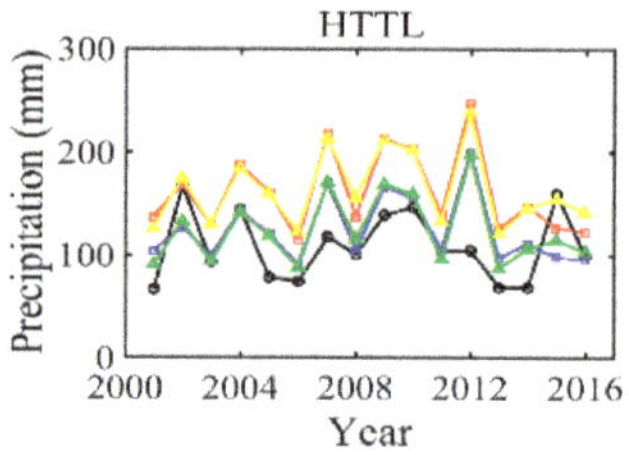

(n)

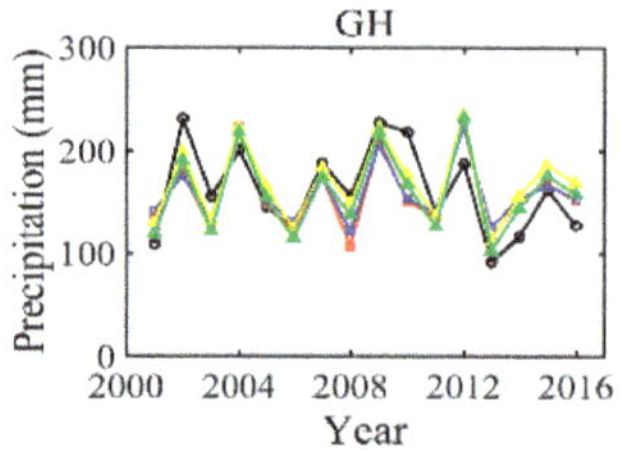

(o)

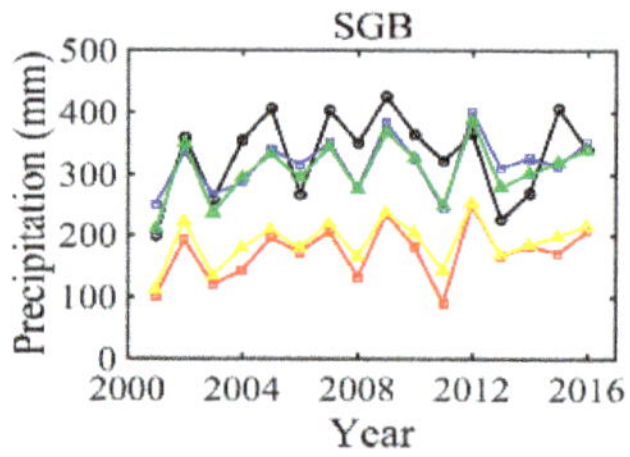

(p)

Figure 8. *Cont.*

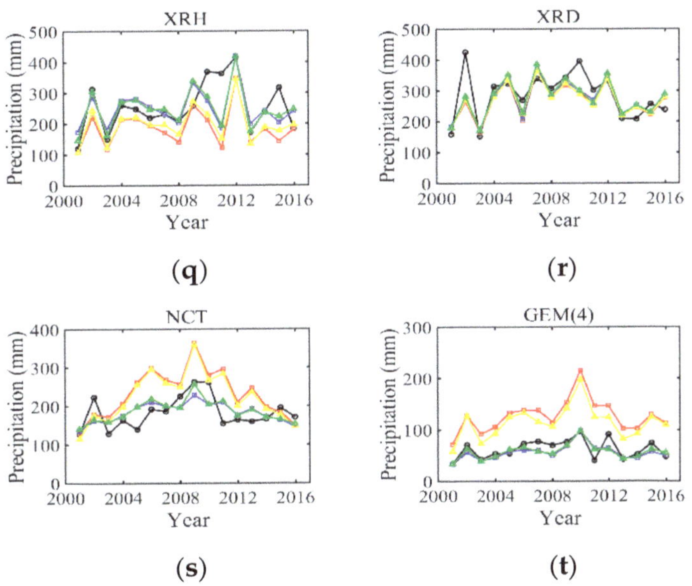

Figure 8. The annual precipitation time series of the gauged precipitation, original and corrected of 3B42RT and 3B43 products at the 20 gauge stations. (**a**) DCD: DaChaDan; (**b**) DLH: DeLingHa; (**c**) DL: DuLan; (**d**) GEM: GeErMu; (**e**) LH: LengHu; (**f**) MY: MangYa; (**g**) NMH: NuoMuHong; (**h**) WL: WuLan; (**i**) XZH: XiaoZaoHuo; (**j**) CHH: ChaHanHe (**k**) CHWS: ChaHanWuSu (**l**) KE: KeEr; (**m**) HX: HeXi; (**n**) HTTL: HuaiTouTaiLa; (**o**) GH: GaHai; (**p**) SGB: ShangGaBa; (**q**) XRH: XiaRiHa; (**r**) XRD: XiangRiDe; (**s**) NCT: NaChiTai; (**t**) GEM(4): GeErMu4.

Table 4. Comparison of the annual precipitation variation trend (2001–2016) between gauged and TRMM products on the grids with the gauge station. LH: LengHu; XZH: XiaoZaoHuo; GEM: GeErMu; HX: HeXi; MY: MangYa; NMH: NuoMuHong; GEM4: GeErMu4; DCD: DaChaDan; HTTL: HuaiTouTaiLa; GH: GaHai; NCT: NaChiTai; WL: WuLan; DLH: DeLingHa; DL: DuLan; CHWS: ChaHanWuSu; XRH: XiaRiHa; XRD: XiangRiDe; KE: KeEr; SGB: ShangGaBa; CHH: ChaHanHe; MAP: Mean Annual Precipitation.

Station Name	Gauged	Original 3B42RT	Corrected 3B42RT	Original 3B43	Corrected 3B43
GH	−2.21	−0.23	−0.18	−0.27	−0.85
XRD	−1.32	0.01	−0.02	−0.06	−0.07
DCD	−0.63	−1.55	−0.75	−1.07	−0.77
KE	−0.62	−0.37	−0.40	−0.47	−0.59
DLH	−0.32	−0.94	−0.91	−0.27	−0.33
HTTL	−0.15	−1.05	−0.63	0.13	−0.10
CHWS	0.09	2.78	2.87	0.58	0.67
LH	0.13	0.91	0.05	1.10	0.26
XZH	0.14	1.44	0.39	1.94	0.57
DL	0.23	0.01	0.02	0.94	1.05
GEM4	0.58	1.87	0.88	1.29	0.58
NMH	0.63	1.91	0.95	2.33	1.29
HX	0.89	2.08	0.69	2.23	0.85
GEM	0.93	1.87	0.84	1.29	0.54
NCT	1.31	0.63	0.32	0.86	0.35
MY	1.45	0.17	0.06	0.41	0.13
SGB	2.43	4.25	3.95	3.22	3.03
WL	4.71	3.14	2.77	2.28	2.69
XRH	3.21	2.21	2.29	2.94	3.06
CHH	10.45	0.19	0.18	1.16	1.78
Average	1.10	0.97	0.67	1.03	0.71

From the average change trends of precipitation at the 20 gauge stations in the three data sources, the annual precipitation of the Qaidam Basin shows an increasing trend. The gauged precipitation at GH, XRD DCD, KE, DLH and HTTL stations showed a decreasing trend, and the others showed an increasing trend in different degrees at the other 14 gauge stations. The change trend direction is basically the same. Therefore, the spatial distribution and temporal variation trend of TRMM products after correction can characterize the spatial–temporal variation characteristics of precipitation in the Qaidam Basin.

According to the corrected 3B42RT and 3B43 precipitation in the Qaidam Basin from 2001 to 2016, the change trend and significance on the grid scale in the study area were calculated, as shown in Figure 9.

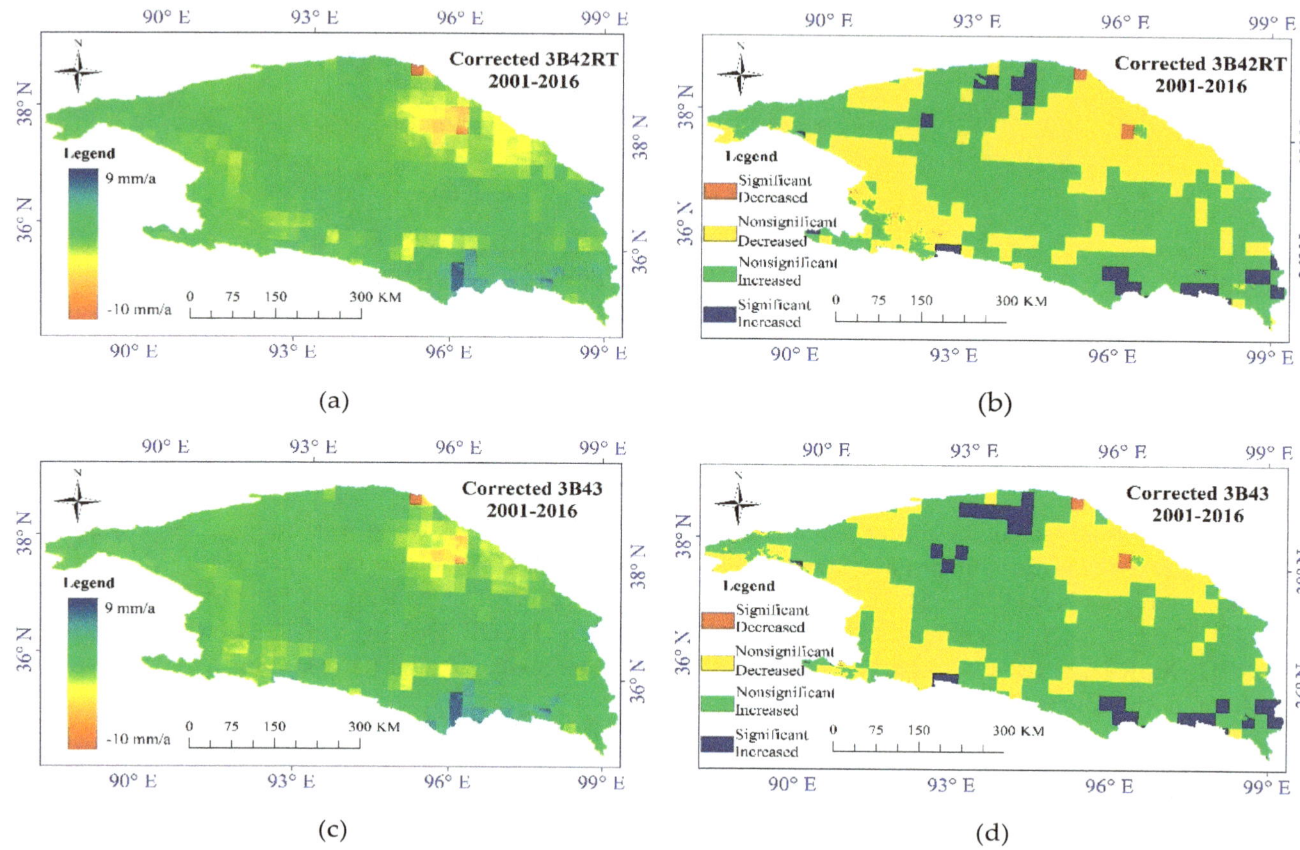

Figure 9. The variation trend and significance of precipitation in the Qaidam Basin. (**a**) Corrected 3B42RT change trend; (**b**) corrected 3B42RT change significance; (**c**) corrected 3B43 change trend; (**d**) corrected 3B43 change significance.

It can be seen from the Figure 9a,b that the precipitation in the northeast of the Qaidam Basin (Qilian Mountain area) showed a decreasing trend, while in the southeastern edge of the basin (east of the Kunlun Mountains) showed an increasing trend based on corrected 3B42RT and 3B43 products. It is obvious the precipitation slightly increased at the center and northwest of the basin. Figure 9b,d showed the significant test results of the precipitation change trend. We know that most areas of the Qaidam Basin were non-significantly increased or decreased, and only a small part of the areas passed the significant change test ($Z \geq 2.58$).

4. Conclusions

(1) The Qaidam Basin is located in an arid and semi-arid region with a dry climate and fragile natural ecological environment. The precipitation process in the Qaidam Basin is significantly different spatially and temporally. Due to the sparse gauge stations and maldistribution, the rainwater resources

are hard to assess. A fusion of remote sensing precipitation of TRMM products and gauged precipitation is helpful.

(2) The average mean annual precipitation is only 104.34 mm by ground gauges interpolation, and 148.45 mm and 146.26 mm by original 3B42RT and 3B43. However, the precision evaluation of TRMM precipitation shows it was overestimated in the LGPA and underestimated in the HGPA. The original TRMM products need a correction.

(3) The correction result shows the average mean annual precipitation is 186.01 mm by 3B42RT and 174.74 mm by 3B43. The average bias of 3B42RT and 3B43 at gauge stations are significantly reduced to −0.08% and 0.11% after being corrected, and the average RMSD is significantly reduced to 37.51 mm and 34.22 mm, respectively. All of those indicated that the precipitation products fusion model based on ANN could effectively work on TRMM products.

(4) The final result shows there are many differences in areal precipitation based on sparse gauge precipitation data, original TRMM data and fusion remote sensing data. The rainwater resources in the study basin have been underestimated in the past year, and both were derived from ground gauge stations and from original TRMM products.

In the future, research needs to test the ANN effectivity on mining the normalized difference vegetation index and rainwater resources in this area.

Author Contributions: Conceptualization, Z.W. and G.X.; methodology, Z.W. and T.X.; formal analysis, G.X.; writing—original draft preparation, G.X.; writing—review and editing, Z.W. All authors read and approved the final manuscript.

Acknowledgments: The authors would like to thank all the colleagues who generously provided their dataset. The authors greatly thank the editors and reviewers for providing thorough and constructive comments to improve the manuscript.

References

1. Bohnenstengel, S.I.; Schlünzen, K.H.; Beyrich, F. Representativity of in situ precipitation measurements—A case study for the LITFASS area in North-Eastern Germany. *J. Hydrol.* **2011**, *400*, 387–395. [CrossRef]
2. Marzano, F.S.; Cimini, D.; Montopoli, M. Investigating precipitation microphysics using ground-based microwave remote sensors and disdrometer data. *Atmos. Res.* **2010**, *97*, 583–600. [CrossRef]
3. Zhuoqi, C.; Xiaogu, Z.; Shupeng, Z.; Tao, L. Mapping Daily Precipitation over China Based on TRMM Multisatellite Precipitation Analysis and Gauge Data. In Proceedings of the International Conference on Remote Sensing, Kyoto, Japan, 29 October–1 November 2012.
4. XIE, P.; Arkin, P.A. Analyses of Global Monthly Precipitation Using Gauge Observations, Satellite Estimates, and Numerical Model Predictions. *J. Clim.* **1996**, *9*, 840–858. [CrossRef]
5. Huffman, G.J.; Adler, R.F.; Rudolf, B.; Schneider, U.; Keehn, P.R. Global Precipitation Estimates Based on a Technique for Combining Satellite-Based Estimates, Rain Gauge Analysis, and NWP Model Precipitation Information. *J. Clim.* **1995**, *8*, 1284–1295. [CrossRef]
6. Vila, D.; Goncalves, L.; Toll, D.; Rozante, J.R. Statistical Evaluation of Combined Daily Gauge Observations and Rainfall Satellite Estimates over Continental South America. *J. Hydrometeorol.* **2009**, *10*, 533–543. [CrossRef]
7. Rozante, J.R.; Moreira, D.S.; de Goncalves, L.G.G.; Vila, D.A. Combining TRMM and Surface Observations of Precipitation: Technique and Validation over South America. *Weather* **2010**, *25*, 885–894. [CrossRef]
8. Shen, Y.; Zhao, P.; Pan, Y.; Yu, J. A high spatiotemporal gauge-satellite merged precipitation analysis over China. *J. Geophys. Res. Atmos.* **2014**, *119*, 3063–3075. [CrossRef]
9. Huffman, G.J.; Robert, F.A.; David, T.B.; Nelkin, E.J.; Acheampong, M. *The TRMM Multi-Satellite Precipitation Analysis (TMPA)*; Springer: Berlin, Germany, 2008.

10. Kubota, T.; Ushio, T.; Shige, S.; Kida, S.; Kachi, M.; Okamoto, K. Verification of High-Resolution Satellite-Based Rainfall Estimates around Japan Using a Gauge-Calibrated Ground-Radar Dataset. *J. Meteorol. Soc. Jpn.* **2009**, *87*, 203–222. [CrossRef]
11. Kidd, C.; Huffman, G. Global precipitation measurement. *Meteorol. Appl.* **2011**, *18*, 334–353. [CrossRef]
12. Tapiador, F.J.; Turk, F.J.; Petersen, W.; Hou, A.Y.; García-Ortega, E.; Machado, L.A.T.; Angelis, C.F.; Salio, P.; Kidd, C.; Huffman, G.J.; et al. Global precipitation measurement: Methods, datasets and applications. *Atmos. Res.* **2012**, *104*, 70–97. [CrossRef]
13. Seto, S.; Iguchi, T.; Oki, T. The Basic Performance of a Precipitation Retrieval Algorithm for the Global Precipitation Measurement Mission's Single/Dual-Frequency Radar Measurements. *IEEE Trans. Geosci. Remote* **2013**, *51*, 5239–5251. [CrossRef]
14. Zhu, Z.; Yong, B.; Ke, L.; Wang, G.; Ren, L.; Chen, X. Tracing the Error Sources of Global Satellite Mapping of Precipitation for GPM (GPM-GSMaP) Over the Tibetan Plateau, China. *IEEE J. Stars* **2018**, *11*, 2181–2191. [CrossRef]
15. Guo, H.; Chen, S.; Bao, A.; Behrangi, A.; Hong, Y.; Ndayisaba, F.; Hu, J.; Stepanian, P.M. Early assessment of Integrated Multi-satellite Retrievals for Global Precipitation Measurement over China. *Atmos. Res.* **2016**, 176–177. [CrossRef]
16. Haile, A.T.; Yan, F.; Habib, E. Accuracy of the CMORPH satellite-rainfall product over Lake Tana Basin in Eastern Africa. *Atmos. Res.* **2015**, *163*, 177–187. [CrossRef]
17. Liu, S.; Yan, D.; Qin, T.; Weng, B.; Li, M. Correction of TRMM 3B42V7 Based on Linear Regression Models over China. *Adv. Meteorol.* **2016**, *2016*, 1–13. [CrossRef]
18. Xia, T.; Wang, Z.; Zheng, H. Topography and Data Mining Based Methods for Improving Satellite Precipitation in Mountainous Areas of China. *Atmosphere* **2015**, *6*, 983–1005. [CrossRef]
19. Zheng, X.; Zhu, J. A methodological approach for spatial downscaling of TRMM precipitation data in North China. *Int. J. Remote Sens.* **2015**, *36*, 144–169. [CrossRef]
20. Tao, Z.; Yuanqing, H.; Jian, M.; Juan, P. Spatial and temporal distribution of precipitation based on corrected TRMM data around the Hexi Corridor, China. *Sci. Cold Arid Reg.* **2014**, *6*, 159–167.
21. Shi, Y.; Song, L.; Xia, Z.; Lin, Y.; Myneni, R.; Choi, S.; Wang, L.; Ni, X.; Lao, C.; Yang, F. Mapping Annual Precipitation across Mainland China in the Period 2001–2010 from TRMM3B43 Product Using Spatial Downscaling Approach. *Remote Sens.* **2015**, *7*, 5849–5878. [CrossRef]
22. Dinku, T.; Chidzambwa, S.; Ceccato, P.; Connor, S.J.; Ropelewski, C.F. Validation of high-resolution satellite rainfall products over complex terrain. *Int. J. Remote Sens.* **2008**, *29*, 4097–4110. [CrossRef]
23. Bitew, M.M.; Gebremichael, M. Assessment of satellite rainfall products for streamflow simulation in medium watersheds of the Ethiopian highlands. *Hydrol. Earth Syst. Sci.* **2011**, *15*, 1147–1155. [CrossRef]
24. Condom, T.; Rau, P.; Espinoza, J.C. Correction of TRMM 3B43 monthly precipitation data over the mountainous areas of Peru during the period 1998–2007. *Hydrol. Process.* **2011**, *25*, 1924–1933. [CrossRef]
25. Darand, M.; Amanollahi, J.; Zandkarimi, S. Evaluation of the performance of TRMM Multi-satellite Precipitation Analysis (TMPA) estimation over Iran. *Atmos. Res.* **2017**, *190*, 121–127. [CrossRef]
26. Romilly, T.G.; Gebremichael, M. Evaluation of satellite rainfall estimates over Ethiopian river basins. *Hydrol. Earth Syst. Sci.* **2011**, *15*, 1505–1514. [CrossRef]
27. Moazami, S.; Golian, S.; Kavianpour, M.R.; Hong, Y. Uncertainty analysis of bias from satellite rainfall estimates using copula method. *Atmos. Res.* **2014**, *137*, 145–166. [CrossRef]
28. Wang, Z.; Zhong, R.; Lai, C.; Chen, J. Evaluation of the GPM IMERG satellite-based precipitation products and the hydrological utility. *Atmos. Res.* **2017**, *196*, 151–163. [CrossRef]
29. Lu, X.; Wei, M.; Tang, G.; Zhang, Y. Evaluation and correction of the TRMM 3B43V7 and GPM 3IMERGM satellite precipitation products by use of ground-based data over Xinjiang, China. *Env. Earth Sci.* **2018**, *77*, 209. [CrossRef]
30. Lekula, M.; Lubczynski, M.W.; Shemang, E.M.; Verhoef, W. Validation of satellite-based rainfall in Kalahari. *Phys. Chem. Earth Parts A/B/C* **2018**, *105*, 84–97. [CrossRef]
31. Duan, Z.; Bastiaanssen, W.G.M. First results from Version 7 TRMM 3B43 precipitation product in combination with a new downscaling–calibration procedure. *Remote Sens. Env.* **2013**, *131*, 1–13. [CrossRef]
32. Guofeng, Z.; Dahe, Q.; Yuanfeng, L.; Fenli, C.; Pengfei, H.; Dongdong, C.; Kai, W. Accuracy of TRMM precipitation data in the southwest monsoon region of China. *Appl. Clim.* **2017**, *129*, 353–362. [CrossRef]

33. Jia, S.; Zhu, W.; Lű, A.; Yan, T. A statistical spatial downscaling algorithm of TRMM precipitation based on NDVI and DEM in the Qaidam Basin of China. *Remote Sens. Env.* **2011**, *115*, 3069–3079. [CrossRef]
34. Liu, S.; Yan, D.; Wang, H.; Li, C.; Qin, T.; Weng, B.; Xing, Z. Evaluation of TRMM 3B42V7 at the basin scale over mainland China. *Adv. Water Sci.* **2016**, *27*, 639–651.
35. Seyyedi, H.; Anagnostou, E.N.; Beighley, E.; McCollum, J. Satellite-driven downscaling of global reanalysis precipitation products for hydrological applications. *Hydrol. Earth Syst. Sci.* **2014**, *18*, 5077–5091. [CrossRef]
36. Agatonovic-Kustrin, S.; Beresford, R. Basic Concepts of Artificial Neural Network (ANN) Modeling and its Application in Pharmaceutical Research. *J. Pharm. Biomed. Anal.* **2000**, *22*, 717–727. [CrossRef]
37. Partridge, D.; Rae, S.; Wang, W.J. Artificial neural networks. *J. Roy. Soc. Med.* **1999**, *92*, 385. [CrossRef]
38. Lyons, W.B.; Flanagan, C.; Lewis, E.; Ewald, H.; Lochmann, S. Interrogation of multipoint optical fibre sensor signals based on artificial neural network pattern recognition techniques. *Sens. Actuators A Phys.* **2004**, *114*, 7–12. [CrossRef]
39. Lyons, W.; Fitzpatrick, C.; Flanagan, C.; Lewis, E. A novel multipoint luminescent coated ultra violet fibre sensor utilising artificial neural network pattern recognition techniques. *Sens. Actuators A Phys.* **2004**, *115*, 267–272. [CrossRef]
40. Nissar, A.I.; Upadhyaya, S.J. Fault Diagnosis of Mixed Signal VLSI Systems Using Artificial Neural Networks. In Proceedings of the Southwest Symposium on Mixed-Signal Design (SSMSD 99), Tucson, AZ, USA, 11–13 April 1999; IEEE: Piscataway, NY, USA, 1999; pp. 93–98.
41. Huesken, D.; Lange, J.; Mickanin, C.; Weiler, J.; Asselbergs, F.; Warner, J.; Meloon, B.; Engel, S.; Rosenberg, A.; Cohen, D.; et al. Design of a genome-wide siRNA library using an artificial neural network. *Nat. Biotechnol.* **2005**, *23*, 995–1001. [CrossRef]
42. Aceto, G.; Ciuonzo, D.; Montieri, A.; Pescapé, A. Mobile Encrypted Traffic Classification Using Deep Learning. In Proceedings of the 2018 Network Traffic Measurement and Analysis Conference (TMA), Vienna, Austria, 26–29 June 2018; IEEE: Piscataway, NJ, USA, 2018; pp. 1–8.
43. Aceto, G.; Ciuonzo, D.; Montieri, A.; Pescape, A. Mobile Encrypted Traffic Classification Using Deep Learning: Experimental Evaluation, Lessons Learned, and Challenges. *IEEE Trans. Netw. Serv. Manag.* **2019**. [CrossRef]
44. Rucco, R.; Sorriso, A.; Liparoti, M.; Ferraioli, G.; Sorrentino, P.; Ambrosanio, M.; Baselice, F. Type and Location of Wearable Sensors for Monitoring Falls during Static and Dynamic Tasks in Healthy Elderly: A Review. *Sensors* **2018**, *18*, 1613. [CrossRef]
45. Gardner, G.G.; Keating, D.; Williamson, T.H.; Elliott, A.T. Automatic detection of diabetic retinopathy using an artificial neural network: A screening tool. *Brit. J. Ophthalmol.* **1996**, *80*, 940–944. [CrossRef]
46. Hsu, K.; Gupta, H.V.; Sorooshian, S. Artificial Neural Network Modeling of the Rainfall-Runoff Process. *Water Resour. Res.* **1995**, *31*, 2517–2530. [CrossRef]
47. Hopfield, J.J. Neural networks and physical systems with emergent collective computational abilities. *Proc. Natl. Acad. Sci. USA* **1982**, *79*, 2554–2558. [CrossRef]
48. AghaKouchak, A.; Nasrollahi, N.; Habib, E. Accounting for Uncertainties of the TRMM Satellite Estimates. *Remote Sens.* **2009**, *1*, 606–619. [CrossRef]
49. Heidinger, H.; Yarlequé, C.; Posadas, A.; Quiroz, R. TRMM rainfall correction over the Andean Plateau using wavelet multi-resolution analysis. *Int. J. Remote Sens.* **2012**, *33*, 4583–4602. [CrossRef]
50. Liu, Z.; Ostrenga, D.; Teng, W.; Kempler, S. Tropical Rainfall Measuring Mission (TRMM) Precipitation Data and Services for Research and Applications. *B Am. Meteorol. Soc.* **2012**, *93*, 1317–1325. [CrossRef]
51. Courault, D.; Seguin, B.; Olioso, A. Review on estimation of evapotranspiration from remote sensing data: From empirical to numerical modeling approaches. *Irrig. Drain. Syst.* **2005**, *19*, 223–249. [CrossRef]
52. Ji, T.; Yang, H.; Liu, R.; He, T.; Wu, J. Applicability analysis of the TRMM precipitation data in the Sichuan-Chongqing region. *Prog. Geogr.* **2014**, 1375–1386.
53. Wang, X.; Liu, H.; Bao, A. Applicability Research on TRMM Precipitation Data in Tianshan Mountains. *J. China Hydrol.* **2014**, 58–64.
54. Yongqing, B.; Juanle, W.; Yujie, W.; Xuehua, H.; Tsydypov, B.Z.; Ochir, A.; Davaasuren, D. Spatio-Temporal Distribution of Drought in the Belt and Road Area During 1998–2015 Based on TRMM Precipitation Data. *J. Resour. Ecol.* **2017**, *8*, 559–570. [CrossRef]
55. Tian, Y.; Peters-Lidard, C.D.; Choudhury, B.J.; Garcia, M. Multitemporal Analysis of TRMM-Based Satellite Precipitation Products for Land Data Assimilation Applications. *J. Hydrometeorol.* **2007**, *8*, 1165–1183. [CrossRef]

56. Bookhagen, B.; Burbank, D.W. Topography, relief, and TRMM-derived rainfall variations along the Himalaya. *Geophys. Res. Lett.* **2006**, *33*.
57. Qu, W.; Lu, J.; Song, W.; Zhang, T.; Tan, Y.; Huang, P. Research on Accuracy Validation and Calibration Methods of TRMM Remote Sensing Precipitation Data in Irrawaddy Basin. *Adv. Earth Sci.* **2014**, 1262–1270.
58. Xu, R.; Tian, F.; Yang, L.; Hu, H.; Lu, H.; Hou, A. Ground validation of GPM IMERG and TRMM 3B42V7 rainfall products over southern Tibetan Plateau based on a high-density rain gauge network. *J. Geophys. Res. Atmos.* **2017**, *122*, 910–924. [CrossRef]
59. Geographical Information Monitoring Cloud Platform. Available online: http://www.dsac.cn/ (accessed on 25 May 2018).
60. National Aeronautics and Space Administration. Available online: https://mirador.gsfc.nasa.gov/ (accessed on 17 March 2018).
61. National Meteorological Information Center. Available online: http://data.cma.cn/ (accessed on 9 May 2018).
62. Qinghai Water Conservancy Bureau. *Annual Hydrological Report P. R. China: Hydrological Data of Inland Rivers and Lakes*, 1st ed.; Journal of Qinghai Water Conservancy Bureau: Xining, China, 2016; pp. 64–73.
63. Marden, J.; Kendall, M.; Gibbons, J. Rank Correlation Methods (5th ed.). *J. Am. Stat. Assoc.* **1992**, *87*, 249. [CrossRef]
64. Mann, H.B. Nonparametric Tests Against Trend. *Econometrica* **1945**, *13*, 245. [CrossRef]
65. Burn, D.H.; Hag Elnur, M.A. Detection of hydrologic trends and variability. *J. Hydrol.* **2002**, *255*, 107–122. [CrossRef]

10

Single-Class Data Descriptors for Mapping *Panax notoginseng* through P-Learning

Fei Deng and Shengliang Pu *

School of Geodesy and Geomatics, Wuhan University, Wuhan 430079, China; fdeng@sgg.whu.edu.cn
* Correspondence: shengliangpu@163.com

Abstract: Machine learning-based remote-sensing techniques have been widely used for the production of specific land cover maps at a fine scale. P-learning is a collection of machine learning techniques for training the class descriptors on the positive samples only. *Panax notoginseng* is a rare medicinal plant, which also has been a highly regarded traditional Chinese medicine resource in China for hundreds of years. Until now, *Panax notoginseng* has scarcely been observed and monitored from space. Remote sensing of natural resources provides us new insights into the resource inventory of Chinese materia medica resources, particularly of *Panax notoginseng*. Generally, land-cover mapping involves focusing on a number of landscape classes. However, sometimes a subset or one of the classes will be the only part of interest. In term of this study, the *Panax notoginseng* field is the right unit class. Such a situation makes single-class data descriptors (SCDDs) especially significant for specific land-cover interpretation. In this paper, we delineated the application such that a stack of SCDDs were trained for remote-sensing mapping of *Panax notoginseng* fields through P-learning. We employed and compared SCDDs, i.e., the simple Gaussian target distribution, the robust Gaussian target distribution, the minimum covariance determinant Gaussian, the mixture of Gaussian, the auto-encoder neural network, the k-means clustering, the self-organizing map, the minimum spanning tree, the k-nearest neighbor, the incremental support vector data description, the Parzen density estimator, and the principal component analysis; as well as three ensemble classifiers, i.e., the mean, median, and voting combiners. Experiments demonstrate that most SCDDs could achieve promising classification performance. Furthermore, this work utilized a set of the elaborate samples manually collected at a pixel-level by experts, which was intended to be a benchmark dataset for the future work. The measuring performance of SCDDs gives us challenging insights to define the selection criteria and scoring proof for choosing a fine SCDD in mapping a specific landscape class. With the increment of remotely sensed satellite data of the study area, the spatial distribution of *Panax notoginseng* could be continuously derived in the local area on the basis of SCDDs.

Keywords: mapping; single-class data descriptors; materia medica resource; *Panax notoginseng*; one-class classifiers; geoherb

---

1. Introduction

Traditional Chinese medicine (TCM) [1] originated in ancient China and has evolved over thousands of years as the only health care and disease healing [2]. A long time before the birth of modern Western medicine, traditional medicinal recipes were handed down orally generation by generation in many parts of the world [3]. Given that TCM is a practical medicine built on experience, and has been mainly practiced and researched in China [4,5], the essence of TCM has always been the most advanced and experienced medicine in the world [6]. Moreover, scientists proved that TCM can coexist with Western medicine [3,7]. Geoherb is a type of Chinese herb with a geographical indication corresponding to a specific geographical location or origin, which has a certification that the product possesses certain qualities, and its production will be protected by intellectual property rights law [8,9]. Compared to the herbal resources produced in other areas, the quality and efficacy of geoherbs are much better [10]. As a highly-regarded TCM resource and a rare kind of geoherb in China [11,12], *Panax notoginseng* (see Figure 1) has been cultivated for more than 400 years in the south-west region of China [13], especially in Wenshan Prefecture, Yunnan province. The conventional methods of TCM resource surveys mainly focus on the qualitative description of species rather than the natural storage or dynamic changes of the planting fields, resulting in a problematic situation that TCM resources appear difficult to monitor over time, which is not conducive to the sustainable development of TCM [14]. In the past few decades, a resource census of TCM has been carried out three times (e.g., 1960–1962, 1969–1973, and 1983–1987). Until 2009, the government of China proposed "*To carry out a nationwide census of TCM resources, strengthen the monitoring of TCM resources and the construction of an information network*" [15]. Furthermore, the State Council of China highlighted "*Strengthening the landscape-scale dynamic monitoring and protection of TCM*" [16] again. From 2011, the government of China planned to conduct the fourth national census of TCM resources, and remote-sensing techniques were regarded as the core-key technologies for surveying and monitoring TCM resources in a large area. The inheritance, innovation, modernization, and internationalization of TCM would be the four basic tasks for a considerable period of time [17,18].

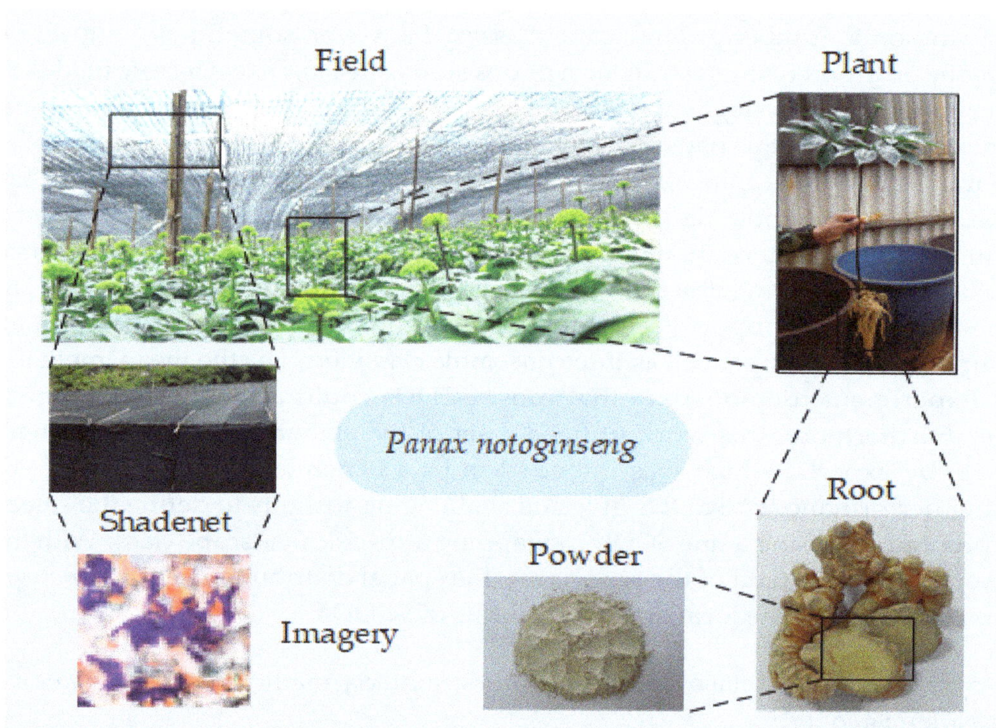

Figure 1. *Panax notoginseng* is a well-known geoherb, which has acquired a very favorable reputation for the treatment of blood disorders, including blood stasis, bleeding, and blood deficiency. Its root can be turned to powder as a medicinal material, and the shadenet cover can be observed from space by means of satellite imagery.

Remote-sensing techniques have been applied to monitor land cover at a range of spatial and temporal scales, in order to satisfy a range of scientific and practical requirements [19–21]. In particular, remote-sensing mapping is an efficient technique to acquire spatial and temporal cropland information repeatedly and consistently [22,23]. In many cases, remotely-sensed data are utilized to derive landscape information on a specific land-cover class of interest [24]. The ability to map and monitor land-cover types and their dynamics for diverse applications has been enhanced by the availability and constantly increased coverage, of satellite images [25,26]. Many problems are encountered in mapping land cover from remotely sensed data by a classification analysis or landscape class description [27] in order to quantify the relationships among all the pixels in an image, such as similarities or differences in spectra signature or spatial texture [22], and extract land cover classes from remote-sensing images [28–30]. The application of remote-sensing techniques plays a significant role in the quantitative resource survey of TCM, particularly in exploring monitoring abilities for the sustainable utilization and bio-diversity protection of Chinese materia medica resources in macrocosm. Under the prerequisite that remote-sensing techniques provide up-to-date landscape surveying at a fine scale [31], one of the motivations for this work comes from the fact that only a single class of interest is involved in a mapping task.

Recently, there has been a number of applications [32], i.e., geological products [33], vegetation indices [34], aerosol products [35], ocean data products [36], dust source identification [37], and crops identification [38] in remote-sensing based on machine learning techniques. In this study, we present an original innovation to apply most of the available single-class data descriptors (SCDDs) through P-learning to conduct mapping of *Panax notoginseng* fields. After that, the work of measuring performance will give us profound insights into defining the selection criteria and scoring proof for choosing a fine SCDD. As such, the introduction of SCDDs for remote-sensing mapping of *Panax notoginseng* fields will be helpful to promoting the development of a *Panax notoginseng* resource inventory and dynamic monitoring towards the quantitative direction. Attributing to the standardized cultivation technique of good agricultural practice (GAP) [39], or so-called controlled-environment agriculture using shade houses provide a distinct image texture to interpret *Panax notoginseng* fields visually [40]. Additionally, due to only *Panax notoginseng* fields being the target class, the task of mapping *Panax notoginseng* fields becomes a specific typology of land-cover classification, which could be regarded as a problem of single-class data description or a special type of one-class classification [41]. SCDDs are the appealing alternatives to the conventional supervised classifiers because they can be trained with only the target training samples [42]. These kinds of algorithms have emerged to only require training samples from the target class, which are referred to as P-learning [43]. Notice that the single class meant no more than one landscape class, and the P-learning based class description may depict the sight that no negative samples are used for training. Such a classification approach aims to identify only one landscape class of interest regardless of the other classes presented in the study area [44]. In the case of single-class data description, we always face an imbalanced binary classification [45,46] including (1) the positive class (i.e., *Panax notoginseng* fields); and (2) the negative class (i.e., the other classes). In this case, the positive class is assumed to be sampled well, while the other classes may be sampled very sparsely or totally absent. When no samples of the other classes are available, most classification errors (e.g., false negative) cannot be estimated [47]. In addition, the procedure that trains an accurate SCDD is challenging, particularly in the face of a large number of unlabeled samples; or say, only a small class or relatively few training samples are available [42]. Therefore, it might be very expensive to collect the negative samples which are so abundant that a good sampling seems elusive. Although this was an extreme case, we carefully designed the training and test sets, which were composed of the qualified positive and negative samples. Note that if we want to improve the overall performance of the numerous classifiers which may differ in complexity, a combination of these classifiers will always be a viable solution [48].

Regarding another motivation of this work with respect to TCM, the quality control of TCM remains a significant issue that affects medicinal herbs, formulations, and even TCM practice. Due more to the lax enforcement of standards [49], resulting in the diminishing popularity of TCM rather than a failure of remedies, particularly, the patchy regulation has led to inconsistent herb quality, unqualified practitioners, unsubstantiated claims for secret formulas, and both deliberate and inadvertent mislabeling and adulteration, sometimes with fatal consequences. Considering *Panax notoginseng* is a vulnerable crop which has a serious succession cropping obstacle [50], consequently the continuous planting of the same crop in the same field will lead to the decrease of yield and quality. In order to promote the quality of production, it is crucial to monitor the spatial planting patterns of *Panax notoginseng* fields, such as crop rotation [51,52]. The planting pattern implies standardized planting with the specific crop structure and spatio-temporal configuration in the same field for a specific region under the particular natural resource and socio-economic conditions [53,54] so as to realize the sustainable utilization of agricultural resources and crop yield. Until now, the concrete planting pattern changes of *Panax notoginseng* are still poorly known. To the best of our knowledge, until now no such work has been done which, on the one hand, enriches the approaches to monitor spatial planting pattern changes of the perennial ginseng from space; on the other hand, employs SCDDs for mapping *Panax notoginseng*.

Our studies on mapping *Panax notoginseng* aim to provide fruitful information for studies on the quality assurance of TCM production, precision farming, the construction of agro-ecosystems, sustainable development, and the protection of the biodiversity of *Panax notoginseng*. Furthermore, determining the planting area of *Panax notoginseng* is an important part of obtaining more accurate information about annual yield and natural storage, except for mapping the spatial distribution. The current study, which involves mapping the planting parcels of *Panax notoginseng* at a 30 m spatial resolution, has three aims: (1) mapping *Panax notoginseng* fields through a stack of SCDDs as the future technical milestone for planting pattern analysis; (2) evaluating the abilities of SCDDs in identifying small *Panax notoginseng* fields in the complex agricultural landscapes; and (3) providing the potential possibilities for monitoring the planting pattern changes of *Panax notoginseng* fields, further giving us new insights into the planting pattern transitions of the perennial ginseng in macrocosm. The case study area is located in Wenshan City of China, which is characterized by a distinctive crop rotation agricultural system. The highlights of this study include: (1) striving for the research of the landscape-scale remote-sensing interpretation of TCM resources for the first time; (2) employing a stack of SCDDs with a comparative perspective to conduct mapping of *Panax notoginseng* fields; (3) defining the selection criteria and scoring proof for choosing a most appropriate SCDD; and (4) evaluating the abilities of SCDDs in identifying the fragmented parcels of *Panax notoginseng* in the complex agricultural landscapes.

The rest of this paper is structured as follows. The description of materials and methods is introduced in Section 2, and the experiments and analysis are presented and discussed in Sections 3 and 4, respectively. Finally, the conclusions of this work are summarized in Section 5.

2. Materials and Methods

2.1. Study Area and Data

Our study area comprises two independent blocks which are situated in Wenshan City, Wenshan Prefecture, Yunnan Province, in the south-west region of China (see Figure 2). These places mainly lie between longitude 103.71° E–104.46° E and latitude 23.07° N–23.73° N. Wenshan Prefecture is on a plateau, where the temperatures are quite constant throughout the year, with more precipitation during the summer months. Due to its low latitude and tempered by its high elevation, Wenshan Prefecture has a mild, humid, and subtropical climate, particularly suitable for planting *Panax notoginseng*. This is the reason why Wenshan Prefecture is the specific geographical location and origin of *Panax notoginseng* (i.e., why it is called a geoherb).

The "sa" and "sb" are two typical square regions. There are 151 × 151 pixels in the image space, respectively, corresponding to an area 20.5209 km^2 or equivalent to 2052.09 ha for both. The two blocks have a typical representation of the dense *Panax notoginseng* fields, upon which mapping *Panax notoginseng* fields will be carried out using multiple SCDDs. Multi-spectral cloud-free images acquired by the Landsat-8 Operational Land Imager (OLI) at a 30 m spatial resolution were utilized in this study. Their acquisition date was 18 March 2015. Since only one scene (path/row: 128/044) was utilized for the analysis in this study, the atmosphere can be considered to be homogeneous, and therefore the atmospheric correction may be not necessary [41,55]. Note that the planting fields of *Panax notoginseng* are rather sparse in most cases, and the cloud-free satellite images are not easy to collect because of the special geographical environment (i.e., a mountainous area is often in heavy clouds, refer to Table 1).

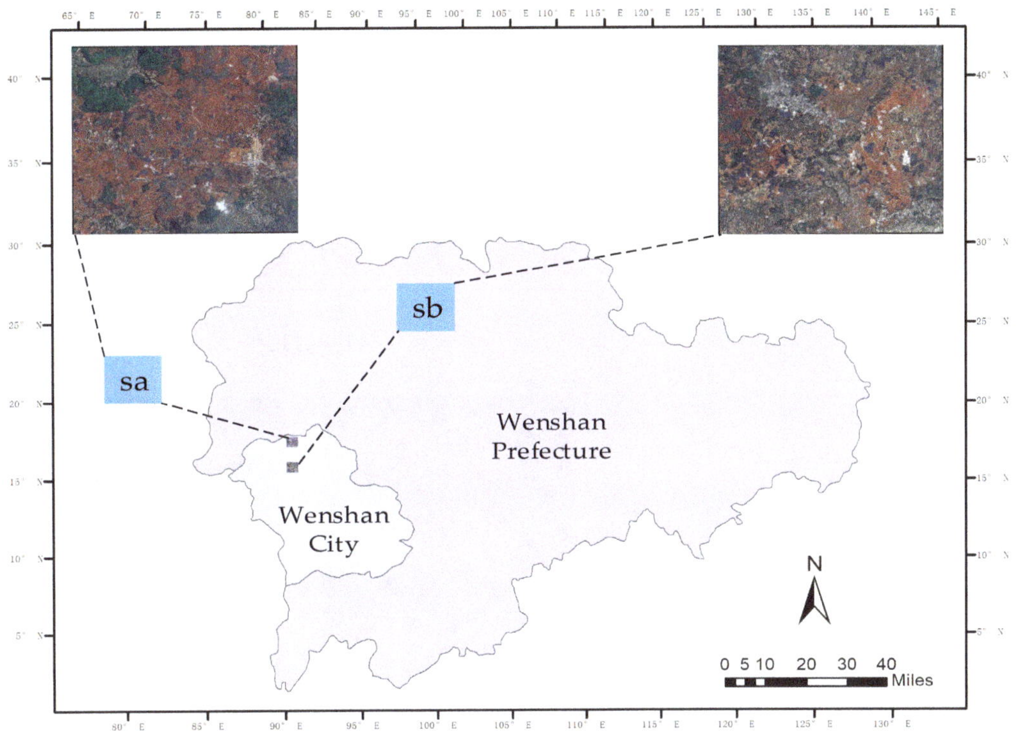

Figure 2. Study area in Wenshan City, Wenshan Prefecture, Yunnan Province, China.

Table 1. Cloud cover statistics of a 16-day revisited Landsat-8 OLI scene (path/row 128/044) until May 2017.

Cloud (%)	Number	Percentages
0–10	3	3.41
10–20	7	7.95
20–40	13	14.77
40–60	17	19.32
60–80	23	26.14
80–100	25	28.41

2.2. Shadenet Structures

Plastic sheets are used as materials to build the shadenet structure which can be regarded as a micro-scale planting environment and are relatively common [56], having unique characteristics [57], i.e., optical transparency, shade percentages, gas-tightness, and high-reflectivity. Agriculturalists have long known the importance of the planting environment for crop growth, always by manipulating the growing environment to provide a more conducive environment for crop growth. As for *Panax notoginseng*, sunlight is often modified by shading to provide the more optimal growing environments so as to enhance their production. Therefore, the production of *Panax notoginseng* takes place within an enclosed growing structure called a shade house. The shade house (see Figure 3) provides the protection and maintains the optimal growing conditions throughout the development of *Panax notoginseng*. The shade house of *Panax notoginseng* is a framed structure often covered with the black plastic nets which are made of the polyethylene thread with different shade percentages. It provides a partially-controlled atmosphere and environment by reducing the light intensity and effective heat during daytime to *Panax notoginseng* grown under the very large plastic sheets.

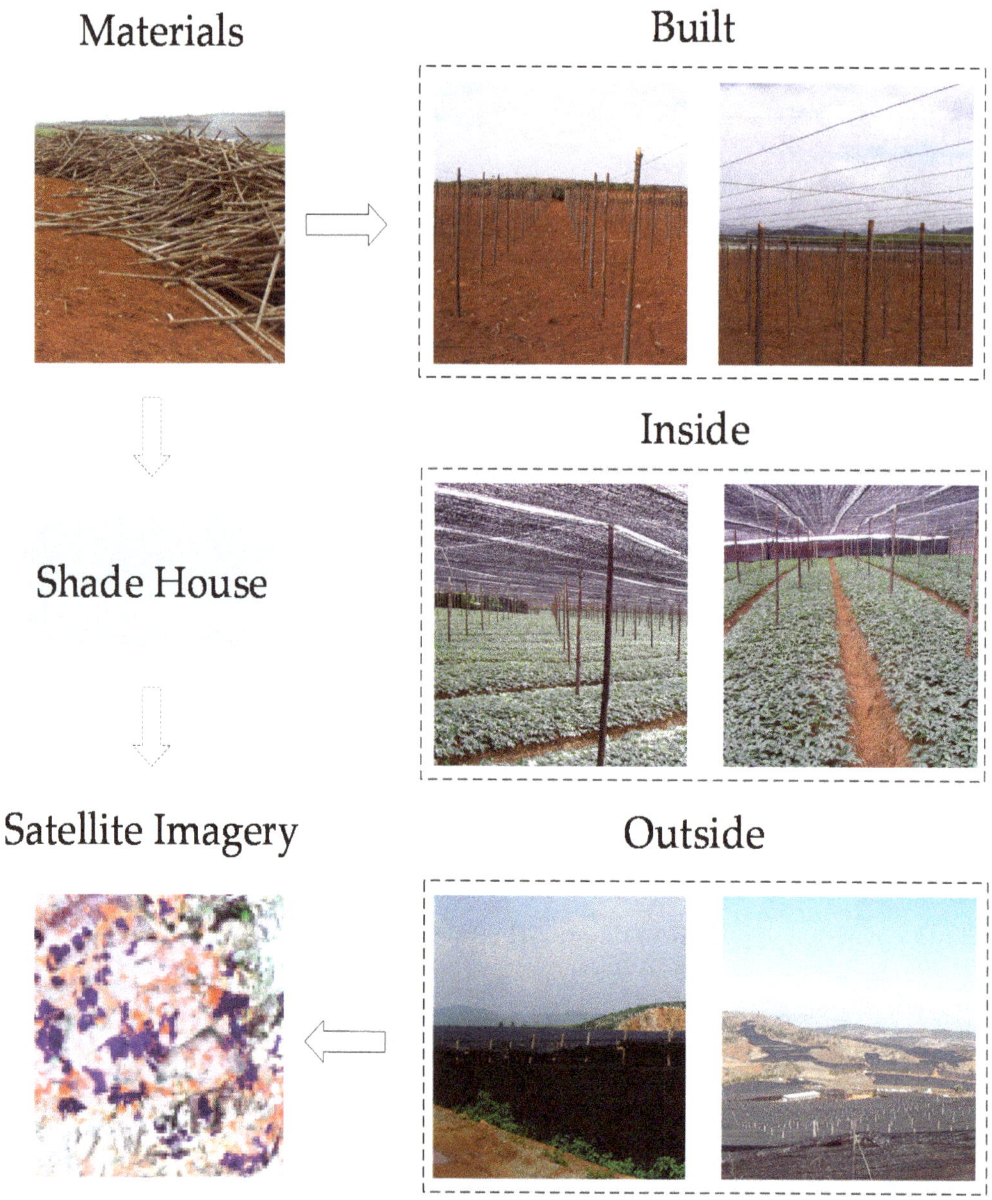

Figure 3. Shade house. Sub-figures are the snapshots and photos in the different phases of the construction of a shade house of *Panax notoginseng* including the materials for building, i.e., sticks (5 c m × 2 m), upright rods, and black plastic net over the poles; viewpoints for the observation, i.e., parallel to bed, perpendicular to bed, at close range, over a long distance, and from a satellite image.

Additionally, attributed to the standardized cultivation technique of GAP, the shade house provides a distinct spatial texture to interpret the spatial distribution of *Panax notoginseng* fields in reference to Google Earth historical images based on a calendar, i.e., there are 20 days when the satellite images were captured from 6 March 2013, to 12 December 2015, which is associated with the different sampling sites (see Figure 4), i.e., there are a total of 123 sites in Wenshan City. Polypropylene fabric shade is the dominant shade for field-cultivated *Panax notoginseng* in Wenshan City. Its black color maintains the proper shade and also forms the distinct texture of the shade net in satellite images, which can be visually interpreted and makes it easier to collect training samples using region of interest (ROI) tools.

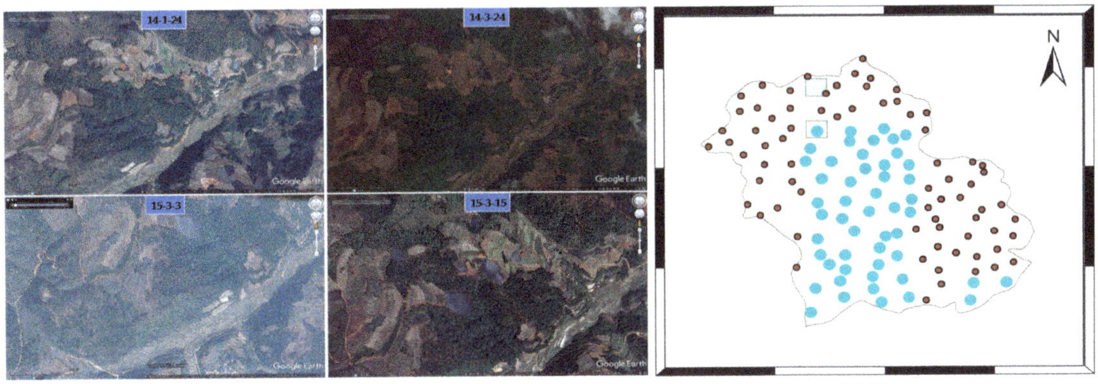

(a) #45 (b) Sites

Figure 4. Sampling sites. (**a**) The site #45; (**b**) the sites in Wenshan City.

2.3. Design Sets

Good classification depends not only on the factors associated with classifiers, but fine design sets also play a significant role in assessing the classification results objectively [58]. There is a non-negligible truth that, if only given positive samples, we cannot estimate most of the classification errors. Therefore, both positive and negative samples are supposed to be prepared for this study. Silva, et al. [41] utilized a manually-collected set of samples of the target class and a random sampling of samples of all classes, to keep the training effort low. In that case, they used the non-pure negative samples under the assumption of which few samples of the target class would be submerged.

In this study, we manually collected the positive (i.e., 1211) and negative (i.e., 8522) samples with a class-wise separability of 1.9918. Subsequently, we prepared three kinds of design sets, i.e., the training set (i.e., 727 positive samples, and 5114 negative samples at a 60% split); test set (484 positive samples, and 3408 negative samples, the remaining 40% split); and validation set (only 51 positive samples for the "sa", and 94 positive samples for the "sb"); as well as additional reference results (see Figure 5) obtained by the expert processing software. The training and test sets are random subsets of the raw collected samples by splitting operation. Note that once determined, they should be fixed so that all SCDDs could be fairly compared. For a good estimate, the test set should be labeled, randomly drawn from the class of interest, independent from the training set, and as large as possible. The validation set (i.e., only comprising true labels of the positive class) is used for validation and a nominal set. Note that the raw samples covering the whole of Wenshan City are specially designed for training and testing SCDDs. Meanwhile, the true labels for validation set are applied to calculate a representative accuracy (i.e., a correct rate) of the classification results as a final validation.

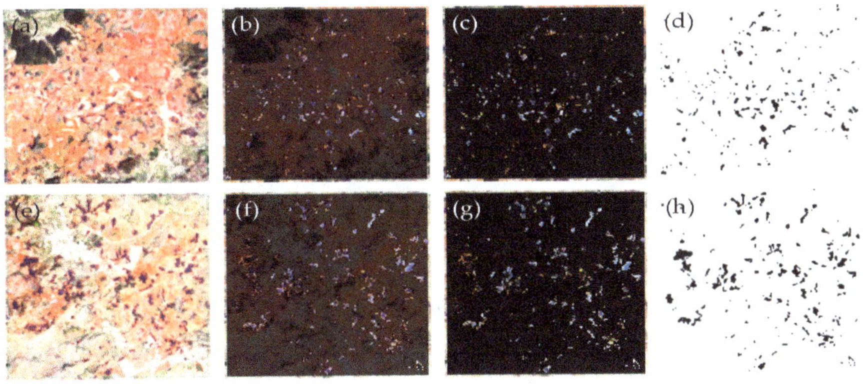

Figure 5. Reference results. (**a**) a regular subset, (**b**) the overlap, (**c**) an irregular subset of Landsat-8 OLI image, and (**d**) the reference result for the "sa"; (**e**) a regular subset, (**f**) the overlap, (**g**) an irregular subset of Landsat-8 OLI image, and (**h**) the reference result for the "sb".

2.4. Single-Class Data Descriptors (SCDDs)

The SCDD [47] is a trained mapping to predict classes. Additionally, they can be divided into several categories depending on the type of the training data and the discrimination function [59]. For example, the positive samples only (i.e., P-learning) or the positive and unlabeled samples (i.e., PU-learning) [31,59]. In general, for certain SCDDs, the corresponding model can be defined as

$$h(x) = \begin{cases} target, \ f(x) \leq \theta \\ other, \ f(x) > \theta \end{cases}, \quad (1)$$

or vice versa in the opposite conditions.

The threshold θ is set according to the target error. Formally, each instance is mapped to one element of the set of the positive and negative class labels. In this study, due to only the target class (i.e., *Panax notoginseng* fields) is the class of interest, and the task of mapping *Panax notoginseng* fields can be regarded as a specific single-class data description problem. Hence, 10 SCDDs [47], i.e., the simple Gaussian target distribution data description (SGTD coded as c1) [60]; the robust Gaussian target distribution data description (RGTD coded as c2) [60]; the minimum covariance determinant Gaussian data description (MCDG coded as c3) [60]; the mixture of Gaussian data description (MoG coded as c4) [60]; the auto-encoder neural network data description (AENN coded as c6) [61]; the k-means clustering data description (k-means coded as c7) [62]; the self-organizing map data description (SOM coded as c10) [63]; the minimum spanning tree data description (MST coded as c11) [64]; the k-nearest neighbor data description (K-NN coded as c13) [65]; the incremental support vector data description (IncSVDD coded as c17) [66]; the Parzen density estimator data description (PDE coded as c5, which is a known underestimated descriptor) [67]; and the principal component analysis data description (PCA coded as c9, which is known as an overestimated descriptor) [68].

In addition to the SCDDs mentioned above, three ensemble classifiers, i.e., the mean combiner (meanc coded as cmea), the median combiner (medianc coded as cmed), and the voting combiner (votec coded as cvot), taking the mean, median, and vote strategies, respectively. The ensemble-based approach refers to the multiple-classifier system in which the outputs of all member classifiers are combined to derive an accurate classification. We combine the SCDDs which should be accurate but different in an ensemble strategy because each of them can focus on a specific feature or characteristic in the feature space [24]. Thus, a much more flexible and outstanding classifier can be obtained by combining all the strong points of the different SCDDs. There are three strategies of combining the different SCDDs, which are referred to as (1) sequential, (2) parallel, and (3) stacked [47]. Here, the above-enumerated SCDDs (i.e., c1–c17) are computed in the same feature space, and which are typically combined in a stacked way. Notice that, the combining procedure is computationally intensive in the face of many different base classifiers. Thus, the action that prunes the base classifiers according to their performance (i.e., underestimated or overestimated) is inevitable sometimes. In this study, ten SCDDs except for the underestimated and overestimated ones, which are regarded as the qualified member classifiers. The abovementioned SCDDs and combiners have been well implemented with a Matlab toolbox for data description developed and distributed by Dr. David Tax.

2.5. Performance Evaluation

In this study, the error computation and performance evaluation involve several accuracy metrics, i.e., the basic errors (see Table 2), F_1 measure, receiver operating characteristics (ROC) curve, area under the ROC curve (AUC) error, cost curve, confusion matrix, and correct rate, which are employed to evaluate the SCDDs in a more comprehensive way. In order to find a good SCDD, four basic errors can be calculated, and two of them have to be minimized, namely the false positives (FP) and false negatives (FN). Hence, we put forward and discuss several representative measures which can reflect the probability that the prediction is informed versus chance. The FN can be estimated on the positive set. In fact, the FN is much harder to estimate when no negative samples are available [47]. If only minimizing the FN will lead to the SCDD which may wrongly label a number of the negative samples as the positive class. In order to avoid such a degenerate solution, the negative samples have to be collected as well. This is the reason why we elaborately prepared the design sets and the reference results for training and testing. Moreover, two other measures, such as precision and recall (i.e., the true positive (TP) rate), are often used in performance evaluation. Finally, a derived performance indicator F_1 score can be computed. Note that a good SCDD should have both small rates of the FN and FP.

Table 2. Binary error matrix.

Types		Predicted Label	
		Target	**Other**
Actual Label	**Target**	true positive (TP)	false positive (FP)
	Other	false negative (FN)	true negative (TN)

Due to the error on the positive class can be estimated relatively well, assuming that, a threshold can be set beforehand on the target error for the SCDDs, and then a ROC curve can be obtained by varying this threshold and measuring the error on the negative samples. The ROC graph [69,70] shows how the FP varies for varying the FN, which is a technique for visualizing and selecting the best classifiers based on their performance. The smaller these fractions are, the more the SCDD is to be preferred. Although the ROC graph shows an intuitive metric of the performance of the SCDD, as one side, it is a bit difficult to compare the ROC curves; for another, we want to reduce the ROC curve to a single scalar value representing the expected performance. Thus, the AUC error [71,72] often is taken, and which is computed from the ROC curve, which integrates the fraction TP over varying thresholds (or by varying fraction FP equivalently). The larger the AUC value, the better the SCDD's performance. That is, the higher values may indicate a better separation between the positive and negative samples. As for the ROC graph [70], there are many thresholds that may be suboptimal. That is, there is another operating point for which at least one of the errors is lower. Nevertheless, this concern can be indicated by a cost curve [73], which will be another method for visualizing the performance of the SCDD. For a varying cost-ratio between both classes, the expected cost is computed. Additionally, once a trained mapping has been determined, we can obtain the sufficient site-specific metrics derived from a confusion matrix (i.e., a binary contingency table for the SCDD). The binary confusion matrix (see Table 2) is a specific table layout that allows the visualization of the performance of the SCDD [74]. Such an error matrix is constructed via classifying a predefined test set or comparing two sets of labels, and which combines the spatial position and quantitative information of the classification results to implement a performance evaluation. What is important is that the very small changes of labels are well reflected in a confusion matrix.

3. Results and Analysis

3.1. Resultant Maps

After the SCDDs have been performed, and then the final outputs associated with mapping *Panax notoginseng* fields can be derived. For one side, we conduct a statistical analysis from a perspective of quantitative evaluation. Besides, we want an intuitive comparison by showing the paired maps which are overlapped in false-color style between the classification and reference maps regarding the spatial distribution of *Panax notoginseng* fields.

As such, the classification map of each SCDD in comparison with the reference map that can be simultaneously graphed for the "sa" and "sb" (see Figures 6 and 7). By visual inspection, the obvious differences have been observed with respect to the underestimated PDE (see Figures 6 and 7) and the overestimated PCA (see Figures 6 and 7). The heavy magenta patches mean seriously underestimated, while a large number of green patches mean severely overestimated. The majority of the rest of the

SCDDs arise from the predominately magenta and slightly green patches, which mean underestimated. Meanwhile, AENN is different (see Figures 6 and 7). Notice that the error threshold on the positive class is set by a default float value of 0.1 for all SCDDs. In a number of clusters a default integer value of 2 will be acceptable because only two classes are available (i.e., the positive and negative class). Additionally, the number of components is set to 5 in terms of PCA.

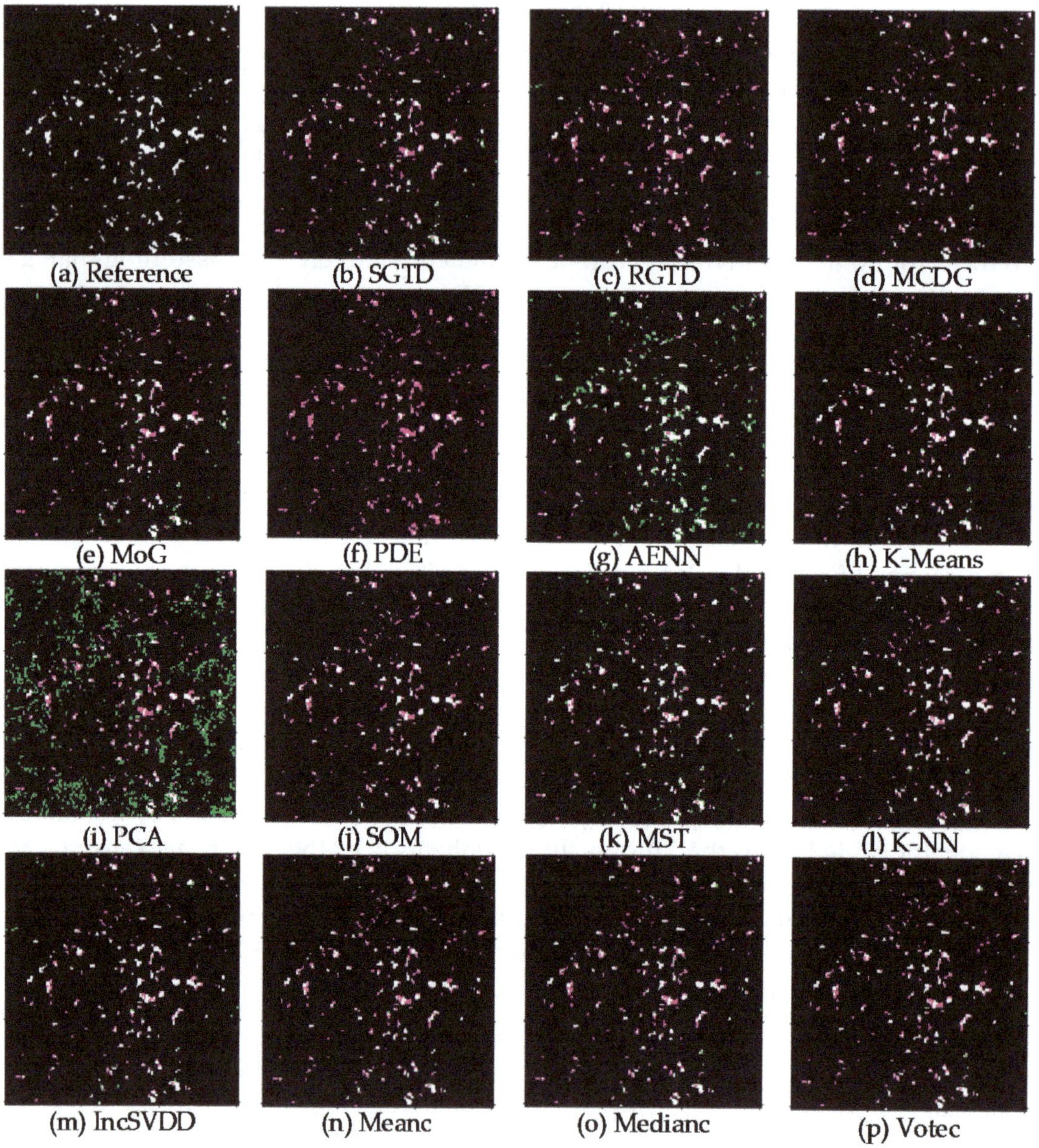

Figure 6. Classification results which are shown in false-color style in comparison with the reference result for the "sa". (**a**) Reference result; (**b**) simple Gaussian target distribution (SGTD); (**c**) robust Gaussian target distribution (RGTD); (**d**) minimum covariance determinant Gaussian (MCDG); (**e**) mixture of Gaussian (MoG); (**f**) Parzen density estimator (PDE); (**g**) auto-encoder neural network (AENN); (**h**) k-means; (**i**) principal component analysis (PCA); (**j**) self-organizing map (SOM); (**k**) minimum spanning tree (MST); (**l**) k-nearest neighbor (K-NN); (**m**) incremental support vector data description (IncSVDD); (**n**) mean combiner (meanc); (**o**) median combiner (medianc); and (**p**) voting combiner (votec). Here, the white pixels means the well estimated, the magenta means underestimated, and the green means overestimated.

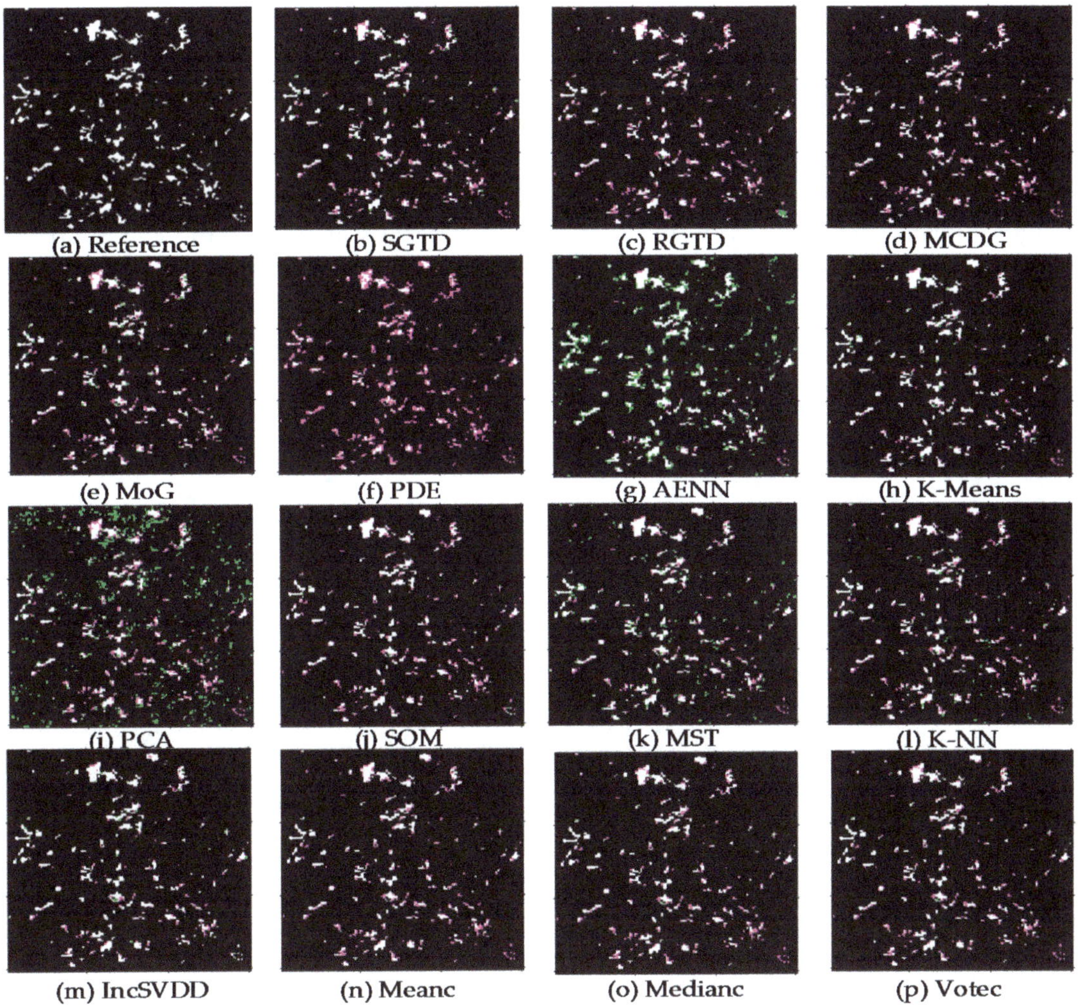

Figure 7. Classification results which are shown in false-color style in comparison with the reference result for the "sb". (**a**) Reference result; (**b**) SGTD; (**c**) RGTD; (**d**) MCDG; (**e**) MoG; (**f**) PDE; (**g**) AENN; (**h**) k-means; (**i**) PCA; (**j**) SOM; (**k**) MST; (**l**) K-NN; (**m**) IncSVDD; (**n**) meanc; (**o**) medianc; and (**p**) votec.

3.2. Measuring Performance

Performance evaluation is crucial in the assessment of a set of SCDDs. For Table 3, the false negative rate (FNR) gives the error on the positive class, while the false positive rate (FPR) shows the error on the negative class. Meanwhile, P is precision, and R denotes recall equivalent with TPR. The statistical metrics can be grouped into (1) a group that the smaller is the better, i.e., the FNR and FPR, and (2) another group that the greater is the better, i.e., P, R, F_1, and AUC. Although it is difficult to describe all SCDDs together, we attempt to make it possible by rating them on a rank table later. Table 3 illustrates that the FNR (e.g., the c5) and FPR (e.g., the c2 and c9) are prominently higher than two known inferior SCDDs (e.g., the c5 and c9), and a slightly inferior one (e.g., c2) is marked as well. Additionally, these unqualified SCDDs are again attention-catching in the second group. For all SCDDs, the precision (e.g., the c2 and c9), recall (e.g., the c5), and F_1 score (e.g., the c2, c5, and c9) support the analysis drawn from the first group, while the AUC error always appears mediocre. However, there are some differences owing to the measuring ability of statistical metrics and the intrinsic characteristics of the diverse SCDDs. The ROC curve (see Figure 8) gives a two-dimensional depiction of the performance of the SCDD. This is due to too many SCDDs with the approximate accuracies so that discriminating the individual ROC curve seems difficult. Thus, we plot them one by one, and the aforementioned inferior SCDDs still can be well reflected. Note that the c5 has the lowest operation point, while the c2 and c9 are found in terms of the ROC curve. In addition, the c6 deserves attention. Here, the ROC graphs are not going to be read significantly, as the purpose is achieved. The cost curve (see Figure 9) is a specific performance visualizer using the expected cost, another technique to measure the performance of SCDDs. Each operating point appears as a line in this plot, while the certain one of them is indicated by the dotted line. The combination of operating points that forms the lower hull is indicated by the thick curve and shows the best operating points over the range of costs. Here, the dotted line of c5, and the thick arcs of c2 and c9, again support the performance analysis drawn from the Figure 8. In particular, c6 has a representation that is the same as the ROC graph.

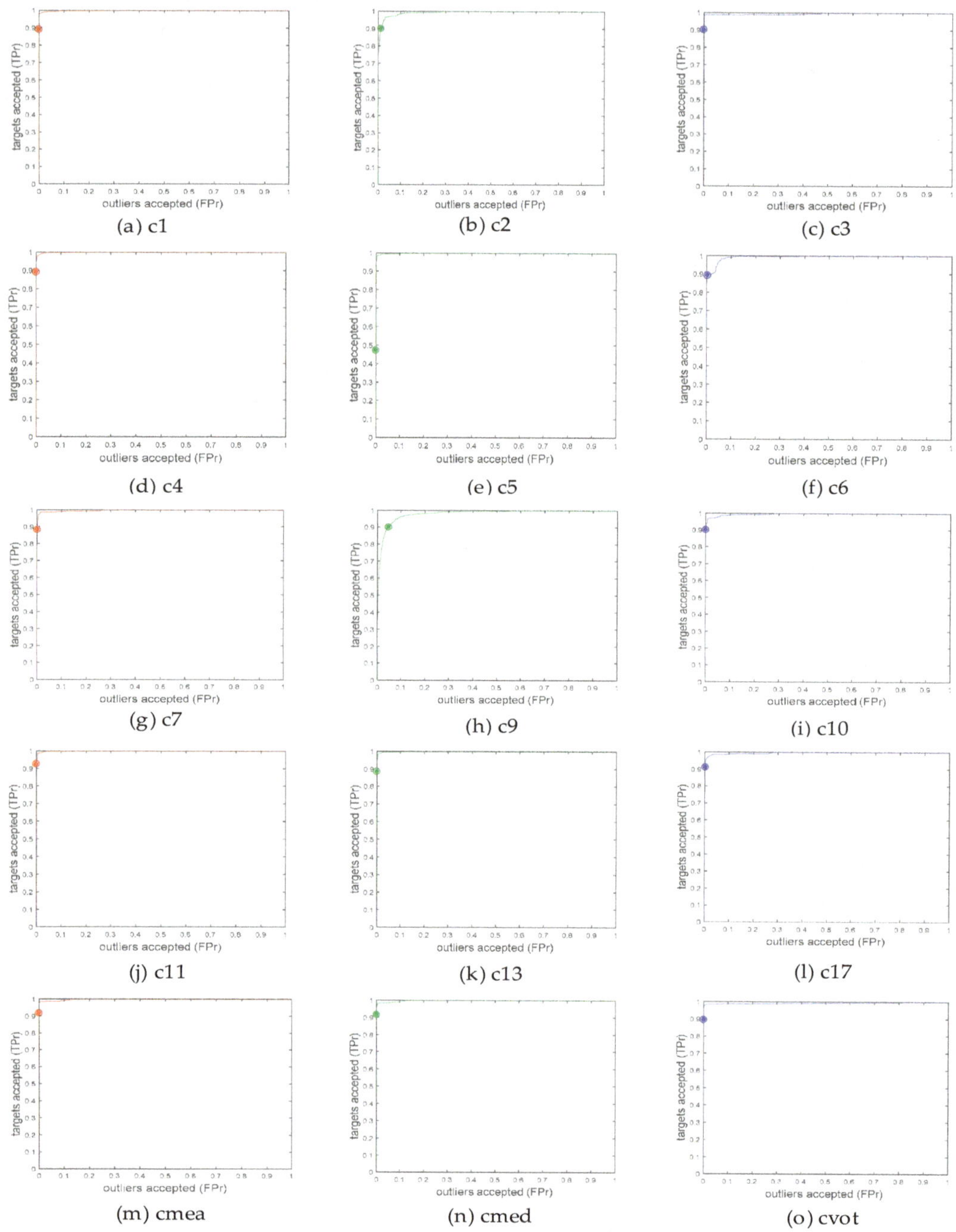

Figure 8. ROC curves. (**a**) c1: SGTD; (**b**) c2: RGTD; (**c**) c3: MCDG; (**d**) c4: MoG; (**e**) c5: PDE; (**f**) c6: AENN; (**g**) c7: k-means; (**h**) c9: PCA; (**i**) c10: SOM; (**j**) c11: MST; (**k**) c13: K-NN; (**l**) c17: IncSVDD; (**m**) cmea: meanc; (**n**) cmed: medianc; and (**o**) cvot: votec.

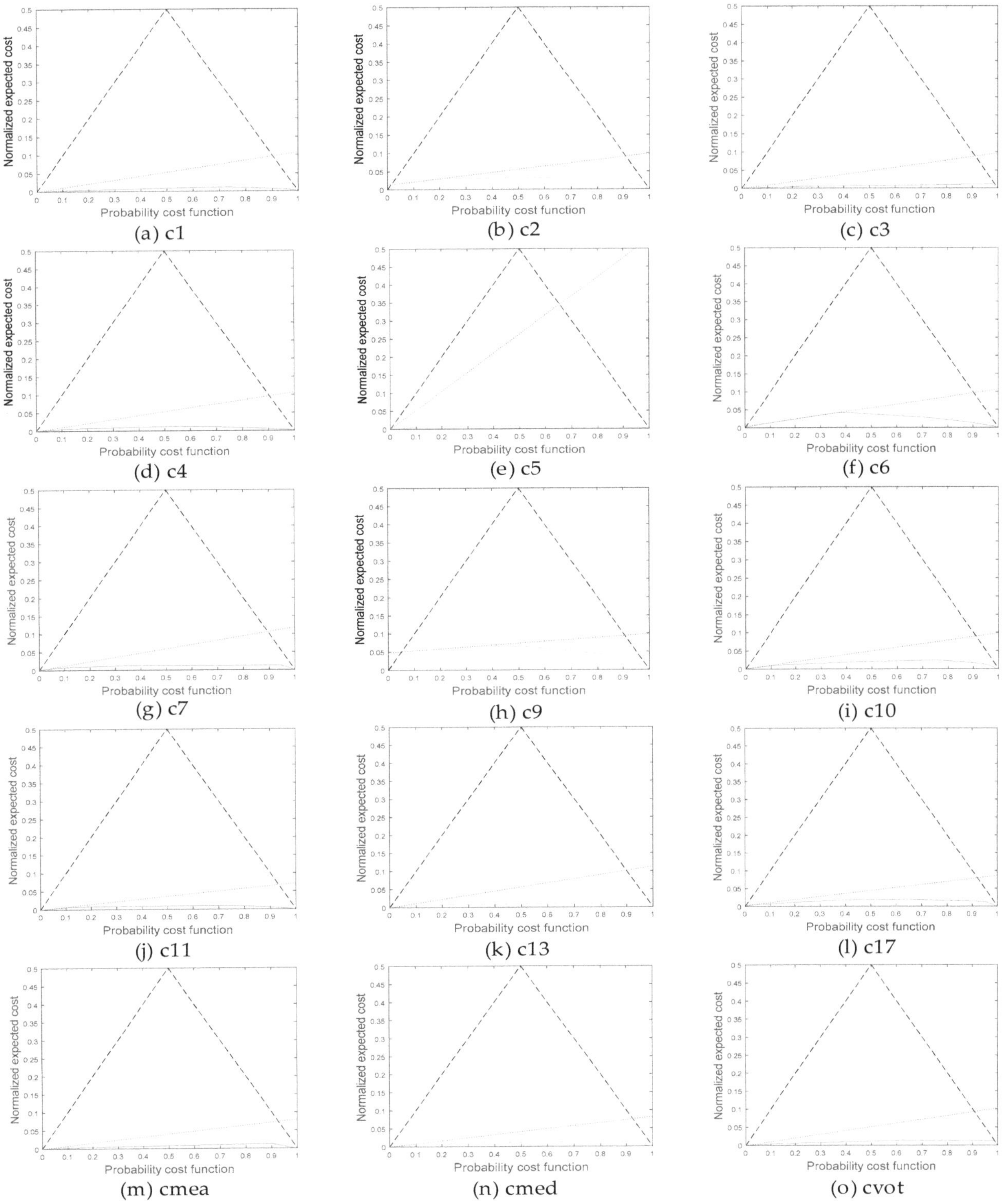

Figure 9. Cost curves. (**a**) c1: SGTD; (**b**) c2: RGTD; (**c**) c3: MCDG; (**d**) c4: MoG; (**e**) c5: PDE; (**f**) c6: AENN; (**g**) c7: k-means; (**h**) c9: PCA; (**i**) c10: SOM; (**j**) c11: MST; (**k**) c13: K-NN; (**l**) c17: IncSVDD; (**m**) cmea: meanc; (**n**) cmed: medianc; and (**o**) cvot: votec.

The confusion matrix is often applied to visualize the performance of the SCDD by a specific table layout. The OA means overall accuracy, K is the Kappa coefficient, PA denotes the producer's accuracy, and UA represents the user's accuracy. It is somewhat intricate that we structure two kinds of confusion matrices together (see Figure 10), which look similar but not identical. That is the reason why we design two types of confusion matrix (1) classifier-dependent (i.e., OAt, Kt, PAt, and UAt), which is generated using the test set; and (2) classifier-independent (i.e., OAa, Ka, PAa, UAa, OAb, Kb, PAb, and UAb), which is produced by utilizing the reference result for "sa" and "sb". For the two classifier-independent error matrices, which give an analogous presentation and a comforting result, though with a slight discrepancy in the amplitude, the inferior SCDDs, i.e., the c5 has a smaller K and PA while the c9 has a smaller K and UA. As for the classifier-dependent confusion matrix, all SCDDs have good performance except for the c5 has a lower K and PA while the c9 has a lower K and UA.

The summary can be drawn as (1) overall accuracy appears mediocre or fails in the face of the SCDD for the imbalanced data; and (2) two classifier-independent error matrices demonstrate that the SCDDs are, indeed, fixed by unchangeable splitting samples. Meanwhile, they seem worse and more unstable compared to the classifier-dependent matrices, which may be disturbed in the presence of more uncertainty. The correct rate (see Figure 10, and Table 4) is a custom performance measure, which is used for validation. Here, the true labels used are a subset of the positive samples drawn from a merged set for the "sa" and "sb". This measure is a simple attempt to repeat and verify the previous analysis.

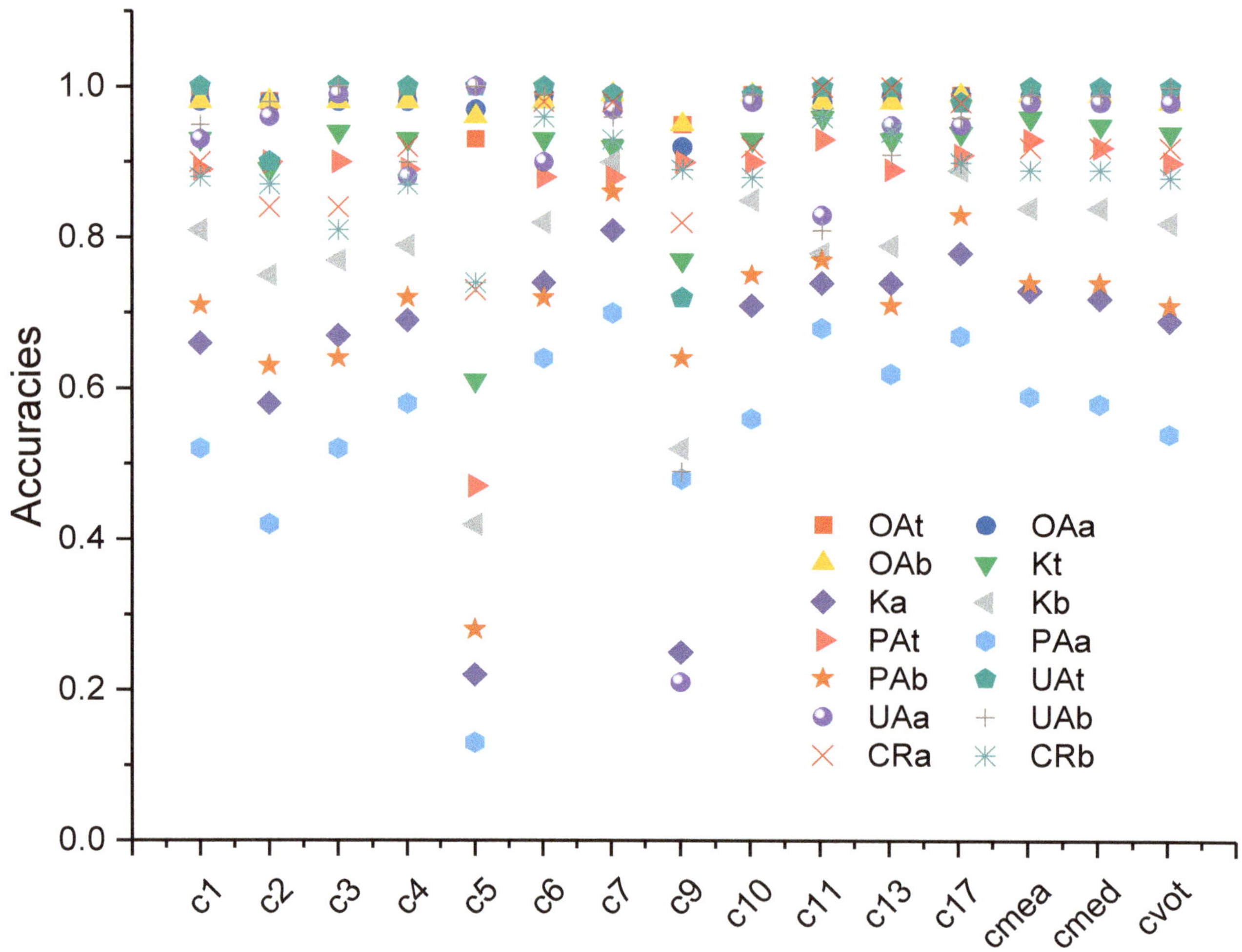

Figure 10. Accuracy metrics derived from the confusion matrix. Here, the postfix "t" means taking the test set as the true labels, while the "a" and "b" mean taking the reference results as the true labels, respectively. The CRa and CRb are the correct rates of the classification results in comparison with the true labels for the "sa" and "sb".

Table 3. Accuracy metrics, i.e., the FNR, FPR, P, R, F_1, and AUC.

	c1	c2	c3	c4	c5	c6	c7	c9	c10	c11	c13	c17	Cmea	Cmed	Cvot
FNR	0.11	0.10	0.10	0.11	0.53	0.12	0.12	0.10	0.10	0.07	0.11	0.09	0.07	0.08	0.10
FPR	0.00	0.01	0.00	0.00	0.00	0.00	0.00	0.05	0.00	0.00	0.00	0.00	0.00	0.00	0.00
P	1.00	0.90	1.00	1.00	1.00	1.00	0.99	0.72	0.99	1.00	1.00	0.98	1.00	1.00	1.00
R	0.89	0.90	0.90	0.89	0.47	0.88	0.88	0.90	0.90	0.93	0.89	0.91	0.93	0.92	0.90
F_1	0.94	0.90	0.95	0.94	0.64	0.94	0.93	0.80	0.94	0.96	0.94	0.94	0.96	0.96	0.95
AUC	1.00	0.99	0.99	1.00	1.00	1.00	1.00	0.98	1.00	1.00	1.00	1.00	1.00	1.00	0.99

Table 4. Rank table. Here, the FN and FP are the false negative rate and false positive rate, respectively. Signs "+" and "−" are the rating directions for judging the ascending or descending order. Here, the bold codes mean the inferior single-class data descriptors (SCDDs). Note that the accuracy metrics derived from confusion matrix for the "sb" have not been included herein.

C.	FN+	FP+	P−	R−	F1−	AUC−	OAt−	Kt−	PAt−	UAt−	OAa−	Ka−	PAa−	UAa−	CRa−	Rank−
c1	**c5**	**c9**	**c9**	**c5**	**c5**	**c9**	**c5**	**c5**	**c5**	**c9**	**c9**	**c5**	**c5**	**c9**	**c5**	15
c2	**c6**	**c2**	**c2**	**c6**	**c9**	cvot	**c9**	**c9**	**c6**	**c2**	**c5**	**c9**	**c2**	c11	**c9**	14
c3	c7	c17	c17	c7	**c2**	**c2**	**c2**	**c2**	c7	c17	**c2**	**c2**	**c9**	c4	**c2**	13
c4	c13	c10	c10	c13	c7	c3	c7	c7	c13	c10	c1	c1	c3	**c6**	c3	12
c5	c4	c7	c7	c4	**c6**	c17	**c6**	**c6**	c4	c7	c4	c3	c1	c1	c1	11
c6	c1	c1	**c6**	c1	c13	c10	c13	c13	c1	**c6**	c3	cvot	cvot	c13	c4	10
c7	c10	c3	c13	c10	c10	c7	c10	c10	c10	c13	cvot	c4	c10	c17	c10	9
c9	**c2**	c4	c1	**c2**	c4	cmea	c4	c4	**c2**	c1	c11	c10	cmed	**c2**	cmea	8
c10	**c9**	**c6**	c4	**c9**	c1	cmed	c1	c1	**c9**	c4	c10	cmed	c4	c7	cmed	7
c11	c3	c11	c3	c3	c17	**c6**	c17	c17	c3	c3	**c6**	cmea	cmea	cmea	cvot	6
c13	cvot	c13	cvot	cvot	c3	c1	c3	c3	cvot	cvot	cmed	**c6**	c13	cmed	**c6**	5
c17	c17	cmea	cmed	c17	cvot	c13	cvot	cvot	c17	cmed	cmea	c13	**c6**	c10	c7	4
cmea	cmed	cmed	c11	cmed	cmed	**c5**	cmed	cmed	cmed	c11	c13	c11	c17	cvot	c17	3
cmed	c11	cvot	cmea	c11	c11	c11	c11	c11	c11	cmea	c17	c17	c11	c3	c11	2
cvot	cmea	**c5**	**c5**	cmea	cmea	c4	cmea	cmea	cmea	**c5**	c7	c7	c7	**c5**	c13	1
C.	FN−	FP−	P+	R+	F1+	AUC+	OAt+	Kt+	PAt+	UAt+	OAa+	Ka+	PAa+	UAa+	CRa+	Rank+

4. Discussion

4.1. Selection Criteria

Although sufficient statistical analyses have been conducted, it is not known how to recognize which SCDD looks good. In fact, it is not easy to determine which is a fine, or even the best in the face of so many SCDDs with multiple performance measures. Therefore, we want to set up a handful of naive selection criteria to achieve such a goal by means of a rank board (see Table 4). For this work, more empirical selection criteria are adopted. Intrinsically, most of the statistical metrics are derived from the basic errors (i.e., the true positive, the true negative, the false positive, and the false negative). The derived measures (i.e., the precision, recall, F_1, AUC, OA, KC, PA, and UA) could be quantitatively analyzed with actions such as rating and scoring. Note that the AUC measure appears mediocre herein. The OA may not be a reliable metric for the real performance of the SCDD in this study, because it yields misleading results supposing the training data are imbalanced when the numbers of observations in different classes vary greatly. The ROC curve and cost graph are used for supporting numerical indicators, which are especially suitable for classification problems in which there are only two classes (i.e., the positive and negative classes). The limitation [70] of both ROC analysis and cost curve is the lack of any effective method to show the performance results obtained from several different data sets in a single plot. This difficulty follows the fact that only two dimensions are used to present the performance of a single data set.

It is important to realize the optimal selection criteria for the hybrid classifiers, such as comparing the performance of an ensemble classifier with a member classifier, which is also presented in this study. The fixed combination strategies or so-called rules (i.e., the mean rule, median rule, and voting rule), are more likely to obtain better classification results, just as the inferior classifiers will reduce the whole performance. In particular, the time taken will be an insufferable problem. It is crucial to address the question of under what criteria does one classifier outperform another. Additionally, a decision needs to be made to determine which classifier should be selected over others. That is if, given the current operating conditions, a set of selection criteria can be derived. It is often easy, by varying the parameter setting, such as a threshold or the variables of the mathematical model, or by varying the class distribution in the training set, to create a whole set of SCDDs. One commonly used selection criterion is to select the SCDD whose parameter settings and training conditions most closely agree with the current operating conditions, which is called the performance-independent criterion [75]. This is the reason why we try to fix all irrelevant conditions prior to developing the performance-dependent selection criteria. A plain criterion is to choose the qualified SCDDs regardless of their training conditions or parameter settings.

4.2. Scoring Model

For SCDDs with multiple performance measures, it would be expected for them to be scored. Then, there is always one possibility that all SCDDs can be quantitatively evaluated and scored. Consequently, a score-oriented method is presented here to clarify this concern. In this study, we put forward a kind of scoreboard on the basis of the rank board to give each SCDD an explicit score so that we can determine which SCDD is optimal.

According to Table 5, we use the rows (M.) to denote the measures and the columns (C.) to represent the different SCDDs. Signs are used to identify the metric belonging to which group: "–" denotes the error metric (i.e., the smaller the better), while "+" is the accuracy metric (i.e., the greater the better). The score variable Sc_j is calculated by the following equation:

$$s_{ij}(x) = \begin{cases} n - \frac{x_{ij} - \mathrm{Min}(x_i)}{(\mathrm{Max}(x_i) - \mathrm{Min}(x_i))/n} & \text{if sign} = \text{"}-\text{"} \\ \frac{x_{ij} - \mathrm{Min}(x_i)}{(\mathrm{Max}(x_i) - \mathrm{Min}(x_i))/n} & \text{if sign} = \text{"}+\text{"} \end{cases}, \quad (2)$$

or

$$s_{ij}(x) = \begin{cases} \frac{\text{Max}(x_i)-x_{ij}}{(\text{Max}(x_i)-\text{Min}(x_i))/n} & \text{if sign} = \text{“}-\text{”} \\ n - \frac{\text{Max}(x_i)-x_{ij}}{(\text{Max}(x_i)-\text{Min}(x_i))/n} & \text{if sign} = \text{“}+\text{”} \end{cases}, \tag{3}$$

and

$$Sc_j = \sum s_j, \tag{4}$$

where x_i represents the measures in the ith row, x_j represents SCDDs in the jth column, and x_{ij} denotes the measured value of the jth SCDD with ith metric. The s_i represents the scores in the ith row, the s_j represents the scores in the jth column, and the s_{ij} denotes the score value of the jth SCDD with the ith metric. Sc_j represents the total score of the jth SCDD. For simplicity, we assume that there are five SCDDs and five measures herein to facilitate the illustration of the scoreboard and the derivation of Equations (2)–(4). The n is a key scale to slice a certain metric for all SCDDs so that each SCDD can be assigned a normalized float value (ranging from 0 to n) associated with this metric. In the end, the gross score of every SCDD can be obtained and plotted by performing the summation by column. Figure 11 illustrates that two inferior SCDDs, i.e., the c5 is underestimated, and the c9 is overestimated, which are prominently identified. Meanwhile, two slightly inferior SCDDs, i.e., the c2 and c6, are apt to be observed again.

Table 5. Score table. Here, we assume five SCDDs with five metrics.

M./C.	j1	j2	j3	j4	j5	Sign–
i1	x11 ǀ s11	x12 ǀ s12	x13 ǀ s13	x14 ǀ s14	x15 ǀ s15	–
i2	x21 ǀ s21	x22 ǀ s22	x23 ǀ s23	x24 ǀ s24	x25 ǀ s25	–
i3	x31 ǀ s31	x32 ǀ s32	x33 ǀ s33	x34 ǀ s34	x35 ǀ s35	+
i4	x41 ǀ s41	x42 ǀ s42	x43 ǀ s43	x44 ǀ s44	x45 ǀ s45	+
i5	x51 ǀ s51	x52 ǀ s52	x53 ǀ s53	x54 ǀ s54	x55 ǀ s55	+
Score	Sc1	Sc2	Sc3	Sc4	Sc5	Sign+

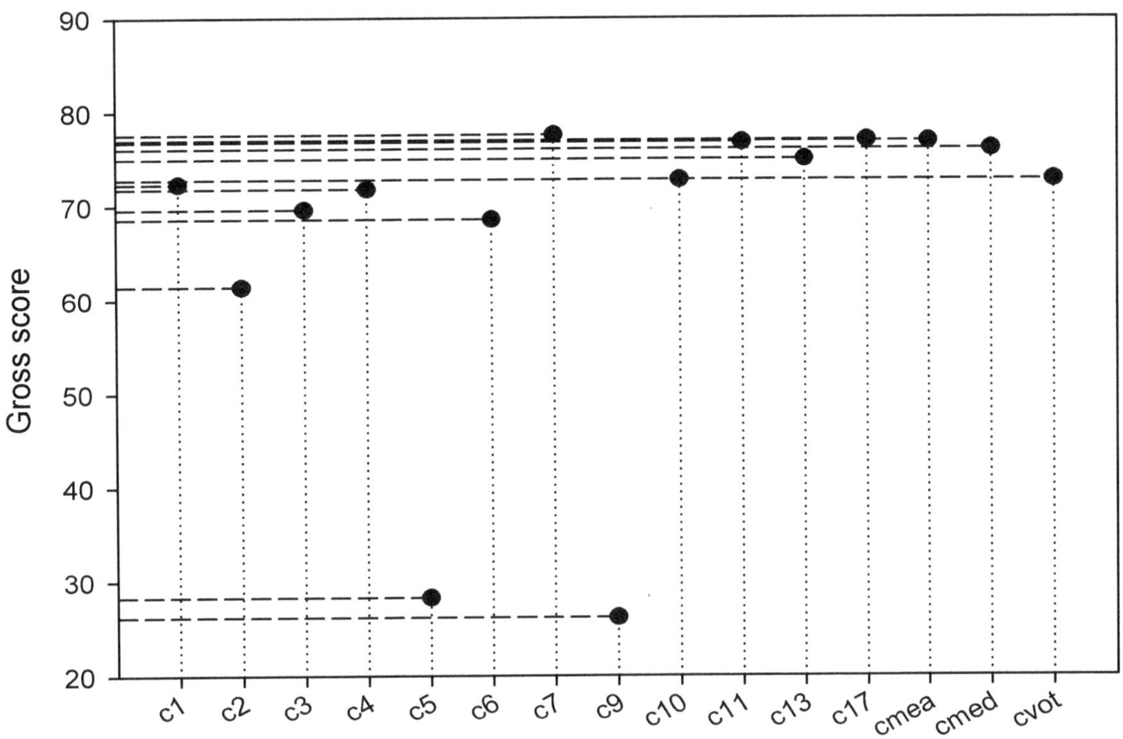

Figure 11. Scoring result, $n = 4$.

4.3. McNemar's Test

The four-cell confusion matrix is very intuitive to show the similarities and differences between the proportions (i.e., the true and false allocated parts) concerning two sets of specific labels. On the basis of the error table, we wish to achieve the statistical significance of the differences between the proportions using McNemar's test [76] so as to assess two allocated results, which are obtained by two SCDDs, or just given, under a position-specific comparison. Here, McNemar's test is specifically useful for comparing paired proportions derived from two sets of samples. In the formulas below, we use the notations:

$$\begin{array}{l} pdisc = p_{F+} + p_{F-} \\ pdiff = p_{F+} - p_{F-} \end{array}, \tag{5}$$

where the p_{F+} is the proportion of testing samples that the first SCDD is true, and the second is false; meanwhile, the p_{F-} denotes the proportion of testing samples that the first SCDD is false while the second is true. Thus, McNemar's test focuses on the proportions of testing samples that one SCDD is true while another is false (and vice versa) [41].

$$SE_p = \sqrt{(pdisc - pdiff^2)/N}, \tag{6}$$

where SE_p represents the standard error derived from the difference between the proportions, and N is the total number of the pairs of objects. McNemar's test will perform the evaluation of the $100(1-\alpha)\%$ confidence interval for comparing the difference between two accuracy values based on the differences (D_α) between the proportions. Assuming a normal distribution z_α, the general expression of the confidence interval [76] can be expressed as:

$$D_a \pm z_\alpha SE_p. \tag{7}$$

For exploring in a straightforward manner, we split Equation (7) into a real image, then the image part may be more crucial to give the proper assessment with regard to a confusion matrix. In this way, the statistical assessment of the differences is carried out to determine if these are significantly different or not [76]. Since three kinds of error matrices are presented in this study, here we name them as CDt (i.e., the classifier-dependent using the test set), CIa (i.e., the classifier-independent using the reference map of the "sa"), and CIb (i.e., the classifier-independent using the reference map of the "sb").

The difference between the accuracies yielded by the paired SCDDs (or two sets of labels) is D_α, ranging from $D_a - |z_\alpha|SE_p$ to $D_a + |z_\alpha|SE_p$ at the $100 \times (1 - 0.05)\%$ confidence interval. In order to make all confidence intervals comparable, the one-sided absolute range ($|z_\alpha|SE_p$) around D_a is exhibited in Table 6. Two inferior SCDDs (i.e., the c5 and c9) have their appearances again. As for two slightly inferior SCDDs (i.e., the c2 and c6), only the c2 can be observed. Such results estimated by McNemar's test provide a useful back-up to previous analysis.

Table 6. Statistical significance ($\times$1000).

	c1	c2	c3	c4	c5	c6	c7	c9	c10	c11	c13	c17	Cmea	Cmed	Cvot
CDt	3	4	3	3	7	3	3	6	3	3	3	3	2	3	3
CIa	1	2	1	1	2	1	1	3	1	1	1	1	1	1	1
CIb	1	2	1	1	2	1	1	3	1	2	1	1	1	1	1

4.4. Special Concerns and Limitations

Panax notoginseng is a rare kind of ginseng, and which also is an antique and endangered medicinal plant (i.e., a traditional Chinese geoherb). This paper aims to explore its potential and provides some insights into the application of SCDDs for the landscape-scale mapping of *Panax notoginseng*. We wish this work could be the referenced technical basis for exploring more novel points that make outstanding contributions and provide the fruitful information for studies on the quality assurance of the production of TCM, precision farming, the construction of agro-ecosystems, sustainable development, and the protection of biodiversity of *Panax notoginseng*. Special concerns and limitations of this study can be summarized as follows:

This work utilizes a manually-collected set of samples of the target class and grid-constraint uniformly-collected negative samples. The uncertainty exists that a few possible land-cover classes may be left out, even though the classification results look rather good.

Thirteen SCDDs are employed and compared, however, there may be many other algorithms and their variants. Anyhow, the available ones have been used in this study.

The comparison with the different SCDDs does not judge them to be good or not. Actually, we wish to extend the ability of SCDDs to achieve the expected experiences in a straightforward way to find the optimal approach to monitoring the plant pattern changes of *Panax notoginseng*.

The class imbalance is a non-negligible problem in terms of a real specific land-cover classification using SCDDs. We strive for trying to observe what influences it would cause, and find two mediocre-appearing measures, i.e., the OA and AUC.

The selection criteria and scoring model are presented to determine the optimal SCDD which is outstanding and deserves attention.

The division of the site-specific error matrices by discriminating if the SCDD is dependent on the training set or not provides a more comprehensive approach to assess the final results.

The combination of SCDDs which are taken as the base classifiers can reduce the error or improve the accuracy. However, lower computational efficiency would be an annoying problem. Additionally, the pruned ensembles can give better performance.

Some classification accuracies may not be the reliable indicators, particularly if the training data are imbalanced [41,77]. As single-class data description is a special type of one-class classification, there are difficulties that may exist when trying to fit a single-class learner using the positive samples only. If SCDDs are trained with the samples of the single target class, then only the sensitivity can be estimated. There is a possibility that using only the sensitivity to fine-tune an algorithm may result in a class descriptor with high sensitivity but low specificity and overestimating the true extension of the class of interest [41]. In terms of the scope of this study, the introduction of single-class data description regarding remote-sensing mapping of *Panax notoginseng* fields based on P-learning, which provides us the new insights to promote the development of the resource inventory and dynamic monitoring of *Panax notoginseng*.

5. Conclusions

Natural TCM resources have seldom been observed and monitored from space before. Due to the promoted GAP technique, the small or fragmented parcels covered by black plastic sheets create the opportunity and probability for us to recognize and analyze the eco-geographic characteristics of *Panax notoginseng* at a landscape-scale. This paper delineates an application whereby a stack of SCDDs is used for remote-sensing mapping of *Panax notoginseng* fields through P-learning. The measuring performance of SCDDs provides us the challenging insights to define the selection criteria and scoring proof for choosing an optimal SCDD for remote-sensing mapping a specific landscape class. Future work would involve (1) developing new algorithms to enrich the approaches of specific land cover mapping; (2) improving the design sets, updating the sampling strategy, and overcoming the imbalance issue; and (3) extending to the state-of-the-art SCDDs published, which have not been presented in this study.

Author Contributions: F.D. and S.P. conceived and designed the experiments; S.P. performed the experiments and analyzed the data; and all relevant co-authors participated in writing and editing the paper.

Acknowledgments: The authors thank the editors and the anonymous reviewers for their insightful comments and helpful suggestions, which highly improved the quality of the manuscript. We are grateful to D. Tax for the development and free use of the dd_tools.

References

1. Tsai, C. A brief introduction to traditional Chinese medicine. In *30 Years' Review of China's Science and Technology*; World Scientific: Singapore, 1981; pp. 125–138.
2. Li, X.; Yang, G.; Li, X.; Zhang, Y.; Yang, J.; Chang, J.; Sun, X.; Zhou, X.; Guo, Y.; Xu, Y.; et al. Traditional Chinese medicine in cancer care: A review of controlled clinical studies published in Chinese. *PLoS ONE* **2013**, *8*, e60338.
3. Stone, R. Lifting the veil on traditional Chinese medicine. *Science* **2008**, *319*, 709–710. [CrossRef] [PubMed]
4. Xiong, X. Integrating traditional Chinese medicine into Western cardiovascular medicine: An evidence-based approach. *Nat. Rev. Cardiol.* **2015**, *12*, 374. [CrossRef] [PubMed]
5. Harvey, A.L.; Edrada-Ebel, R.; Quinn, R.J. The re-emergence of natural products for drug discovery in the genomics era. *Nat. Rev. Drug Discov.* **2015**, *14*, 111–129. [CrossRef] [PubMed]
6. Dong, J. The relationship between traditional Chinese medicine and modern medicine. *Evid.-Based Complement. Altern.* **2013**, *2013*, 153148. [CrossRef] [PubMed]
7. Xue, T.; Roy, R. Studying traditional Chinese medicine. *Science* **2003**, *300*, 740–741. [CrossRef] [PubMed]
8. General Administration of Quality Supervision, Inspection and Quarantine of the People's Republic of China. Provisions for the Protection of Products of Geographical Indication. Available online: http://www.wipo.int/edocs/lexdocs/laws/en/cn/cn041en.pdf (accessed on 10 August 2018).
9. Addor, F.; Grazioli, A. Geographical indications beyond wines and spirits. *J. World Intellect. Prop.* **2002**, *5*, 865–897. [CrossRef]
10. Standing Committee of the National People's Congress. Law of the People's Republic of China on Traditional Chinese Medicine. Available online: http://www.gov.cn/xinwen/2016-12/26/content_5152773.htm (accessed on 10 August 2018).
11. Fan, Z.; Miao, C.; Qiao, X.; Zheng, Y.; Chen, H.; Chen, Y.; Xu, L.; Zhao, L.; Guan, H. Diversity, distribution, and antagonistic activities of rhizobacteria of *Panax notoginseng*. *J. Ginseng Res.* **2016**, *40*, 97–104. [CrossRef] [PubMed]
12. Park, H.J.; Kim, D.H.; Park, S.J.; Kim, J.M.; Ryu, J.H. Ginseng in traditional herbal prescriptions. *J. Ginseng Res.* **2012**, *36*, 225–241. [CrossRef] [PubMed]
13. Wei, J.X.; Du, Y.C. *Modern Science Research and Application of Panax Notoginseng*; Yunnan Science and Technology Press: Kunming, China, 1996.
14. Zhou, Y.Q.; Chen, S.L.; Zhang, B.G.; Zhang, J.S.; Zhang, J.; Chen, Z.J.; Cun, X.M. Studies on the resources survey methods of Panax notogingseng based on remote sensing. *China J. Chin. Mater. Med.* **2005**, *30*, 1902–1905.
15. The State Council of the People's Republic of China. Several Opinions of the State Council on Supporting and Promoting the Development of Traditional Chinese Medicine. Available online: http://www.gov.cn/zwgk/2009-05/07/content_1307145.htm (accessed on 10 August 2018).
16. The State Council Information Office of the Peoples Republic of China. Health Service Development Plan of Traditional Chinese Medicine (2015–2020). Available online: http://www.gov.cn/zhengce/content/2015-05/07/content_9704.htm (accessed on 10 August 2018).
17. The Ministry of Science and Technology of the People's Republic of China. Outline of Traditional Chinese Medicine Innovation and Development Plan (2006–2020). Available online: http://www.most.gov.cn/tztg/200703/t20070320_42240.htm (accessed on 10 August 2018).
18. Sun, X.; Lin, D.; Wu, W.; Lv, Z. Translational Chinese medicine: A way for development of traditional Chinese medicine. *Chin. Med.* **2011**, *2*, 186–190. [CrossRef]
19. Sanchez-Hernandez, C.; Boyd, D.S.; Foody, G.M. One-class classification for mapping a specific land-cover class: SVDD classification of fenland. *IEEE Trans. Geosci. Remote Sens.* **2007**, *45*, 1061–1073. [CrossRef]

20. Boyd, D.S.; Foody, G.M. *Changing Land Cover*; Global Environmental Issues; John Wiley & Sons: Hoboken, NJ, USA, 2004; pp. 65–94.
21. Cihlar, J. Land cover mapping of large areas from satellites: Status and research priorities. *Int. J. Remote Sens.* **2000**, *21*, 1093–1114. [CrossRef]
22. Wang, J.; Xiao, X.; Qin, Y.; Dong, J.; Zhang, G.; Kou, W.; Jin, C.; Zhou, Y.; Zhang, Y. Mapping paddy rice planting area in wheat-rice double-cropped areas through integration of Landsat-8 OLI, MODIS, and PALSAR images. *Sci. Rep.* **2015**, *5*, 10088. [CrossRef] [PubMed]
23. Thenkabail, P.S.; Knox, J.W.; Ozdogan, M.; Gumma, M.K.; Congalton, R.G.; Wu, Z.; Milesi, C.; Finkral, A.; Marshall, M.; Mariotto, I.; et al. Assessing future risks to agricultural productivity, water resources and food security: How can remote sensing help? *Photogramm. Eng. Rem. Sens.* **2012**, *78*, 773–782.
24. Foody, G.M.; Mathur, A.; Sanchez-Hernandez, C.; Boyd, D.S. Training set size requirements for the classification of a specific class. *Remote Sens. Environ.* **2006**, *104*, 1–14. [CrossRef]
25. Song, B.; Li, P.; Li, J.; Plaza, A. One-class classification of remote sensing images using kernel sparse representation. *IEEE J-STARS* **2016**, *9*, 1613–1623. [CrossRef]
26. Chen, C.H. An overview of recent progress on information processing for remote sensing. In *Information Processing for Remote Sensing*; World Scientific: Singapore, 1999; pp. 39–49.
27. Foody, G.M. Status of land cover classification accuracy assessment. *Remote Sens. Environ.* **2002**, *80*, 185–201. [CrossRef]
28. Wilkinson, G.G. Results and implications of a study of fifteen years of satellite image classification experiments. *IEEE Trans. Geosci. Remote Sens.* **2005**, *43*, 433–440. [CrossRef]
29. Chen, C.H. *Frontiers of Remote Sensing Information Processing*; World Scientific: Singapore, 2003.
30. Cristianini, N.; Shawe-Taylor, J. *An Introduction to Support Vector Machines and Other Kernel-Based Learning Methods*; Cambridge University Press: Cambridge, UK, 2000.
31. Wan, B.; Guo, Q.; Fang, F.; Su, Y.; Wang, R. Mapping US urban extents from MODIS data using one-class classification method. *Remote Sens.* **2015**, *7*, 10143–10163. [CrossRef]
32. Mathieu, P.P.; Aubrecht, C. *Earth Observation Open Science and Innovation*; Springer Open: Cham, Switzerland, 2018; pp. 165–218.
33. Tse, C.H.; Lam, E.Y. Geological applications of machine learning on hyperspectral remote sensing data. *Proc. SPIE Int. Soc. Opt. Eng.* **2015**, *9405*, 940512.
34. Brown, M.E.; Lary, D.J.; Vrieling, A.; Stathakis, D.; Mussa, H. Neural networks as a tool for constructing continuous NDVI time series from AVHRR and MODIS. *Int. J. Remote Sens.* **2008**, *29*, 7141–7158. [CrossRef]
35. Lary, D.J.; Remer, L.A.; MacNeill, D.; Roscoe, B.; Paradise, S. Machine Learning and Bias Correction of MODIS Aerosol Optical Depth. *IEEE Geosci. Remote Sens. Lett.* **2009**, *6*, 694–698. [CrossRef]
36. Aurin, D.A.; Mannino, A. A Database for Developing Global Ocean Color Algorithms for Colored Dissolved Organic Material, CDOM Spectral Slope, and Dissolved Organic Carbon. In Proceedings of the Ocean Optics XXI, Glasgow, Scotland, UK, 8–12 October 2012.
37. Lary, D.J.; Alavi, A.H.; Gandomi, A.H.; Walker, A.L. Machine learning in geosciences and remote sensing. *Geosci. Front.* **2016**, *7*, 3–10. [CrossRef]
38. Khobragade, A.N.; Raghuwanshi, M.M. *Contextual Soft Classification Approaches for Crops Identification Using Multi-sensory Remote Sensing Data: Machine Learning Perspective for Satellite Images*; Springer International Publishing: Cham, Switzerland, 2015; pp. 333–346.
39. FAO. Development of a Framework for Good Agricultural Practices. Available online: http://www.fao.org/docrep/meeting/006/y8704e.htm (accessed on 10 August 2018).
40. Davis, N. Controlled-environment agriculture-past, present and future. *Food Technol.* **1985**, *39*, 124–126.
41. Silva, J.; Bacao, F.; Caetano, M. Specific Land Cover Class Mapping by Semi-Supervised Weighted Support Vector Machines. *Remote Sens.* **2017**, *9*, 181. [CrossRef]
42. Mack, B.; Roscher, R.; Stenzel, S.; Feilhauer, H.; Schmidtlein, S.; Waske, B. Mapping raised bogs with an iterative one-class classification approach. *ISPRS J. Photogramm.* **2016**, *120*, 53–64. [CrossRef]
43. Liu, X.; Liu, H.; Gong, H.; Lin, Z.; Lv, S. Appling the one-class classification method of maxent to detect an invasive plant *Spartina alterniflora* with time-series analysis. *Remote Sens.* **2017**, *9*, 1120. [CrossRef]
44. Marconcini, M.; Fernández-Prieto, D.; Buchholz, T. Targeted land-cover classification. *IEEE Trans. Geosci. Remote Sens.* **2014**, *52*, 4173–4193. [CrossRef]

45. Sahare, M.; Gupta, H. A review of multi-class classification for imbalanced data. *Int. J. Adv. Comput. Res.* **2012**, *2*, 160–164.
46. Sun, Y.; Wong, A.K.C.; Kamel, M.S. Classification of imbalanced data: A review. *Int. J. Pattern Recogn. Artif. Intell.* **2009**, *23*, 687–719. [CrossRef]
47. Tax, D.M.J. One-Class Classification: Concept-Learning in the Absence of Counter-Examples. Ph.D. Thesis, Delft University of Technology, Delft, The Netherlands, 2001; p. 65.
48. Khan, S.S.; Madden, M.G. One-class classification: Taxonomy of study and review of techniques. *Knowl. Eng. Rev.* **2014**, *29*, 345–374. [CrossRef]
49. Normile, D. The new face of traditional Chinese medicine. *Science* **2003**, *299*, 188–190. [CrossRef] [PubMed]
50. Chen, L.; Dong, K.; Yang, Z.; Dong, Y.; Tang, L.; Zheng, Y. Allelopathy autotoxcity effect of successive cropping obstacle and its alleviate mechanism by intercropping. *Chin. Agric. Sci. Bull.* **2017**, *33*, 91–98.
51. Fu, G.; Zhang, Q.; Liang, C.; Cheng, Z. Stereoscopic planting pattern of kernel-used apricot and medicinal plants in the loess drought hilly region in West Henan Province. *Med. Plant* **2011**, *2*, 5–11.
52. Panigrahy, S.; Sharma, S.A. Mapping of crop rotation using multidate Indian remote rensing satellite digital data. *ISPRS J. Photogramm.* **1997**, *52*, 85–91. [CrossRef]
53. Pirkouhi, M.G.; Nobahar, A.; Dadashi, M.A. Effects of variety, planting pattern and density of plant phenology traits basil plants (*Ocimum basilicum* L.). *Int. J. Agric. Crop Sci.* **2012**, *4*, 1221–1227.
54. Yunusa, I.A.M. Effects of planting density and plant arrangement pattern on growth and yields of maize (*Zea mays* L.) and soya bean (*Glycine max* (L.) Merr.) grown in mixtures. *J. Agric. Sci.* **1989**, *112*, 1–8. [CrossRef]
55. Song, C.; Woodcock, C.E.; Seto, K.C.; Lenney, M.P.; Macomber, S.A. Classification and change detection using Landsat TM data: When and how to correct atmospheric effects? *Remote Sens. Environ.* **2001**, *75*, 230–244. [CrossRef]
56. Yang, D.; Chen, J.; Zhou, Y.; Chen, X.; Chen, X.; Cao, X. Mapping plastic greenhouse with medium spatial resolution satellite data: Development of a new spectral index. *ISPRS J. Photogramm.* **2017**, *128*, 47–60. [CrossRef]
57. Von Elsner, B.; Briassoulis, D.; Waaijenberg, D.; Mistriotis, A.; Von Zabeltitz, C.; Gratraud, J.; Russo, G.; Suay-Cortes, R. Review of structural and functional characteristics of greenhouses in European Union countries: Part I, design requirements. *J. Agric. Eng. Res.* **2000**, *75*, 1–16. [CrossRef]
58. Foody, G.M.; Boyd, D.S.; Sanchez-Hernandez, C. Mapping a specific class with an ensemble of classifiers. *Int. J. Remote Sens.* **2007**, *28*, 1733–1746. [CrossRef]
59. Mack, B.; Roscher, R.; Waske, B. Can I trust my one-class classification? *Remote Sens.* **2014**, *6*, 8779–8802. [CrossRef]
60. McLachlan, G.; Peel, D. *Finite Mixture Models*; John Wiley & Sons: Hoboken, NJ, USA, 2004.
61. Møller, M.F. A scaled conjugate gradient algorithm for fast supervised learning. *Neural Netw.* **1993**, *6*, 525–533. [CrossRef]
62. Arthur, D.; Vassilvitskii, S. In k-means++: The advantages of careful seeding. In Proceedings of the Eighteenth Annual ACM-SIAM Symposium on Discrete Algorithms, New Orleans, LA, USA, 7–9 January 2007; pp. 1027–1035.
63. Kohonen, T. The self-organizing map. *Neurocomputing* **1998**, *21*, 1–6. [CrossRef]
64. Gallager, R.G.; Humblet, P.A.; Spira, P.M. A distributed algorithm for minimum-weight spanning trees. *ACM Trans. Program. Lang. Syst. (TOPLAS)* **1983**, *5*, 66–77. [CrossRef]
65. Altman, N.S. An introduction to kernel and nearest-neighbor nonparametric regression. *Am. Stat.* **1992**, *46*, 175–185.
66. Tax, D.; Ypma, A.; Duin, R. Support vector data description applied to machine vibration analysis. In Proceedings of the 5th Annual Conference of the Advanced School for Computing and Imaging, Heijen, The Netherlands, 15–17 June 1999; pp. 15–23.
67. Parzen, E. On estimation of a probability density function and mode. *Ann. Math. Stat.* **1962**, *33*, 1065–1076. [CrossRef]
68. Jolliffe, I.T. Principal component analysis and factor analysis. In *Principal Component Analysis*; Springer: New York, NY, USA, 2002; pp. 150–166.
69. Powers, D.M. Evaluation: From precision, recall and F-measure to ROC, informedness, markedness and correlation. *J. Mach. Learn. Technol.* **2011**, *2*, 37–63.
70. Fawcett, T. An introduction to ROC analysis. *Pattern Recogn. Lett.* **2006**, *27*, 861–874. [CrossRef]

71. Bradley, A.P. The use of the area under the ROC curve in the evaluation of machine learning algorithms. *Pattern Recogn.* **1997**, *30*, 1145–1159. [CrossRef]
72. Hanley, J.A.; McNeil, B.J. The meaning and use of the area under a receiver operating characteristic (ROC) curve. *Radiology* **1982**, *143*, 29–36. [CrossRef] [PubMed]
73. Drummond, C.; Holte, R.C. Explicitly representing expected cost: An alternative to ROC representation. In Proceedings of the 6th ACM SIGKDD International Conference on Knowledge Discovery and Data Mining, Boston, MA, USA, 20–23 August 2000; pp. 198–207.
74. Stehman, S.V. Selecting and interpreting measures of thematic classification accuracy. *Remote Sens. Environ.* **1997**, *62*, 77–89. [CrossRef]
75. Ting, K.M. Matching model versus single model: A study of the requirement to match class distribution using decision trees. In *European Conference on Machine Learning*; Springer: Berlin/Heidelberg, Germany, 2004; pp. 429–440.
76. Foody, G.M. Classification accuracy comparison: Hypothesis tests and the use of confidence intervals in evaluations of difference, equivalence and non-inferiority. *Remote Sens. Environ.* **2009**, *113*, 1658–1663. [CrossRef]
77. Hwang, J.P. A new weighted approach to imbalanced data classification problem via support vector machine with quadratic cost function. *Expert Syst. Appl.* **2011**, *38*, 8580–8585. [CrossRef]

11

Landslide Susceptibility Mapping Based on Random Forest and Boosted Regression Tree Models and a Comparison of their Performance

Soyoung Park [1] and Jinsoo Kim [2,*]

1 BK21 Plus Project of the Graduate School of Earth Environmental Hazard System, Pukyong National University, Busan 48513, Korea; yac100@pknu.ac.kr
2 Department of Spatial Information Engineering, Pukyong National University, Busan 48513, Korea
* Correspondence: jinsookim@pknu.ac.kr

Abstract: This study aims to analyze and compare landslide susceptibility at Woomyeon Mountain, South Korea, based on the random forest (RF) model and the boosted regression tree (BRT) model. Through the construction of a landslide inventory map, 140 landslide locations were found. Among these, 42 (30%) were reserved to validate the model after 98 (70%) had been selected at random for model training. Fourteen landslide explanatory variables related to topography, hydrology, and forestry factors were considered and selected, based on the results of information gain for the modeling. The results were evaluated and compared using the receiver operating characteristic curve and statistical indices. The analysis showed that the RF model was better than the BRT model. The RF model yielded higher specificity, overall accuracy, and kappa index than the BRT model. In addition, the RF model, with a prediction rate of 0.865, performed slightly better than the BRT model, which had a prediction rate of 0.851. These results indicate that the landslide susceptibility maps (LSMs) produced in this study had good performance for predicting the spatial landslide distribution in the study area. These LSMs could be helpful for establishing mitigation strategies and for land use planning.

Keywords: landslide susceptibility; random forest; boosted regression tree; information gain; landslide susceptibility map

---

1. Introduction

A landslide is defined as a natural disaster that occurs when gravity causes a mass of debris, soil, or rock to move on a downward slope [1]. The majority of landslides occur as a result of hydroclimatic events, such as prolonged or intensive rain. Furthermore, mechanisms such as seismic triggers, wind, and freeze–thaw cycles are known to initiate landslides [2].

Mountains with shallow layers of soil that have formed in place from weathered gneiss and granite make up roughly 70% of the Korean peninsula [3]. Such terrain is vulnerable to weakening during heavy rainfall. Most of the annual precipitation occurs during the summer, when heavy rain and typhoons frequently occur. In particular, the heavy rain associated with typhoons has the potential to cause landslides in South Korea [4]. The year 2011 was a particularly devastating year, with 43 landslide-related casualties in Chuncheon and at Woomyeon Mountain in the area surrounding Seoul City. This is the largest number of landslide-related casualties since 2000.

South Korea has not been alone in experiencing an increase in such natural disasters. Other regions around the world have also experienced more frequent landslides on a larger scale and with more severe damage. In future decades, this trend will probably continue because of ongoing deforestation, increased urbanization, and an increase in regional precipitation in landslide-prone areas due to climate change [5]. It is essential that both susceptible and stable areas be identified to mitigate property damage, environmental degradation, and loss of life. Consequently, landslide susceptibility assessments, i.e., assessments of the spatial probability of a landslide occurring, are a huge step forward in the comprehensive hazard management of landslides [6,7]. The landslide susceptibility map (LSM) produced by a landslide susceptibility assessment can be a useful tool for authorities with decision-making capabilities.

Many methods and techniques have been proposed to evaluate landslide susceptibility. In the past few decades, statistical approaches have become popular in the use of remote sensing (RS) with a geographic information system (GIS). There are many statistical approaches used in landslide susceptibility assessment, including a frequency ratio (FR) [8,9], certainty factor (CF) [10], statistical index (SI) [11,12], as well as weight of evidence (WoE) [7,13,14] and logistic regression (LR) [15,16] approaches.

Recently, machine learning techniques have become popular in various fields. Machine learning, a branch of artificial intelligence, uses computer algorithms to analyze and predict information based on learning from training data [17,18]. Due to its robustness and high generalization capability, the use of machine learning has increased in landslide susceptibility analysis. Among the machine learning methods, artificial neural network [19,20], fuzzy logic [21,22], neuro-fuzzy [23], support vector machine [24,25], random forest [26,27], and naïve Bayes tree [17,28] methods have been popularly applied.

More recently, ensemble machine learning techniques have been used to enhance the prediction power and robustness of landslide susceptibility assessment. The ensemble methods, formed by a combination of variously based classifiers, have typically demonstrated significant improvement [17,24,29,30]. Ensemble techniques, which are relatively new approaches for producing a landslide susceptibility map, have been rarely used in the field. Therefore, the main objective of this research was to analyze and compare the performance of different ensemble models—namely, the random forest (RF) and boosted regression tree (BRT) models—for landslide susceptibility analysis. The RF and BRT models are very popular ensemble methods. Both are tree-based algorithms that predict the results by combining individual trees. However, the RF and BRT models build trees in different ways. Considering these characteristics, these models are appropriate for producing LSMs and for comparing LSM results. The results of the models were compared using the receiver operating characteristic (ROC) curve and statistical indices to determine the more robust model.

2. Study Area and Data Used

2.1. Study Area

The study area, Woomyeon Mountain, is located in the Seocho district of Seoul City, South Korea. This area lies within 37°27′00″–37°28′55″ N and 126°59′02″–127°01′41″ E (Figure 1). The average elevation is 293 m above sea level, and the slope is approximately 30°–35°. The bedrock is Precambrian banded biotite gneiss, which is believed to be highly susceptible to landslides because of severe weathering and abundant faults (Figure A1). In addition, granite gneiss with relatively poor compositional differentiation is excavated en masse, and there is partial distribution of an embedded dike. The gneiss outcrop is poor, as a result of severe weathering in the overall area, and its foliation structure is irregular, due to several folding events [31].

This area experienced concentrated precipitation from 26–29 July 2011. The maximum precipitation, which occurred during 2 h one morning, was 164 mm. This exceeded the 156-mm, 100-year return period. This heavy precipitation led to a debris flow landslide in the area near Woomyeon Mountain, and 1–1.5 m of stratum flowed over areas near the mountain. Seven locations in the study area, including two locations in the valley area damaged primarily from flooding and five locations damaged by debris flow, were affected by the landslide. The total area damaged by the debris was approximately 276,683 m^2, and the maximum length of damage from the upper part of the steep-slope disaster area to diffuse areas was approximately 764 m. The event caused 16 deaths and 10 building collapses [31].

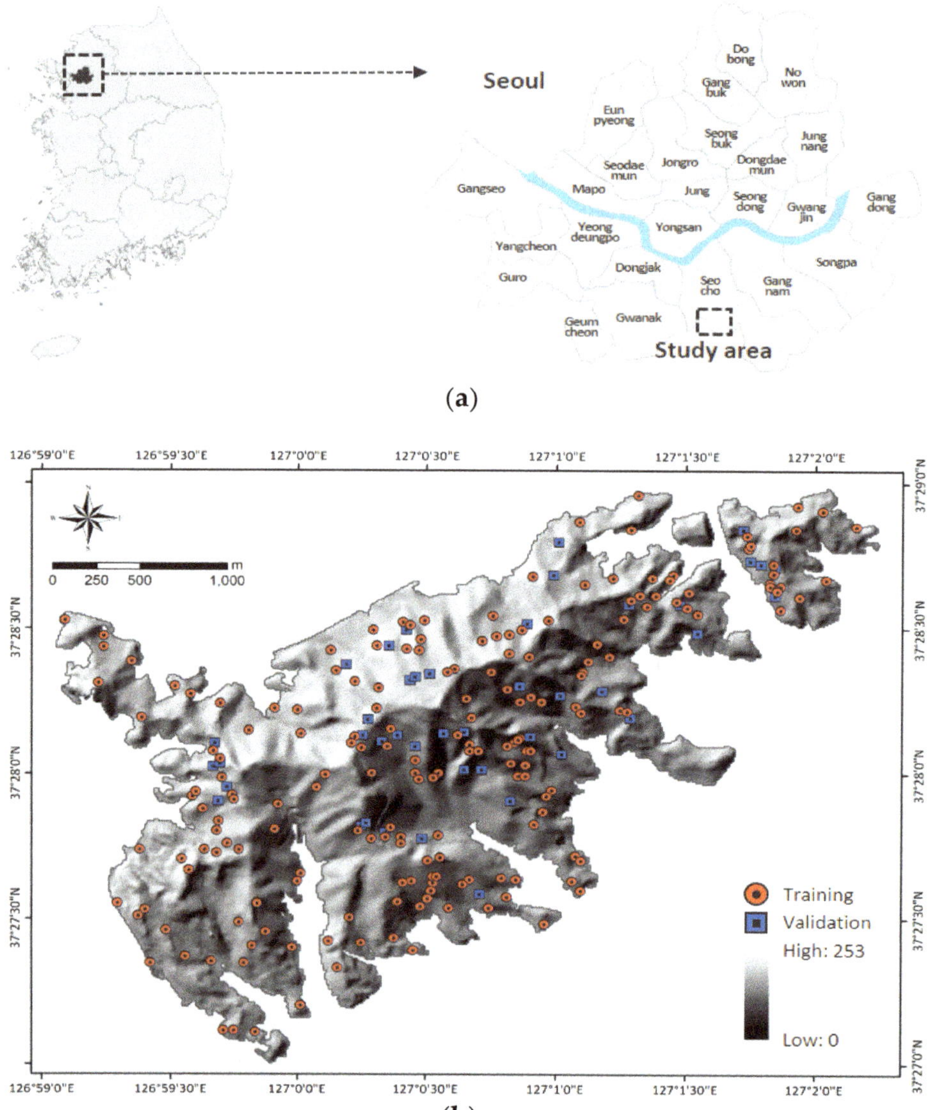

Figure 1. Location of study area (**a**) and landslide inventory map with hill shading (**b**).

2.2. Landslide Inventory

Landslide locations were identified using 32 aerial photographs of the study area, taken after the occurrence of the landslides. These aerial photographs were taken by a digital mapping camera with a spatial resolution of 10 cm. The orthorectified photographs were produced using the Leica Photogrammetry Suite (LPS) mounted on ERDAS Imagine 2011 (Erdas, Inc., Norcross, GA, United States). Landslide locations were digitized by visual interpretation using ArcGIS 10.2 (ESRI, Inc., Redlands, CA, USA). Among the digitized landslide locations, landslide locations belonging to rupture zones were converted to point data using a centroid technique. The point data representing the landslide locations were converted to a pixel format, with resolution of 10 m. From the 140 identified landslides, 42 (30%) were reserved to validate the model, after 98 (70%) had been chosen at random for model training. Additionally, non-landslide pixels were selected randomly from the non-landslide area: 98 non-landslide pixels were used for the training dataset, and 42 non-landslide pixels were used to build the validation dataset. This generating and splitting process was performed repeatedly more than 10 times. Finally, the combination utilized was found through the area under the receiver operating characteristic (ROC) curve (AUC) method.

2.3. Landslide Explanatory Variables

Landslides usually occur by complex interactions among various explanatory variables, and there is no consensus about which landslide explanatory variables to use. In this study, 14 explanatory variables were selected, based on a literature review and data availability. These factors were divided into the following three categories: topography, hydrology, and forestry (Table 1, Figure A2). These factors were produced in raster format with a cell size of 10 × 10 m, considering the scale of the input data, using ArcGIS 10.2 and ERDAS Imagine 2011; the total number of cells in the study area was 67,005. For the next process, the continuous variables among the explanatory variables were reclassified into seven classes, using ArcGIS 10.2. ArcGIS 10.2 provides various classification schemes, such as equal interval, standard deviation, natural break, quantile, etc. Natural break classification groups the classes based on break points that are relatively large jumps in data values. This classification method can be used to maximize the variance between classes. In addition, Cao et al. (2016) [32] indicated that natural break classification is more appropriate for the classification of variables, because their results showed that the LSM produced had higher accuracy compared to that using a different classification method. Therefore, natural break classification was used in this study.

Table 1. Information and sources of data used for the landslide susceptibility assessment at Woomyeon Mountain.

Category	Factor	Source	Scale (Resolution)	GIS and Data Type
-	Landslide inventory	Aerial photographs	1:5000	Raster
Topography	Altitude Slope degree Slope aspect Profile curvature Plan curvature	Topographic maps Digital elevation map	1:5000 10 × 10 m	Vector Raster
Hydrology	Distance to streams Topographic wetness index Stream power index Sediment transport index Terrain roughness index	Digital elevation map	10 × 10 m	Raster
Forestry	Timber type Timber diameter Timber age Timber density	Forest map	1:5000	Vector

2.3.1. Topography Factors

Topography factors include altitude, slope degree, slope aspect, profile curvature, and plan curvature. Altitude is an influential factor among the various landslide explanatory variables, because it is affected by several geomorphologic and geological processes. Slope, which can be described as the form between any section of the surface and a horizontal datum, has considerable influence on slope stability [33]. The degree of vulnerability to landslides may differ based on slope direction, because the water content of the surface, vegetation type, and soil strength may be different. In addition, both the profile and plan curvatures can be classified as flat, concave, or convex. During the rainy season, concave slopes may contain more moisture than convex slopes or flat slopes, so the concave slopes may be more vulnerable to landslides. All of these variables were extracted from the 10-m digital elevation model (DEM), using the spatial analyst tool of ArcGIS. The DEM was produced from 1:5000 topographic maps provided by the Korean National Geographic Information Institute.

2.3.2. Hydrology Factors

The hydrology factors were distance to streams, topographic wetness index (TWI), stream power index (SPI), sediment transport index (STI), and terrain roughness index (TRI). The streams were delineated by flow accumulation and converted to a vector format. The distance to streams was calculated using the Euclidean distance function in ArcGIS. Beven and Kirby (1979) [34] developed a TWI that reflects water's tendency to accumulate anywhere within the catchment area, accumulations that will then tend to move downslope as a result of gravity [35]. The water flow's power to erode is measured by the SPI, based on the assumption of proportionality of discharge to a catchment's specific area [36]. The STI is also often used to reflect the overland flow's power to erode [37]. The TWI, SPI, and STI were calculated with their base in specific catchment areas (A_s) and slope maps, using the following:

$$\mathrm{TWI} = \ln\left(\frac{A_s}{tan\beta}\right) \tag{1}$$

$$\mathrm{SPI} = A_s \times tan\beta \tag{2}$$

$$\mathrm{STI} = \left(\frac{A_s}{22.13}\right)^{0.6}\left(\frac{sin\beta}{0.0896}\right)^{1.3} \tag{3}$$

where A_s represents the specific catchment area (m^2/m), and β represents the local slope gradient (degrees).

In addition, the TRI, which represents the concave and convex upward slopes [38], was calculated as

$$\mathrm{TRI} = \sqrt{|x|(max^2 - min^2)}, \tag{4}$$

where max and min represent the maximum and minimum values of altitude among the nine rectangular neighbor pixels, respectively.

2.3.3. Forestry Factors

Vegetation prevents erosion on a slope by buffering the impact of rain falling on the slope, and vegetation roots increase the shear strength of the slope by increasing the shear strength of the soil. The forestry factors include timber type, timber diameter, timber age, and timber density. Here, timber type and timber age mean the species and average age of planted trees, respectively. In addition, timber diameter represents the size of the diameter at chest height. Timber density refers to the degree of closure of the crown canopy. These values were obtained from a 1:5000 scale forest map produced by the Korea Forest Research Institute.

3. Methodology

This study was performed using the following main steps: (1) collection and construction of database of landslides and landslide explanatory variables, (2) preparation of the training and test datasets through repeated random sampling, (3) feature selection using information gain (IG), (4) landslide susceptibility mapping using RF and BRT models, and (5) validation and comparison of performance among landslide susceptibility maps (LSMs) (Figure 2). The IG, RF, and BRT models were implemented in R (Foundation for Statistical Computing, Vienna, Austria) using the "FSelector," "randomForest," and "gbm" packages, respectively. These algorithms were performed employing a 10-fold cross-validation approach, to reduce the variability of the model results.

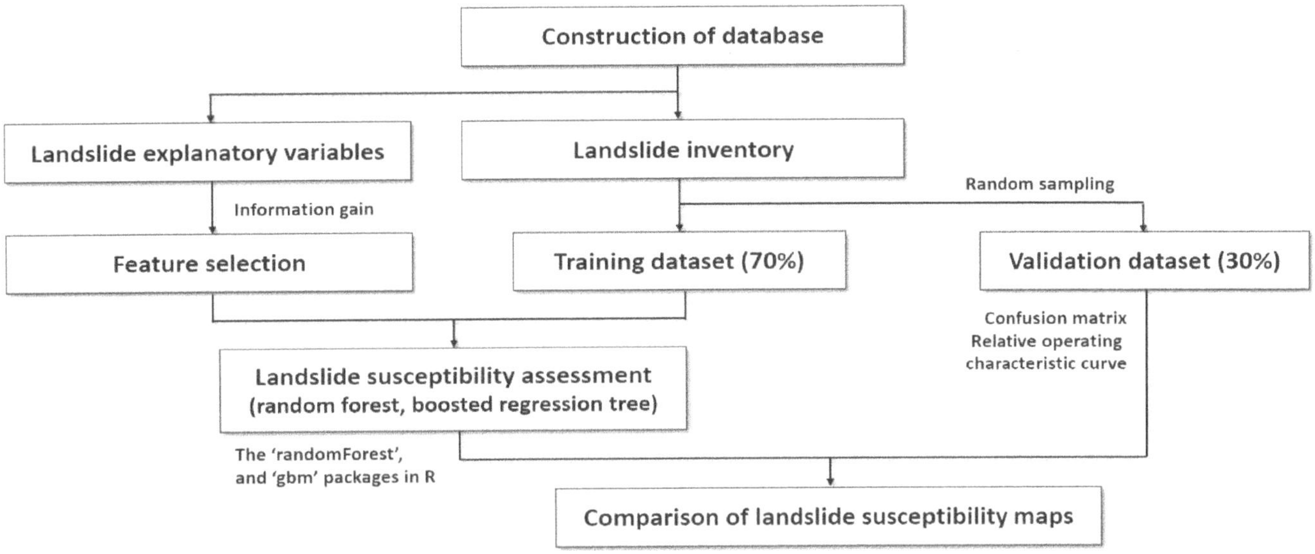

Figure 2. Flow chart of the overall methodology.

3.1. Landslide Dataset Preparation

In this study, correlations between landslide and landslide explanatory variables were analyzed using the FR (Table A1). The FR is the ratio of the area where landslides occurred to the total study area. The FR was calculated by dividing the ratio of landslide occurrence, for the class or type of each factor, into an area ratio, for the class or type of each factor to the total area. The calculated FRs pertaining to each landslide explanatory variable were normalized from 0 to 1. The normalized FRs were extracted for the landslide and non-landslide dataset. Subsequently, these data were used for the training and validation datasets, to run the models and evaluate the prediction capabilities of the models, respectively.

3.2. Information Gain

The landslide explanatory variables have a crucial role in producing LSM. Some landslide explanatory variables might be associated with reductions in model performance, overfitting, model training time, and predictive capability [39]. Therefore, it is necessary to recognize and choose proper landslide explanatory variables.

Various methods such as IG [17], chi-square statistics [30], and Relief-F [29] have been proposed for feature selection in landslide modeling. In this study, the IG, proposed by Quinlan (1993) [40], was used to determine irrelevant and unimportant variables. The IG evaluates an attribute by determining the overall information gain in terms of the class. Consequently, the result can determine the ranking of importance, based on the normalized average merit contributed by each attribute [41,42].

The IG value of landslide explanatory variable C_i belonging to class L (landslide and non-landslide) is calculated as [24,30]

$$IG(L,\ C_i) = IF(L) - IF(L|C_i), \tag{5}$$

where $IF(L)$ is the entropy value of L, and $IF(L|C_i)$ is the entropy of L after integrating the values of landslide explanatory variable C_i. These values are calculated using Equation (6) and Equation (7), respectively:

$$IF(L) = -\sum_i P(L_i)\log_2(P(L_i)), \tag{6}$$

$$IF(L|C_i) = -\sum_i P(L_i)\sum_j P(L_i|C_i))\log_2(P(L_i|C_i)), \tag{7}$$

where $P(L_i)$ is the prior probability of the class L, and $P(L_i|C_i)$ is the posterior probabilities of L given the values of explanatory variable C_i. An explanatory variable with a higher IG value has higher rank, meaning that it is more important to landslide models. By contrast, an explanatory variable with an IG value of zero must be removed from the dataset, because that factor does not make a contribution.

3.3. Landslide Susceptibility Analysis

3.3.1. Random Forest

RF, developed by Breiman (2001) [43], is a popular ensemble learning method that has been used widely for classification, regression, clustering, and interaction detection. A single decision tree is a weak classification, because of its high variance and bias. However, RF tends to produce robust models, because it can mitigate these problems by using ensemble trees [44].

RF generates thousands of random binary trees to form a forest. Each tree is grown based on a bootstrap sample, using a classification and regression trees (CART) procedure with a random subset of variables selected at each node [26,45]. For each tree grown on a bootstrap sample, the "out-of-bag" (OOB) error rate is calculated using observations left out of the bootstrap sample. The final decisions of class membership and model construction (output) are determined by the majority vote among all trees [46].

Two types of error rate—the mean decrease in accuracy and the mean decrease in the Gini coefficient—were calculated. These measures have been widely used to rank and select variables [26,47]. To run the RF model, the user should optimize two priori parameters, the number of trees in the forest (*ntree*) and the number of variables tested at each node (*mtry*), to minimize the OOB error and obtain good model performance [44,45].

3.3.2. Boosted Regression Tree Model

The BRT model is a combination of statistical and machine learning techniques. The BRT model fits different techniques and combines them to improve the performance of a single model [48,49]. Two different algorithms, namely boosting and regression, are used in the model, and the strengths of these algorithms are combined to improve model accuracy and decrease model variance [45,50]. Boosting is one of the most powerful learning methods for improving model accuracy, by iteratively fitting new trees to the residual errors (RE) of the existing tree assemblage [45,51]. In addition to boosting, the BRT model uses regression trees in the modeling process. Regression trees are categorized from the classification and regression tree approaches from the decision tree group of models [52].

In the model, among the various parameters, the number of trees is automatically set through internal cross-validation. In addition, the learning rate, the number of nodes in a single tree, and bag fraction were determined through a trial-and-error approach [53]. The complexity of the model and the contribution of each tree to the model are controlled by a shrinkage parameter and the learning rate, respectively. The bag fraction and shrinkage parameter determine the number of trees required to reach the optimal solution [54].

3.4. Model Performance Assessment and Comparison

3.4.1. Confusion Matrix

The confusion matrix includes true positive (TP), false positive (FP), true negative (TN), and false negative (FN) categories. Using these values, various statistical indices, such as accuracy, sensitivity, specificity, threat score, equitable threat score, Pierce's skill score, odds ratio, and odds ratio skill score can be calculated [55]. The value calculated from the confusion matrix provides useful information on model performance and classification accuracy.

In this study, the sensitivity, specificity, overall accuracy, and kappa statistic were used to validate the performance of the LSMs. The percentages of landslide and non-landslide pixels classified correctly into those two categories enable the calculation of sensitivity and specificity, and the overall percentage classified correctly (in both categories together) indicates the accuracy of the LSMs [56]. In addition, the kappa statistic is used to evaluate the reliability of the landslide models. Its value ranges from −1 (non-reliable) to 1 (reliable) [57].

3.4.2. Receiver Operating Characteristic

The receiver operating characteristic (ROC) curve has been commonly used to validate the quality of a probabilistic model. The ROC curve is plotted by statistical index value pairs, with the false positive rate (sensitivity) on the x-axis and the "100−false negative rate" (100−specificity) on the y-axis. The ROC curve can be classified as a success rate curve or prediction rate curve, depending on the dataset used. The success rate curve, calculated using the training dataset, represents how well the LSMs fit the data. The prediction rate curve, calculated using the validation dataset, represents how well the model and landslide explanatory variables predict a landslide [11]. The ROC curve can be verified quantitatively when the area under the ROC curve (AUC) is calculated. AUC values range from 0.5 to 1.0. AUC values closer to 1 indicate a more accurate model.

4. Results

4.1. Selection of Landslide Explanatory Variables

The average information gain (AIG) value, and its standard deviation for each landslide explanatory variable, were calculated and ranked (Table 2). All landslide explanatory variables used in this study contributed to the landslide models, because the AIG values of these variables were more than 0. According to the results, the TRI had the highest AIG value (0.086), which means that this factor made the greatest contribution to the landslide models in this study area. By contrast, timber diameter made the smallest contribution to the landslide models, as indicated by the lowest AIG value (0.005).

Table 2. Information gain values for the landslide explanatory variables used in this study.

No.	Landslide Explanatory Variable	Average Merit	Standard Deviation
1	Terrain roughness index	0.086	±0.010
2	Slope aspect	0.071	±0.012
3	Distance to streams	0.06	±0.010
4	Altitude	0.049	±0.008
5	Timber type	0.049	±0.008
6	Stream power index	0.041	±0.008
7	Slope degree	0.038	±0.008
8	Sediment transport index	0.037	±0.006
9	Topographic wetness index	0.033	±0.011
10	Plan curvature	0.025	±0.006
11	Profile curvature	0.013	±0.005
12	Timber age	0.012	±0.002
13	Timber density	0.008	±0.002
14	Timber diameter	0.005	±0.002

4.2. Training the Random Forest and Boosted Regression Tree Models

The training dataset was used to train the RF and BRT models for landslide susceptibility assessment. During the training process, the optimum values of the parameters for the models were applied to obtain high model predictive capability. The optimized values for the RF model were 300 for *ntree* and 2 for *mtry*. In the case of the BRT model, the optimized values for *n.trees*, *interaction.depth*, *shrinkage*, and *n.minobsinnode* were 500, 1, 0.01, and 10, respectively. Subsequently, the RF and BRT models were constructed using the optimized parameters, based on the training dataset.

After their construction, the RF and BRT models were applied throughout the whole study area to produce LSMs.

4.3. Model Validation and Comparison

The performance of each model was analyzed using the training dataset. The RF model showed a higher sensitivity value (98.00%) than did the BRT model (79.57%). This result showed that the RF model classified more correctly than the BRT model in the landslide class. The specificity results also indicated that the RF model had higher specificity (100.00%) in the non-landslide class, indicating that the non-landslide pixels were more correctly classified. The specificity value of the BRT model was 76.70%. Because of the lower sensitivity and specificity values of the BRT model, the overall accuracy and kappa index values were lower, with values of 78.16% and 0.561, respectively. In the case of the RF model, the overall accuracy and kappa index were 98.98% and 0.980, respectively.

In addition, the success rate and the prediction rate were analyzed using the training dataset and the validation dataset, respectively (Figure 3). In the case of success rate, the RF and BRT models had values of 0.999 and 0.887, respectively. The prediction rate curve also showed that the RF model had a higher AUC (0.865) than the BRT model (0.851). Overall, the AUC values of all models were greater than 0.8. These results show that the LSMs constructed in this study have good accuracy in the spatial prediction of landslide susceptibility.

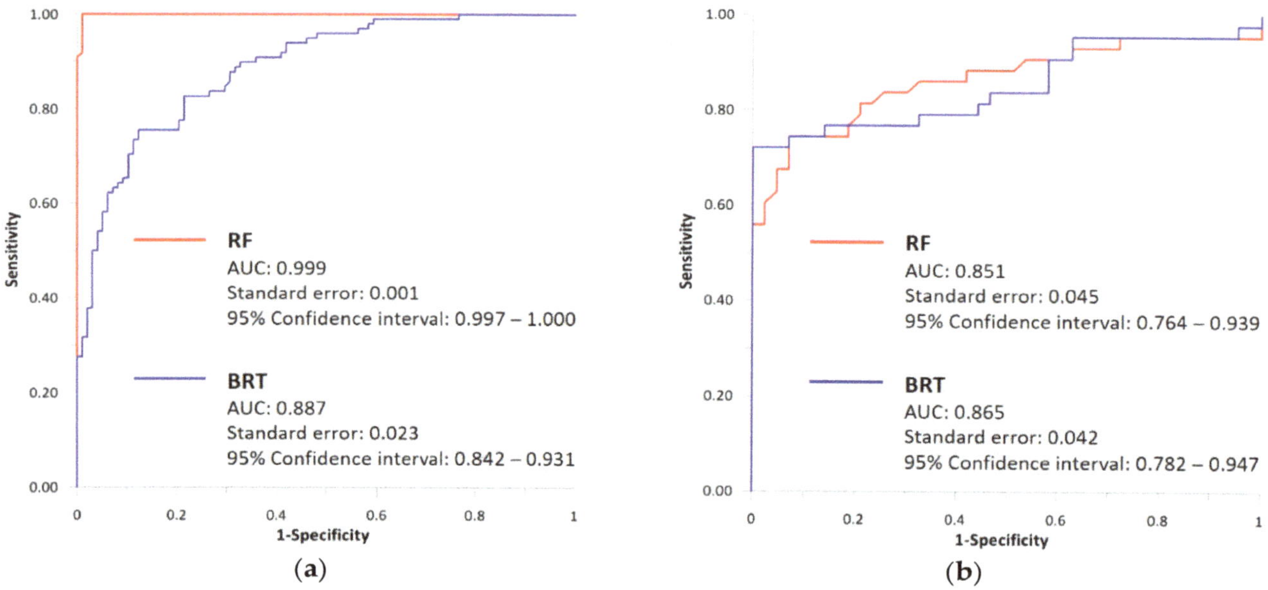

Figure 3. Analysis of the receiver operating characteristic (ROC) curve for the two landslide susceptibility maps: (**a**) success rate curve using the training dataset and (**b**) prediction rate curve using the validation dataset.

4.4. Generating Landslide Susceptibility Maps

The RF and BRT models were used to develop LSMs in the study area. The LSMs were prepared by generating landslide susceptibility indices (LSIs) and reclassifying the class. The LSIs were calculated based on the trained RF and BRT models. Using the natural breaks method, the LSMs were reclassified into five susceptibility classes: very high, high, moderate, low, and very low (Figure 4). Overall, the distribution of LSI for each susceptibility class was similar between the LSM produced by RF (RF LSM) and that produced by BRT (BRT LSM). The "high" and "very high" susceptibility classes covered about 30% of the total area. The RF model had a value of 34.69%, and the BRT model had the lower value of 31.11%.

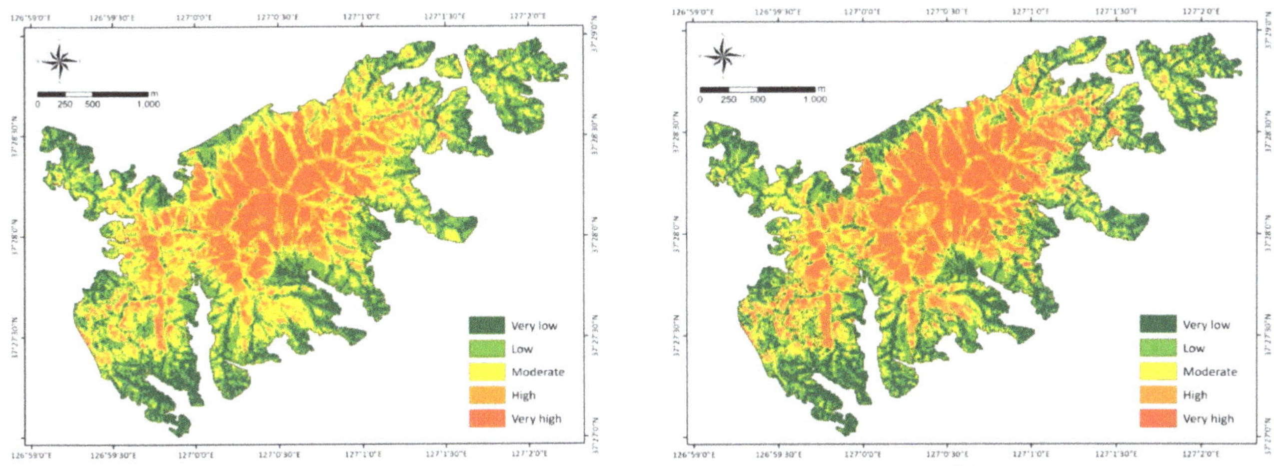

(a) (b)

Figure 4. Landslide susceptibility maps produced by random forest (RF) (**a**) and boosted regression tree (BRT) (**b**) models.

The LSMs produced from the two models were validated based on the landslide density (LD) of each susceptibility class on the LSMs. The LD is the ratio of the percentage of landslide pixels to the percentage of all pixels for each susceptibility class shown on the map [56]. LD was calculated by overlaying the five LSMs and the landslide inventory map. Generally, for the study area, the value of LD increased gradually, from very low to very high susceptibility (Table 3). At the "very high" class, the RF and BRT models had LD values of 3.799 and 2.721, respectively. Overall, the models used in this study are suitable for LSM.

Table 3. Landslide density on landslide susceptibility maps produced from the different models.

	Random Forest			Boosted Regression Tree		
	Pixels of Class	Pixels of Landslide	Landslide Density	Pixels of Class	Pixels of Landslide	Landslide Density
Very low	13,034	2	0.073	10,671	3	0.135
Low	16,871	4	0.113	18,629	16	0.411
Moderate	13,854	16	0.553	16,860	35	0.994
High	12,917	36	1.334	12,930	41	1.518
Very high	10,329	82	3.799	7915	45	2.721
Total	67,005	140		67,005	140	

4.5. Discussion

The LSMs produced using the models were evaluated by statistical indices and ROC curves. The RF model had better sensitivity, specificity, overall accuracy, and kappa values. The AUC values of the LSMs used in this study were about 80%, indicating reasonable accuracy. The RF model had higher AUC values for the success rate and prediction rate curves than the BRT model. Thus, these models had very high predictive performance. Furthermore, the LSMs would be produced differently depending on the methods used and the landslide explanatory variables selected. The landslide explanatory variables may not make equal contributions, which can affect prediction ability. In this study, the landslide explanatory variables used made different contributions to the models. Table 4 illustrates the importance of each explanatory variable, calculated and normalized in the RF and BRT models. In general, TRI had the highest importance to the models, whereas timber diameter, timber age, and timber density had lower predictive capability.

Table 4. Relative importance of each landslide explanatory variable calculated in the random forest and boosted regression tree model.

	Random Forest		Boosted Regression Tree	
	Importance	**Rank**	**Importance**	**Rank**
Terrain roughness index	1.000	1	1.000	1
Distance to streams	0.857	2	0.556	3
Altitude	0.766	3	0.277	5
Sediment transport index	0.654	4	0.562	2
Timber type	0.484	5	0.158	7
Slope degree	0.469	6	0.000	-
Stream power index	0.449	7	0.084	9
Topographic wetness index	0.440	8	0.378	4
Slope aspect	0.408	9	0.242	6
Plan curvature	0.214	10	0.099	8
Profile curvature	0.118	11	0.016	10
Timber diameter	0.026	12	0.000	-
Timber age	0.009	13	0.000	-
Timber density	0.000	14	0.000	-

From the results, ensemble classification, such as that done by the model used in this study, can improve the performance of single (weak) classifiers and the prediction accuracy of LSM [56]. However, the models had overfitting problems, as indicated by the AUC values calculated using the training and validation datasets. The AUC values of the success rate curve were very high, almost reaching a value of 1, but the AUC values of the prediction rate curve were lower. Especially in the case of the RF model, the AUC value of the prediction rate was decreased by about 20%. This result showed that the RF model was trained excessively by the training data. This can be associated with poor generalization from training data and increased error for real data. Overfitting is a common problem affecting researchers performing machine learning and data mining. There can be many reasons for overfitting. However, in this study, the landslide explanatory variables used still included noise, despite the feature selection process. In addition, because the landslide area is very small compared to the non-landslide area, the model could not learn and predict the non-landslide area.

5. Conclusions

This study compared and analyzed landslide susceptibility at Woomyeon Mountain using different models. For this purpose, landslide-related spatial data consisting of a landslide inventory, and landslide explanatory variables were collected and prepared. The landslide inventory map was built using aerial photographs. The 14 landslide explanatory variables were constructed from spatial data collected by government organizations. These factors included altitude, slope degree, slope aspect, profile curvature, plan curvature, distance to streams, TWI, SPI, STI, TRI, timber type, timber diameter, timber age, and timber density.

The contribution of each landslide explanatory variable was evaluated using the average IG value with a 10-fold cross-validation approach. All of the landslide explanatory variables contributed to the models, because the IG values of all factors were greater than zero. Therefore, the landslide susceptibility analysis and mapping were performed with all landslide explanatory variables using the RF and BRT models. The RF and BRT models were implemented in R. A popular open-source software, R is helpful for statistical computing and data visualization [58]. The models were constructed using optimized parameters, and LSI was predicted over the study area.

The LSMs produced in this study may prove useful for decision makers, planners, and engineers in disaster planning to minimize economic losses and casualties. In a future study, the accuracy of the LSMs of this study could be enhanced by selecting more optimal landslide explanatory variables and solving the problem of overfitting.

Author Contributions: S.P. analyzed the data and wrote the paper. J.K. suggested the idea.

Appendix A

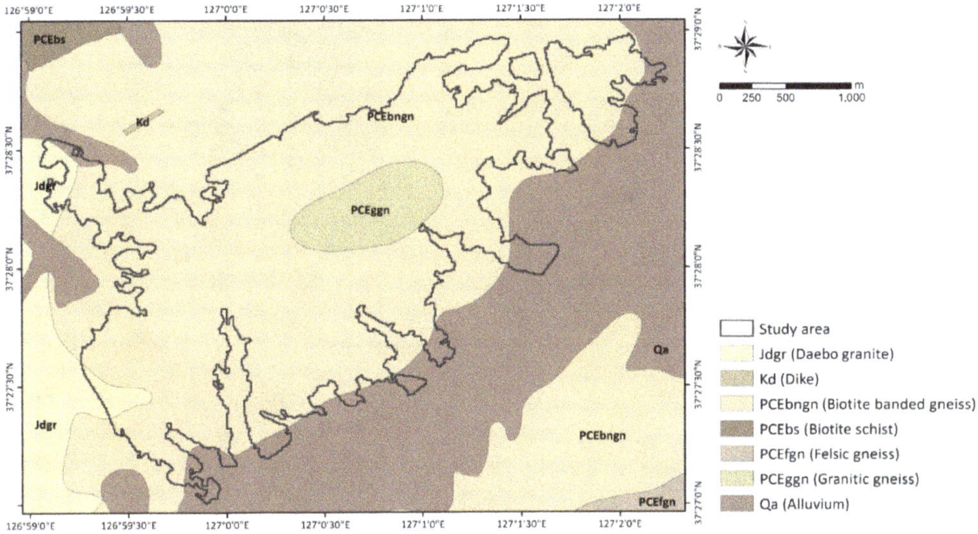

Figure A1. Geological features of the study area produced from 1:50,000 geological maps provided by the Korea Institute of Geoscience and Mineral Resources.

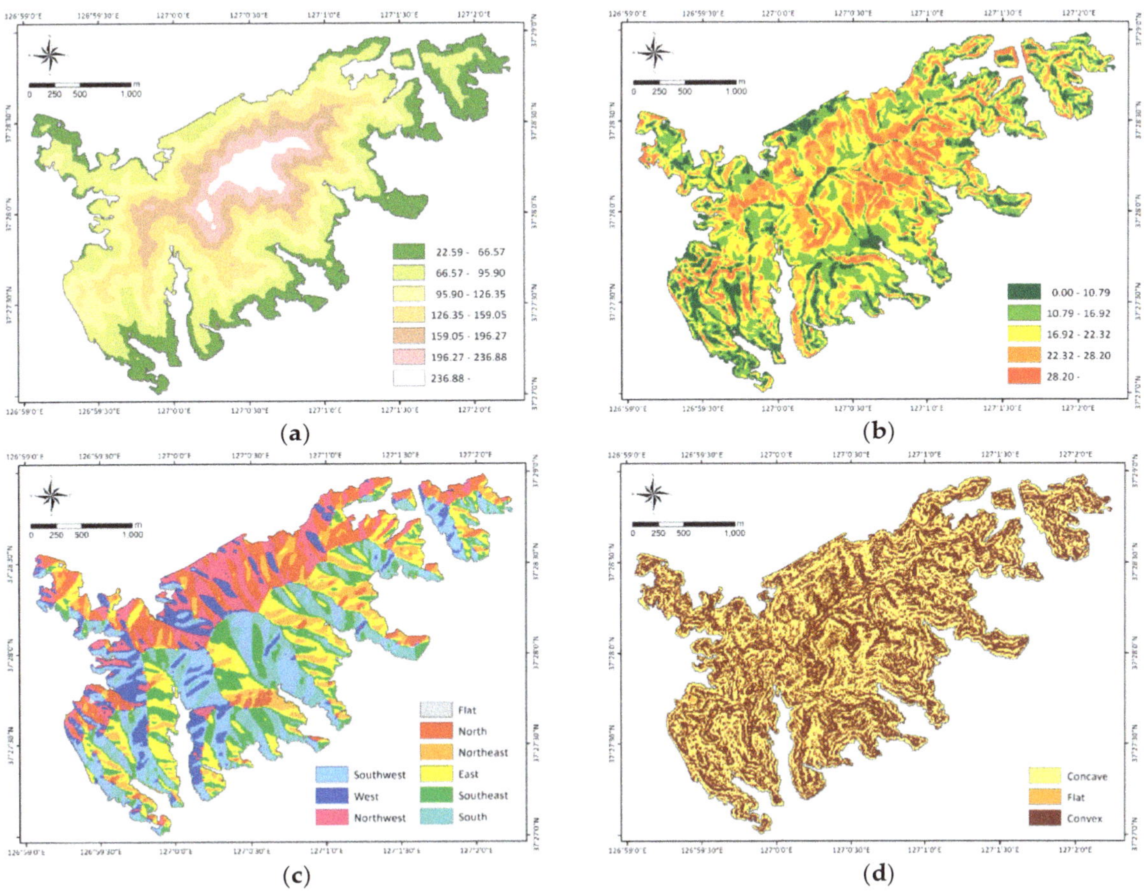

Figure A2. *Cont.*

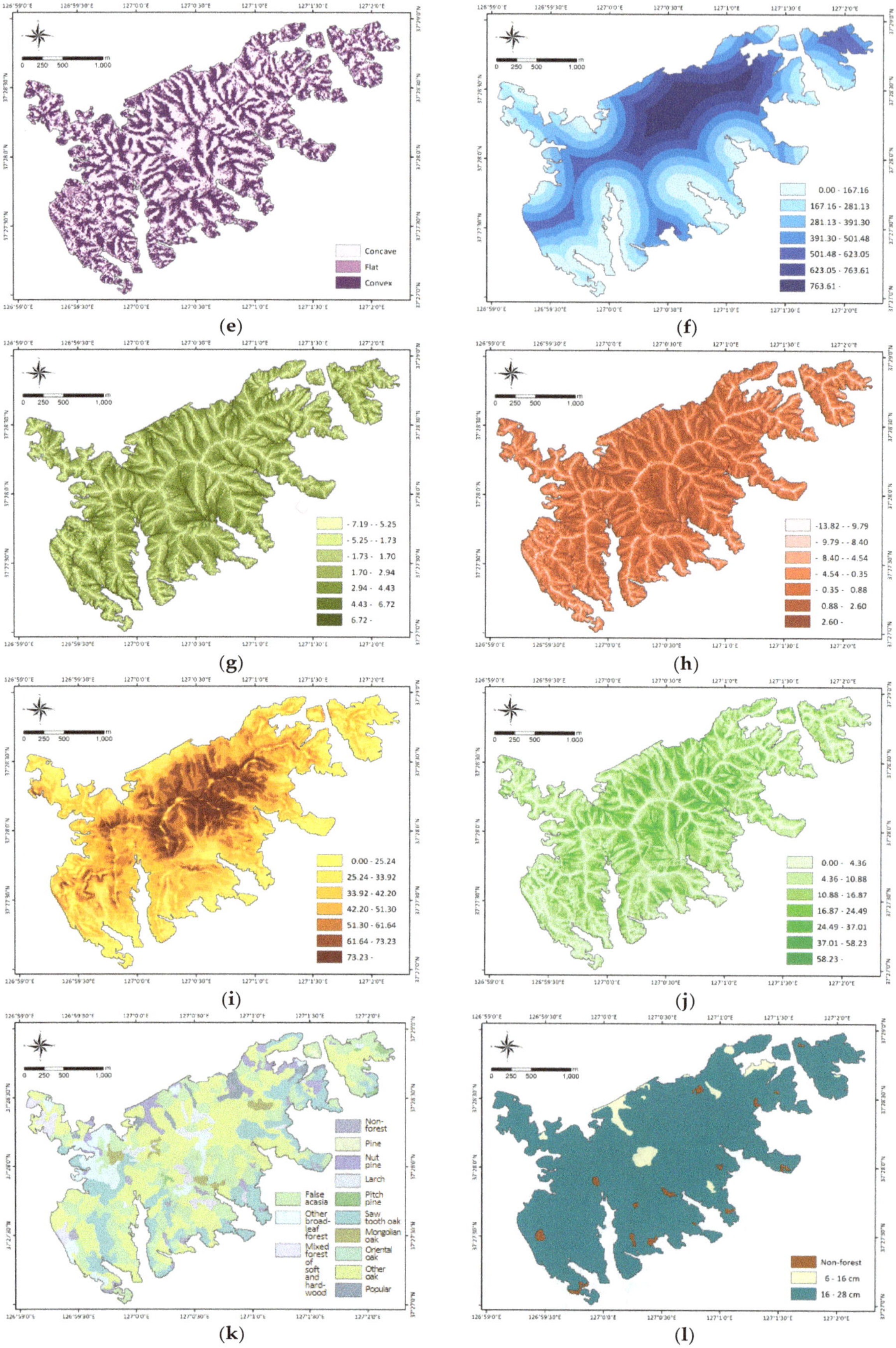

Figure A2. *Cont.*

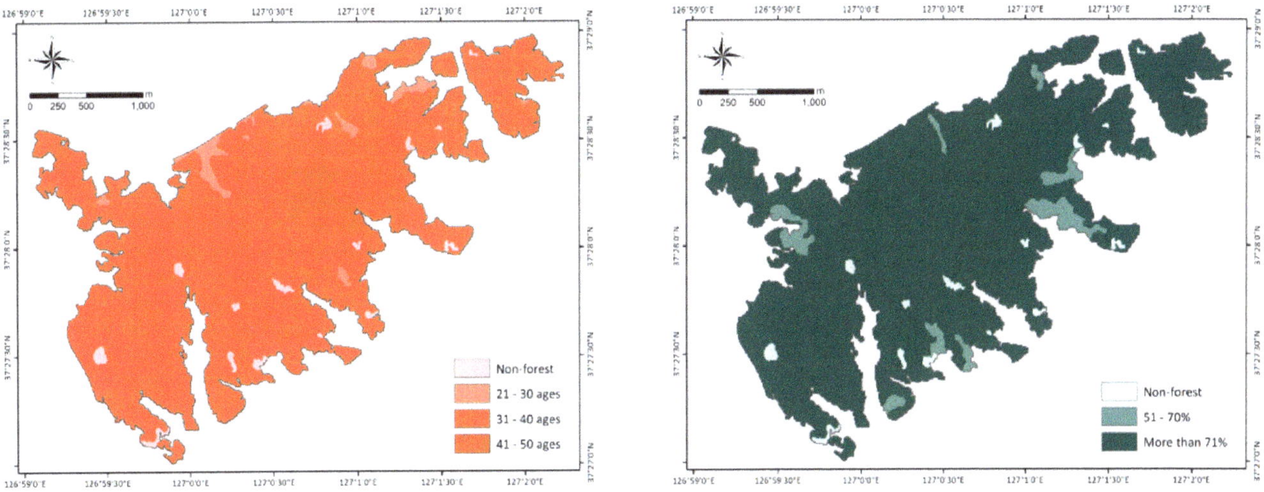

(m) (n)

Figure A2. Landslide explanatory variables used to analyze landslide susceptibility: (**a**) altitude, (**b**) slope degree, (**c**) slope aspect, (**d**) profile curvature, (**e**) plan curvature, (**f**) distance to streams, (**g**) topographic wetness index, (**h**) stream power index, (**i**) sediment transport index, (**j**) terrain roughness index, (**k**) timber type, (**l**) timber diameter, (**m**) timber age, and (**n**) timber density.

Table A1. Correlations between landslide and landslide explanatory variables using the frequency ratio.

Factor	Class	No. of Pixels in Domains	No. of Landslide Pixels	Frequency Ratio	Normalized Frequency Ratio
Altitude (m)	[22.59, 66.57]	12,477	10	0.38	0.00
	[66.57, 95.90]	17,042	20	0.56	0.11
	[95.90, 126.35]	13,614	31	1.09	0.42
	[126.35, 159.05]	10,060	33	1.57	0.71
	[159.05, 196.27]	6766	17	1.20	0.49
	[196.27, 236.88]	4409	19	2.06	1.00
	[>236.88]	2637	10	1.81	0.85
Slope degree (°)	[0.00, 8.83]	4704	3	0.31	0.00
	[8.83, 13.98]	11,554	11	0.46	0.11
	[13.98, 18.15]	15,716	33	1.00	0.52
	[18.15, 22.07]	14,692	27	0.88	0.43
	[22.07, 26.24]	11,362	36	1.52	0.91
	[26.24, 31.39]	6708	23	1.64	1.00
	[>31.39]	2269	7	1.48	0.88
Slope aspect	Flat	69	0	0.00	0.00
	North	8483	12	0.68	0.50
	Northeast	7928	9	0.54	0.40
	East	9395	18	0.92	0.67
	Southeast	8837	20	1.08	0.80
	South	9522	26	1.31	0.96
	Southwest	8970	19	1.01	0.75
	West	6333	18	1.36	1.00
	Northwest	7468	18	1.15	0.85
Profile curvature	Concave	33,763	79	1.12	1.00
	Flat	247	0	0.00	0.00
	Convex	32,995	61	0.88	0.79
Plan curvature	Concave	31,592	82	1.24	1.00
	Flat	772	0	0.00	0.00
	Convex	34,641	58	0.80	0.65

Table A1. *Cont.*

Factor	Class	No. of Pixels in Domains	No. of Landslide Pixels	Frequency Ratio	Normalized Frequency Ratio
Distance to streams (m)	[0.00, 167.16]	8947	3	0.61	0.28
	[167.16, 281.13]	12,553	16	1.46	0.81
	[281.13, 391.30]	13,430	41	0.77	0.38
	[391.30, 501.48]	12,372	20	1.28	0.69
	[504.48, 623.05]	9369	25	1.77	1.00
	[623.05, 763.61]	6205	23	1.39	0.76
	[>763.61]	4129	12	0.63	0.50
Topographic wetness index	[−7.19, −5.25]	8350	11	0.63	0.50
	[−5.25, −1.73]	3148	2	0.30	0.24
	[−1.73, 1.70]	13,316	24	0.86	0.69
	[1.70, 2.94]	21,415	56	1.25	1.00
	[2.94, 4.43]	14,843	38	1.23	0.98
	[4.43, 6.72]	4652	9	0.93	0.74
	[>6.72]	1281	0	0.00	0.00
Stream power index	[−13.82, −9.79]	487	0	0.00	0.00
	[−9.79, −8.40]	3850	2	0.25	0.17
	[−8.40, −4.54]	7251	11	0.73	0.48
	[−4.54, −0.35]	13,013	16	0.59	0.39
Stream power index	[−0.35, 0.88]	22,507	51	1.08	0.72
	[0.88, 2.60]	16,613	52	1.50	1.00
	[>2.60]	3284	8	1.17	0.78
Sediment transport index	[0.00, 4.35]	15,486	17	0.53	0.20
	[4.35, 10.88]	21,160	31	0.70	0.27
	[10.88, 16.87]	16,659	44	1.26	0.48
	[16.87, 24.49]	9054	29	1.53	0.58
	[24.49, 37.01]	3267	18	2.64	1.00
	[37.01, 58.23]	1065	0	0.00	0.00
	[>58.23]	314	1	1.52	0.58
Terrain roughness index	[0.00, 25.24]	6302	0	0.00	0.00
	[25.24, 33.92]	13,887	14	0.48	0.20
	[33.92, 42.20]	14,614	23	0.75	0.31
	[42.20, 51.30]	12,675	33	1.25	0.52
	[51.30, 61.64]	9415	27	1.37	0.57
	[61.64, 73.23]	6747	34	2.41	1.00
	[>73.23]	3365	9	1.28	0.53
Timber type	Non-forest	1027	0	0.00	0.00
	Pine	143	0	0.00	0.00
	Nut pine	2319	3	0.62	0.24
	Larch	1389	2	0.69	0.27
	Pitch pine	431	0	0.00	0.00
	Sawtooth oak	11,482	18	0.75	0.30
	Mongolian oak	1183	2	0.81	0.32
	Oriental oak	565	3	2.54	1.00
	Other oak	25,845	63	1.17	0.46
	Poplar	1419	6	2.02	0.80
	False acasia	15,809	30	0.91	0.36
	Other broadleaf forest	3707	12	1.55	0.61
	Mixed forest of soft and hardwood	1686	1	0.28	0.11
Timber diameter (cm)	Non-forest	1027	0	0.00	0.00
	[6, 16]	2124	3	0.68	0.66
	[16, 28]	63,854	137	1.03	1.00

Table A1. *Cont.*

Factor	Class	No. of Pixels in Domains	No. of Landslide Pixels	Frequency Ratio	Normalized Frequency Ratio
Timber age (ages)	Non-forest	1027	0	0.00	0.00
	[21, 30]	1508	1	0.32	0.16
	[31, 40]	62,234	130	1.00	0.52
	[41, 50]	2236	9	1.93	1.00
Timber density (%)	Non-forest	1027	0	0.00	0.00
	[51, 70]	2994	5	0.80	0.78
	[>71]	62,984	135	1.03	1.00

References

1. Das, I.; Stein, A.; Kerle, N.; Dadhwal, V.K. Landslide susceptibility mapping along road corridors in the Indian Himalayas using Bayesian logistic regression models. *Geomorphology* **2012**, *179*, 116–125. [CrossRef]
2. Jakob, M.; Lambert, S. Climate change effects on landslides along the southwest coast of British Columbia. *Geomorphology* **2009**, *107*, 275–284. [CrossRef]
3. Vasu, N.N.; Lee, S.R. A hybrid feature selection algorithm integrating an extreme learning machine for landslide susceptibility modeling of Mt Woomyeon, South Korea. *Geomorphology* **2016**, *263*, 50–70. [CrossRef]
4. Park, S.; Choi, C.; Kim, B.; Kim, J. Landslide susceptibility mapping using frequency ratio, analytic hierarchy process, logistic regression, and artificial neural network methods at the Inje area, Korea. *Environ. Earth Sci.* **2013**, *68*, 1443–1464. [CrossRef]
5. Schuster, R.L. *Socioeconomic Significance of Landslides: Investigation and Mitigation*; National Academy Press Transportation Research Board Special Report; National Academy Press: Washington, DC, USA, 1996; Volume 247, pp. 12–35.
6. Guzzetti, F.; Reichenbach, P.; Ardizzone, F.; Cardinali, M.; Galli, M. Estimating the quality of landslide susceptibility models. *Geomorphology* **2006**, *81*, 166–184. [CrossRef]
7. Mohammady, M.; Pourghasemi, H.R.; Pradhan, B. Landslide susceptibility mapping at Golestan Province, Iran: A comparison between frequency ratio, Dempster–Shafer, and weights-of-evidence models. *J. Asian Earth Sci.* **2012**, *61*, 221–236. [CrossRef]
8. Akgun, A.; Dag, S.; Bulut, F. Landslide susceptibility mapping for a landslide-prone area (Findikli, NE of Turkey) by likelihood-frequency ratio and weighted linear combination models. *Environ. Geol.* **2008**, *54*, 1127–1143. [CrossRef]
9. Yilmaz, I. Landslide susceptibility mapping using frequency ratio, logistic regression, artificial neural networks and their comparison: A case study from Kat landslides (Tokat—Turkey). *Comput. Geosci.* **2009**, *35*, 1125–1138. [CrossRef]
10. Kanungo, D.P.; Sarkar, S.; Sharma, S. Combining neural network with fuzzy, certainty factor and likelihood ratio concepts for spatial prediction of landslides. *Nat. Hazards* **2011**, *59*, 1491–1512. [CrossRef]
11. Bui, D.T.; Lofman, O.; Revhaug, I.; Dick, O. Landslide susceptibility analysis in the Hoa Binh Province of Vietnam using statistical index and logistic regression. *Nat. Hazards* **2011**, *59*, 1413–1444. [CrossRef]
12. Zhang, G.; Cai, Y.; Zheng, Z.; Zhen, J.; Liu, Y.; Huang, K. Integration of the statistical index method and the analytic hierarchy process technique for the assessment of landslide susceptibility in Huizhou, China. *Catena* **2016**, *142*, 233–244. [CrossRef]
13. Ozdemir, A.; Altural, T. A comparative study of frequency ratio, weights of evidence and logistic regression methods for landslide susceptibility mapping: Sultan Mountains, SW Turkey. *J. Asian Earth Sci.* **2013**, *64*, 180–197. [CrossRef]
14. Xu, C.; Xu, X.; Lee, Y.H.; Tan, X.; Yu, G.; Dai, F. The 2010 Yushu earthquake triggered landslide hazard mapping using GIS and weight of evidence modeling. *Environ. Earth Sci.* **2012**, *66*, 1603–1616. [CrossRef]
15. Pourghasemi, H.R.; Moradi, H.R.; Aghda, S.F. Landslide susceptibility mapping by binary logistic regression, analytical hierarchy process, and statistical index models and assessment of their performances. *Nat. Hazards* **2013**, *69*, 749–779. [CrossRef]

16. Yalcin, A.; Reis, S.; Aydinoglu, A.C.; Yomralioglu, T. A GIS-based comparative study of frequency ratio, analytical hierarchy process, bivariate statistics and logistics regression methods for landslide susceptibility mapping in Trabzon, NE Turkey. *Catena* **2011**, *85*, 274–287. [CrossRef]
17. Chen, W.; Shirzadi, A.; Shahabi, H.; Ahmad, B.B.; Zhang, S.; Hong, H.; Zhang, N. A novel hybrid artificial intelligence approach based on the rotation forest ensemble and naïve Bayes tree classifiers for a landslide susceptibility assessment in Langao County, China. *Geomat. Nat. Hazards* **2017**, *8*, 1955–1977. [CrossRef]
18. Jordan, M.; Mitchell, T. Machine learning: Trend, perspectives, and prospects. *Science* **2015**, *349*, 255–260. [CrossRef] [PubMed]
19. Gomez, H.; Kavzoglu, T. Assessment of shallow landslide susceptibility using artificial neural networks in Jabonosa River Basin, Venezuela. *Eng. Geol.* **2005**, *78*, 11–27. [CrossRef]
20. Nefeslioglu, H.A.; Gokceoglu, C.; Sonmez, H. An assessment on the use of logistic regression and artificial neural networks with different sampling strategies for the preparation of landslide susceptibility maps. *Eng. Geol.* **2008**, *97*, 171–191. [CrossRef]
21. Shahabi, H.; Hashim, M.; Ahmad, B.B. Remote sensing and GIS-based landslide susceptibility mapping using frequency ratio, logistic regression, and fuzzy logic methods at the central Zab basin, Iran. *Environ. Earth Sci.* **2015**, *73*, 8647–8668. [CrossRef]
22. Zhu, A.X.; Wang, R.; Qiao, J.; Qin, C.Z.; Chen, Y.; Liu, J.; Zhu, T. An expert knowledge-based approach to landslide susceptibility mapping using GIS and fuzzy logic. *Geomorphology* **2014**, *214*, 128–138. [CrossRef]
23. Vahidnia, M.H.; Alesheikh, A.A.; Alimohammadi, A.; Hosseinali, F. A GIS-based neuro-fuzzy procedure for integrating knowledge and data in landslide susceptibility mapping. *Comput. Geosci.* **2010**, *36*, 1101–1114. [CrossRef]
24. Bui, D.T.; Ho, T.C.; Pradhan, B.; Pham, B.T.; Nhu, V.H.; Revhaug, I. GIS-based modeling of rainfall-induced landslides using data mining-based functional trees classifier with AdaBoost, Bagging, and MultiBoost ensemble frameworks. *Environ. Earth Sci.* **2016**, *75*, 1101.
25. Kavzoglu, T.; Sahin, E.K.; Colkesen, I. An assessment of multivariate and bivariate approaches in landslide susceptibility mapping: A case study of Duzkoy district. *Nat. Hazards* **2015**, *76*, 471–496. [CrossRef]
26. Pourghasemi, H.R.; Kerle, N. Random forests and evidential belief function-based landslide susceptibility assessment in Western Mazandaran Province, Iran. *Environ. Earth Sci.* **2016**, *75*, 1–17. [CrossRef]
27. Youssef, A.M.; Pourghasemi, H.R.; Pourtaghi, Z.S.; Al-Katheeri, M.M. Landslide susceptibility mapping using random forest, boosted regression tree, classification and regression tree, and general linear models and comparison of their performance at Wadi Tayyah Basin, Asir Region, Saudi Arabia. *Landslides* **2015**, *13*, 1–18.
28. Tsangaratos, P.; Ilia, I. Comparison of a logistic regression and Naïve Bayes classifier in landslide susceptibility assessments: The influence of models complexity and training dataset size. *Catena* **2016**, *145*, 164–179. [CrossRef]
29. Hong, H.; Liu, J.; Bui, D.T.; Pradhan, B.; Acharya, T.D.; Pham, B.T.; Ahmad, B.B. Landslide susceptibility mapping using J48 Decision Tree with AdaBoost, Bagging and Rotation Forest ensembles in the Guangchang area (China). *Catena* **2018**, *163*, 399–413. [CrossRef]
30. Pham, B.T.; Bui, D.T.; Dholakia, M.B.; Prakash, I.; Pham, H.V.; Mehmood, K.; Le, H.Q. A novel ensemble classifier of rotation forest and Naïve Bayer for landslide susceptibility assessment at the Luc Yen district, Yen Bai Province (Viet Nam) using GIS. *Geomat. Nat. Hazards Risk* **2017**, *8*, 649–671. [CrossRef]
31. Korean Geotechnical Society (KGS). *The Study on Investigation of Cause and Development of Restoration Policy about Landslide in Wumyon Area*; Korean Geotechnical Society: Seoul, Korea, 2011. (In Korean)
32. Cao, C.; Xu, P.; Wang, Y.; Chen, J.; Zheng, L.; Niu, C. Flash flood hazard susceptibility mapping using frequency ratio and statistical index methods in coalmine subsidence areas. *Sustainability* **2016**, *8*, 948. [CrossRef]
33. Kalantar, B.; Pradhan, B.; Naghibi, S.A.; Motevalli, A.; Mansor, S. Assessment of the effects of training data selection on the landslide susceptibility mapping: A comparison between support vector machine (SVM), logistic regression (LR) and artificial neural networks (ANN). *Geomat. Nat. Hazards Risk* **2018**, *9*, 49–69. [CrossRef]
34. Beven, K.J.; Kirkby, M.J. A physically based, variable contributing area model of basin hydrology/Un modèle à base physique de zone d'appel variable de l'hydrologie du bassin versant. *Hydrol. Sci. J.* **1979**, *24*, 43–69. [CrossRef]

35. Poudyal, C.P.; Chang, C.; Oh, H.J.; Lee, S. Landslide susceptibility maps comparing frequency ratio and artificial neural networks: A case study from the Nepal Himalaya. *Environ. Earth Sci.* **2010**, *61*, 1049–1064. [CrossRef]
36. Moore, I.D.; Grayson, R.B.; Ladson, A.R. Digital terrain modelling: A review of hydrological, geomorphological, and biological applications. *Hydrol. Process.* **1991**, *5*, 3–30. [CrossRef]
37. Moore, I.D.; Burch, G.J. Sediment transport capacity of sheet and rill flow: Application of unit stream power theory. *Water Resour. Res.* **1986**, *22*, 1350–1360. [CrossRef]
38. Althuwaynee, O.F.; Pradhan, B.; Park, H.J.; Lee, J.H. A novel ensemble bivariate statistical evidential belief function with knowledge-based analytical hierarchy process and multivariate statistical logistic regression for landslide susceptibility mapping. *Catena* **2014**, *114*, 21–36. [CrossRef]
39. Liu, H.; Motoda, H. Ensemble-based variable selection using independent probes. In *Computational Methods of Feature Selection*; Chapman and Hall/CRC: London, UK, 2007; pp. 147–162.
40. Quinlan, J.T. *C4.5: Programs for Machine Learning*; Morgan Kaufmann Publishers: San Francisco, CA, USA, 1993.
41. Oommen, T.; Cobin, P.F.; Gierke, J.S.; Sajinkumar, K.S. Significance of variable selection and scaling issues for probabilistic modeling of rainfall-induced landslide susceptibility. *Spat. Inf. Res.* **2018**, *26*, 21–31. [CrossRef]
42. Witten, I.G.; Frank, E. *Data Mining: Practical Machine Learning Tools and Techniques*, 2nd ed.; Morgan Kaufmann Publishers: San Francisco, CA, USA, 2005.
43. Breiman, L. Random forests. *Mach. Learn.* **2001**, *45*, 5–32. [CrossRef]
44. Taalab, K.; Cheng, T.; Zhang, Y. Mapping landslide susceptibility and types using Random Forest. *Big Earth Data* **2018**, *2*, 1–20. [CrossRef]
45. Rahmati, O.; Naghibi, S.A.; Shahabi, H.; Bui, D.T.; Pradhan, B.; Azareh, A.; Melesse, A.M. Groundwater spring potential modelling: Comprising the capability and robustness of three different modeling approaches. *J. Hydrol.* **2018**, *565*, 248–261. [CrossRef]
46. Micheletti, N.; Foresti, L.; Robert, S.; Leuenberger, M.; Pedrazzini, A.; Jaboyedoff, M.; Kanevski, M. Machine learning feature selection methods for landslide susceptibility mapping. *Math. Geosci.* **2014**, *46*, 33–57. [CrossRef]
47. Calle, M.L.; Urrea, V. Letter to the editor: Stability of random forest importance measures. *Brief Bioinform.* **2010**, *12*, 86–89. [CrossRef] [PubMed]
48. Schapire, R.E. The boosting approach to machine learning: An overview. In *Nonlinear Estimation and Classification*; Springer: New York, NY, USA, 2003; pp. 149–171.
49. Elith, J.; Leathwick, J.R.; Hastie, T. A working guide to boosted regression trees. *J. Anim. Ecol.* **2008**, *77*, 802–813. [CrossRef] [PubMed]
50. Aertsen, W.; Kint, V.; Van Orshoven, J.; Özkan, K.; Muys, B. Comparison and ranking of different modelling techniques for prediction of site index in Mediterranean mountain forests. *Ecol. Model.* **2010**, *221*, 1119–1130. [CrossRef]
51. Cao, D.S.; Xu, Q.S.; Liang, Y.Z.; Zhang, L.X.; Li, H.D. The boosting: A new idea of building models. *Chemom. Intell. Lab.* **2010**, *100*, 1–11. [CrossRef]
52. Naghibi, S.A.; Pourghasemi, H.R.; Abbaspour, K. A comparison between ten advanced and soft computing models for groundwater qanat potential assessment in Iran using R and GIS. *Theor. Appl. Climatol.* **2018**, *131*, 967–984. [CrossRef]
53. França, S.; Cabral, H.N. Predicting fish species richness in estuaries: Which modelling technique to use? *Environ. Model. Softw.* **2015**, *66*, 17–26. [CrossRef]
54. Al-Abadi, A.M. A novel geographical information system-based Ant Miner algorithm model for delineating groundwater flowing artesian well boundary: A case study from Iraqi southern and western deserts. *Environ. Earth Sci.* **2017**, *76*, 534. [CrossRef]
55. Frattini, P.; Crosta, G.; Carrara, A. Techniques for evaluating the performance of landslide susceptibility models. *Eng. Geol.* **2010**, *111*, 62–72. [CrossRef]
56. Pham, B.T.; Bui, D.T.; Dholakia, M.B.; Prakash, I.; Pham, H.V. A comparative study of least square support vector machines and multiclass alternating decision trees for spatial prediction of rainfall-induced landslides in a tropical cyclones area. *Geotech. Geol. Eng.* **2016**, *34*, 1807–1824. [CrossRef]

57. Bennett, N.D.; Croke, B.F.; Guariso, G.; Guillaume, J.H.; Hamilton, S.H.; Jakeman, A.J.; Pierce, S.A. Characterising performance of environmental models. *Environ. Model. Softw.* **2014**, *40*, 1–20. [CrossRef]
58. Zhou, J.; Li, X.; Mitri, H.S. Classification of rockburst in underground projects: Comparison of ten supervised learning methods. *J. Comput. Civ. Eng.* **2016**, *30*, 04016003. [CrossRef]

12

Landslide Prediction with Model Switching †

Darmawan Utomo, Shi-Feng Chen and Pao-Ann Hsiung *

Computer Science and Information Engineering, National Chung Cheng University, No. 168, Sec. 1, University Rd., Minhsiung, Chiayi 62102, Taiwan; du88@yahoo.com (D.U.); vincet0809@gmail.com (S.-F.C.)

* Correspondence: pahsiung@cs.ccu.edu.tw

† This paper is an extended version of paper published in the 2017 IEEE Conference on Dependable and Secure Computing, held in Taipei, Taiwan, 7–10 August 2017.

Abstract: Landslides could cause huge damages to properties and severe loss of lives. Landslides can be detected by analyzing the environmental data collected by wireless sensor networks (WSNs). However, environmental data are usually complex and undergo rapid changes. Thus, if landslides can be predicted, people can leave the hazardous areas earlier. A good prediction mechanism is, thus, critical. Currently, a widely-used method is Artificial Neural Networks (ANNs), which give accurate predictions and exhibit high learning ability. Through training, the ANN weight coefficients can be made precise enough such that the network works in analogy to a human brain. However, when there is an imbalanced distribution of data, an ANN will not be able to learn the pattern of the minority class; that is, the class having very few data samples. As a result, the predictions could be inaccurate. To overcome this shortcoming of ANNs, this work proposes a model switching strategy that can choose between different predictors, according to environmental states. In addition, ANN-based error models have also been designed to predict future errors from prediction models and to compensate for these errors in the prediction phase. As a result, our proposed method can improve prediction performance, and the landslide prediction system can give warnings, on average, 44.2 min prior to the occurrence of a landslide.

Keywords: landslide prediction; machine learning; neural networks; model switching

---

1. Introduction

Landslides are natural disasters which can cause huge damage to properties and severe loss of lives. Many studies have focused on how to detect and monitor landslides. Though landslide detections could be performed in real-time, there might not be enough time to react, so as to save human lives and properties. In order to minimize the losses caused by landslides, an early prediction mechanism, with pre-warnings, is necessary. Once the system can give an alarm in advance, people would have more response time to evacuate before the landslide occurs.

There are several problems in landslide prediction. First, just as in most safety-critical applications, landslide prediction also exhibits the same data imbalance problem, where the class of stable data has much more data than the class of unstable data. Stable data, here, refers to the normal conditions (where there is no landslide), while unstable data represents landslide-related information. Second, a low true-positive rate (TPR) problem is often found in safety-critical applications, because of the interference in learning between two or more classes in the datasets. For example, learning from the normal stable conditions often affects the learning from the unstable (landslide) conditions, thus resulting in a low TPR. Third, predictive applications are often faced with the problem of determining an appropriate prediction horizon; that is, the size of the time window of past history to be used for predictions. Finally, real-time applications face the problem of determining when to re-train the models.

To address the above-mentioned four problems existing in landslide prediction, this work provides a total solution in the form of an early warning system, called the Model Switch-based Landslide Prediction System (MoSLaPS). To address the data imbalance problem, we adapt the popular Adaptive Synthetic Sampling (ADASYN) method [1] to landslide prediction. To address the low TPR problem, we propose a novel event-class model switch predictor design that significantly improves TPR. To address the problem of customizing the prediction horizon, we also propose a novel dynamic tuning method for the prediction horizon, in order to achieve the goal of early warning. To address the problem of determining model re-training time, we propose a novel learning-based re-training method, based on an error model which considers both the long-term and short-term accumulated errors. Errors are also predicted, so re-training can be done earlier in preparation for future data changes; as a result of which, our proposed system can achieve the goal of early warning.

Section 2 introduces some related work. Section 3 presents the proposed model switching method. Section 5 presents and analyzes the experimental results. In Section 6, the conclusions and future work are described.

2. Related Work

Landslide prediction methods can be classified into three types: Image analysis, machine learning, and mathematical evaluation models. Table 1 shows a comparison among these types of methods. First, image analysis uses Geographic Information Systems (GIS), which can collect, store, manage, and analyze geographical data. By analyzing disaster data, such as history of landslides and data on land development for agriculture, the risk of landslides can be predicted. The probability of landslides is variable, as it is based on the number of layers of data used for analysis. Second, machine learning techniques, such as Bayesian networks [2], neural networks [3], or genetic algorithm [4], use computational intelligence to calculate the probability of landslides. These methods incorporate different factors that might cause landslides to evaluate the probability of landslide occurrence. They are not real-time, because they require huge computational times for prediction. Finally, mathematical evaluation models use a single evaluate equation, such as Factor of Safety (FS) [5]. A hazard model is combined with the physical concepts of mechanics and hydrographic data for the stability of slopes. It is easy for simulation and fits a wide range of environments, but it is difficult to obtain the whole hydrographic data as groundwater elevation is difficult to measure.

Table 1. Comparison of Landslide Detection/Prediction Methods.

Types	Methods	Advantages	Accuracies
Image Analysis	Geographical Information System [4]	Suitable for Large Area	Accuracy based on number of layers
Machine Learning	Bayesian Network [2]	Simple Network	75%
	Neural Network [3]	Simple Network	67%
	Genetic Algorithm [4]	Optimal Solution	90%
Mathematical Evaluation	SHALSTAB * [6]	High Accuracy	>90%

* Shallow Landsliding Stability Model.

Landslides occur when the down-slope shear stress is large. As shown in Equation (1) [5], the Factor of Safety (FS) refers to the stability of the soils. It takes physical properties, including rainfall, slope, and soil properties, into consideration. It can easily predict landslides with the trend of each parameter. Therefore, to predict landslides, a FS equation is matched with these attributes. Three regions of the FS value, based on the SHALSTAB model [6], are defined to distinguish between the dangerous levels of a slope, as shown in Table 2. The Stable Region is classified as the stable class and the Marginally Stable and Actively Unstable Regions are classified as the unstable class. Based on the classes, different training samples are used to train multiple neural network predictors.

$$FS = \frac{C + (1 - \frac{R}{T}\frac{\alpha}{sin\theta}\frac{\rho_w}{\rho_s})\rho_s g Z cos^2\theta tan\phi}{\rho_s g Z cos\theta sin\theta}, \quad (1)$$

with

- C: The effective coefficient (kPa);
- R: The rainfall intensity (mm/hr);
- T: The soil transmissivity (mm/hr);
- Z: The soil depth (m);
- ρ_w: The density of water (kg/m^2);
- ρ_s: The density of soil (kg/m^2);
- ϕ: The internal friction angle of the slope material (degree);
- θ: The slope gradient (degree); and
- α: The specific contributing area [5].

Table 2. Classification of Factor of Safety (FS) Levels.

Stable Region	**FS** $\geq$ 1.3
Marginally Stable Region	1.3 > **FS** $\geq$ 1
Actively Unstable Region	**FS**<1

Of particular mention is the work done by Lian et al. [7,8] on landslide displacement prediction using Prediction Intervals (PIs) and an ANN switched prediction method. The authors employed K-means clustering for dividing the landslide data into two classes; namely, a majority class (stationary points) and a minority class (mutational points). Then, a weighted Extreme Learning Machine (ELM) classifier was used to construct the switch rules. Finally, bootstrap and kernel-based ELMs were applied to construct the PIs. This work was concerned with how the displacements are predicted accurately and early. In contrast, our work is focused on how landslides can be actually predicted accurately and early. Not only is the goal different, the methods or techniques used or proposed are also quite different. We employ a very popular mathematical estimation model for landslide prediction, as defined above (namely, the SHALSTAB model and the factor of safety). We use ANN models for the model switching, as well as for the predictions. We adapt ADASYN for resolving the data imbalance problem. We also propose novel methods for model retraining and prediction horizon tuning. Details are given in the next section.

3. Model Switched Landslide Prediction System

A total solution for landslide prediction with early warnings is proposed in this work. The design of the proposed Model Switched Landslide Prediction System (MoSLaPS) is shown in Figure 1. It consists of two parts; namely, Physical Entities and Computation Elements. In the Physical Entities, environmental data, such as rainfall, soil moisture, and slope, are collected by sensor nodes. Coordinator nodes integrate the sensed data and transmit them to the Computation Elements through Zigbee transmitters. The Computation Elements consists of a SHALSTAB Model, a Switch-based Prediction Model, and an Accurate Early Warning System, as described in the following.

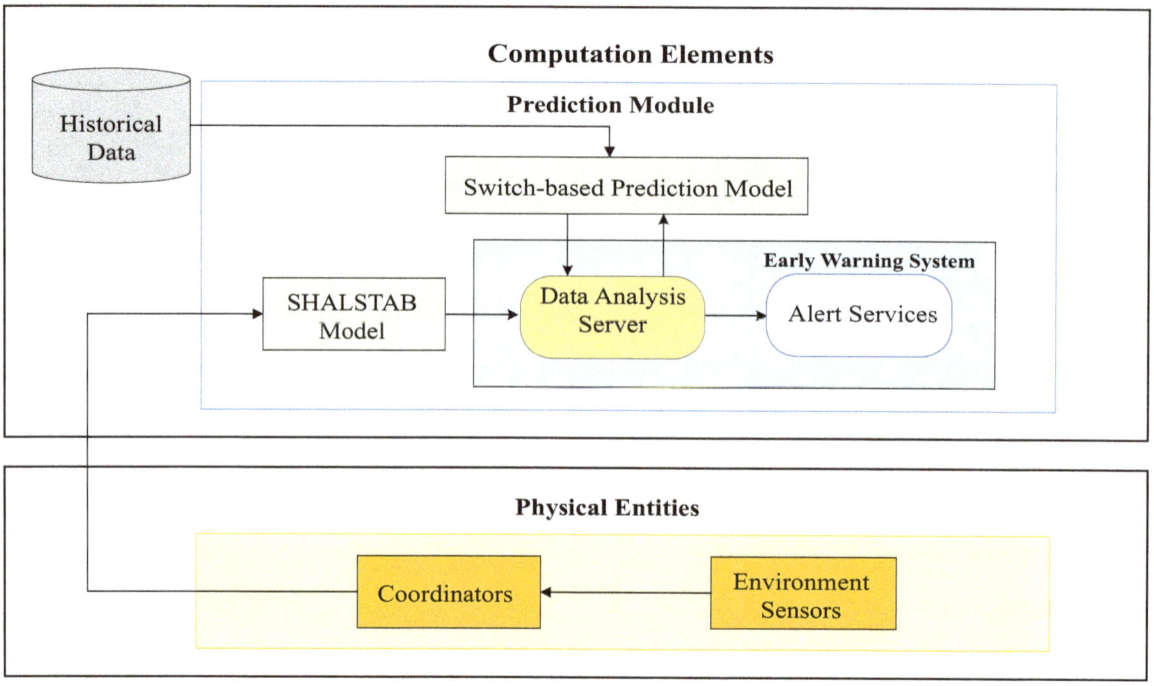

Figure 1. Model switched landslide prediction system architecture.

- SHALSTAB Model
  The SHALSTAB model takes the sensed environmental data, including rainfall, soil moisture, and slope, from the Physical Entities and evaluates the Factor of Safety, using Equation (1). Over time, the FS values are recorded as $FS_{actual} = \{FS_0, FS_1, \ldots, FS_t\}$, which is the input to the Accurate Early Warning System to predict the occurrence of landslides.

- Switch-based Prediction Model
  From historical environmental data, the proposed system consists of two prediction models to learn two different data patterns; namely, the stable pattern and the unstable pattern. To switch between the different prediction models, a neural network classifier is designed to predict the future class. The Switch-based Prediction Model can improve the prediction accuracy when the neural network classifier switches the prediction models accurately. The detailed technique is described in Section 3.3.

- Accurate Early Warning System
  The accurate early warning system consists of a data analysis server and alert services. The data analysis server uses the above-described switch-based prediction model to predict landslides. The input data, FS_{actual}, is used to predict future FS values, denoted as $FS_{predict} = \{FS_{t+1}, FS_{t+2}, \ldots, FS_{t+n}\}$. For each predicted FS value, there is a difference between the predicted FS and the actual FS calculated using Equation (1). This difference is called prediction error. The data analysis server will assess the applicability of the prediction model, according to the trend of prediction errors. If the error exceeds a pre-defined threshold, it means the prediction model is not suitable for the environment at that time and the predicted results have large prediction errors. Thus, based on the error measurement and a given error tolerance threshold, the prediction model is re-trained. The entire process will be described in Section 4. If a predicted FS value, FS_{t+k}, is smaller than 1, then it is estimated that a landslide will occur after k time slots [9]. Thus, alert services can send an alert in advance.

The details of the prediction models, model switching, and early warning system will be introduced in the rest of this section.

3.1. Prediction Model Design

To predict the future FS, a feed-forward Back-Propagation Neural Network (BPNN) is employed as the prediction model. A BPNN is a powerful computation system, created by generalizing the Widrow-Hoff learning rules [10] into a multi-layer with a non-linear differential transfer function network. The complex network connections imply that the high learning and reasoning ability of BPNN can be applied to deal with problems with high complexity. Figure 2 shows the framework of a BPNN-based prediction model.

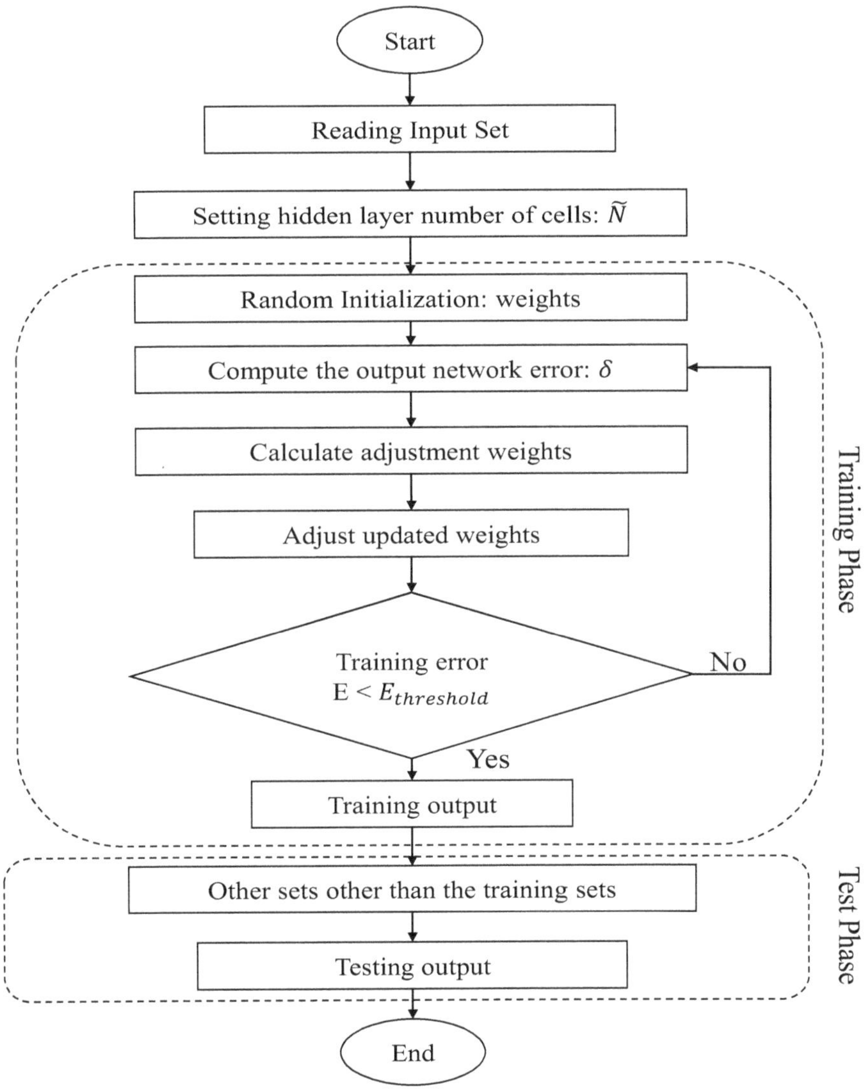

Figure 2. Back-Propagation Neural Network (BPNN) prediction model framework design.

Training Method for Time-Series Based BPNN

The basic computational procedure of a BPNN is explained to provide a basic description of the type of ANN that is implemented here. Figure 3 shows the basic structure of a time-series based BPNN. There are three types of layers: input, hidden, and output layers. In the input layer, time-series input for time slots $t - n$ to t, corresponding to a specific feature, such as factor of safety, rainfall, and soil moisture, is taken as input data. Each pair of nodes in the adjacent layers are linked by a weight. The values in all nodes in a previous layer and the weights are multiplied and accumulated as the input for a node of the next layer. The inputs are, then, given to an activation function to calculate the output value of the node. Repeating the above operations, layer by layer, from input layer to output layer, the final output can be derived.

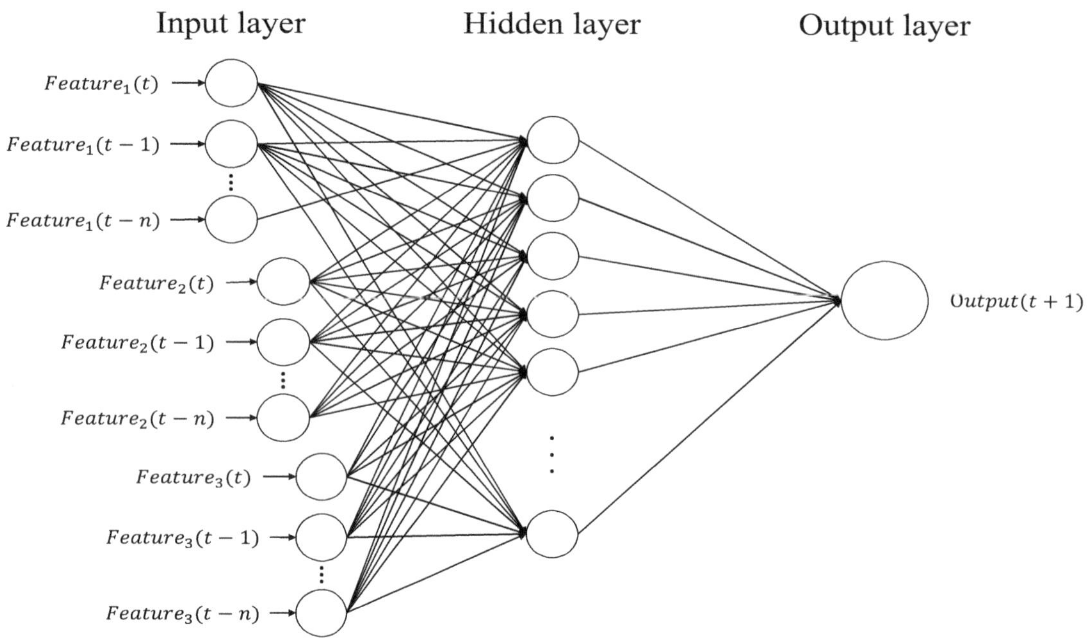

Figure 3. A fully connected feed-forward Back-Propagation Neural Network with time-series.

Algorithm 1 has a complete description as a prediction model; the following equations explain the functions that are used in this algorithm. At first, the previous FS is used to predict the future FS by a BPNN model. To dynamically determine suitable weights for different FS, a back-propagation method is applied to train and update the weights before prediction, in order to completely illustrate the details of back propagation method. Given the training sample, $T_{data} = (x_i, p_i), i = 1, ..., N$, where $x_i = [x_{i1}, x_{i2}, ...x_{in}] \in R^n$ is the impact factor and $p_i = [p_{i1}, p_{i2}, ...p_{ik}] \in R^k$ is the training target, the general mathematical model of a standard single hidden layer feed-forward network with $\tilde{N}$ hidden neurons is shown in Equation (2).

$$o_j = \sum_{i=1}^{\tilde{N}} g(w_i \cdot x_i), \qquad j = 1, ..., N, \tag{2}$$

where $o_j = [o_{j1}, o_{j2}, ...o_{jk}]$ is the jth output of the BPNN, w_i is the weight of the connection from the input neurons to the ith hidden neuron, and $g(x)$ is an activation function that represents how much adjustment the output should be from the neuron, based on the sum of the input. The activation function used in our BPNN model is depicted in Equation (3); namely, the sigmoid function. The sigmoid function, also called the logistic function, is a commonly-used activation function which has an output range from 0 to 1.

$$g(x) = \frac{1}{1 + e^{-x}}. \tag{3}$$

The difference between the prediction result and the actual result is called the prediction error. In order to reduce prediction error, the weights need to be updated. The Levenberg-Marquardt (LMA) [11] method has the fastest convergence and the lowest mean square error. Therefore, LMA is selected as a training function to calculate the output network error δ and adjustment weight W. It is depicted in Equation (4):

$$W_{k+1} = W_k - [J^T J + uI]^{-1} J^T \delta, \tag{4}$$

where W_k represents the weight matrix in the kth iteration, J is the Jacobian matrix [12] that contains the network error for weight and the first-order differential of weight, I is the unit matrix, δ is the network output error, and u is a constant. LMA can dynamically adjust the constant u to reduce the network output error δ.

Algorithm 1: Prediction Model Algorithm.

```
Input:
T_data: Data of FS values used for training BPNN;
P_input: Inputs of FS values for prediction;
E_threshold: Training error threshold;
Output:
P_output: Prediction output of FS values;
Variable:
W: Weights of BPNN;
E: Training error between training outputs and target;
δ: Error value for adjusting weights;
c_sat: 1: Training cycle is complete, 0: Training cycle is incomplete;

Set hidden layer number of cells N̂ ; // Equation (6)
Set the maximum iteration number for training Epochs;
Randomly initialize weight W;
// Training model
i = 0;
while (E > E_threshold)&&(i <= Epochs) do
    c_sat = 0;
    while c_sat = 0 do
        CalculateBPNN(T_data, W) ; // Equations (2) and (3)
        Calculate output network error δ ; // Equation (4)
        Calculate adjustment weight W ; // Equation (4)
        if all samples are trained then
            c_sat = 1;
    Calculate training error E ; // Equation (5)
    i++;
// Do prediction
P_output = CalculateBPNN(P_input, W);
return P_output;
```

After the training phase, the training error E is calculated by Equation (5), to determine whether the training step has reached convergence. If it is not less than the training error threshold, $E_{threshold}$, the training phase is restarted.

$$E = \frac{1}{N}\sum_{i=1}^{N}(T_i - O_i)^2, \tag{5}$$

where:

- T_i: Target FS value of training sample i;
- O_i: BPNN output FS value of training sample i; and
- N: Number of training samples.

Considering the training time in our proposed re-training process, the number of outputs, Num_{output}, is set to 1 to avoid a long training time. This means that only one prediction result per iteration will be obtained, by taking the past FS value as input.

The number of neurons in the hidden layer, denoted as Num_{neuron}, is determined based on the following experience rule [13]:

- The number of neurons in the first hidden layer is calculated using (6):

$$Num_{neuron} = \sqrt{Num_{input} * Num_{output}}. \quad (6)$$

3.2. Imbalanced-Class Prediction Design

Class balance enhancements are needed to handle training samples with an unbalanced class distribution [14]. The Adaptive Synthetic Sampling (ADASYN) method [1] is used here for balancing the imbalanced data (i.e., data is pre-processed using ADASYN), and then the processed dataset are used to train the event-class predictor, which is also a BPNN model. In the following, the ADASYN algorithm is described as follows.

Data Pre-Processing Using ADASYN Algorithm

The ADASYN algorithm can improve the data imbalance problem by synthetically creating new samples from the unstable class by linear interpolation between existing unstable class samples. This approach, by itself, is known as the Synthetic Minority Over-sampling Technique (SMOTE) method [15]. ADASYN is an extension of SMOTE, creating more samples in the vicinity of the boundary between the two classes than in the interior of the unstable class.

To create more synthetic data for the unstable class, FS, rainfall, and soil moisture are used as the training data and the corresponding class label, y_i, is constructed according the classification region. Given the training samples $(X_i, y_i), i = 1, \ldots, N$, where $X_i =< x_{i,fs}, x_{i,r}, x_{i,sm}, x_{i,slope} >\in R^n$, $x_{i,fs}$ is the FS value, $x_{i,r}$ is the rainfall, $x_{i,sm}$ is the soil moisture, $x_{i,slope}$ is the slope gradient, and y_i is the class label. The training samples, X_i, are classified by Equation (7). For $x_{i,fs} \geq 1.3$, x_i classifies as stable class. On the other hand, if $x_{i,fs} < 1.3$, then x_i classifies as unstable class. After X_i is classified, the class label, y_i, is set by Equation (8). If $X_i \in StableClass$, y_i is set as 0. If $X_i \in UnstableClass$, y_i is set as 1.

$$\begin{cases} \text{if } x_{i,fs} \geq 1.3, X_i \in Stable\ Class \\ \text{if } x_{i,fs} < 1.3, X_i \in Unstable\ Class \end{cases}, \quad (7)$$

$$y_i = \begin{cases} 0 & \text{if } X_i \in Stable\ Class \\ 1 & \text{if } X_i \in Unstable\ Class \end{cases}. \quad (8)$$

To adjust class balance, the degree of class imbalance is needed to be calculated by Equation (9):

$$d = N_m / N_s, \quad (9)$$

where

- N_m: The size of unstable class examples; and
- N_s: The size of stable class examples.

Furthermore, the default level, $d_{default}$, which is the threshold for the level of maximum class imbalance tolerated, also needs to be determined beforehand. If the current d is smaller than the threshold degree, $d_{default}$, then the number of synthetic data samples that need to be generated for the unstable class is calculated using Equation (10):

$$R = (N_s - N_m) \times \beta, \quad (10)$$

where $\beta \in [0,1]$ is a parameter used to specify the desired balance level after generation of the synthetic data: $\beta = 1$ means a fully balanced data set is created after the generalization process.

For each $X_i \in Unstable\ Class$, K nearest neighbours can be found by using the Euclidean distance. The ratio r_i, defined in Equation (11), which represents the number of stable-classified in the K nearest neighbours, and its normal form, $\hat{r}_i$, defined in Equation (12), are calculated, where $\hat{r}_i$ is called a density distribution of r_i, ($\sum_i \hat{r}_i = 1$).

$$r_i = h_i/K, \qquad i = 1, ..., N_m, \tag{11}$$

$$\hat{r}_i = r_i / \sum_{i=1}^{N_m} r_i, \tag{12}$$

where h_i is the number of samples in the K nearest neighbours of x_i that belong to the stable class; therefore, $r_i \in [0, 1]$.

Thus, the number of synthetic data samples that need to be generated for each unstable class sample, $X_{unstable}$, can be calculated by Equation (13).

$$g_i = \hat{r}_i \times R, \tag{13}$$

where R is the total number of synthetic data samples that need to be generated for the unstable class, as defined in Equation (10).

By random, the program chooses one unstable data sample, X_{zi}, from the K nearest neighbours for $X_{unstable}$ to generate new synthetic samples, sd_{new}, by Equation (14). This procedure is repeated g_i times to produce new synthetic samples.

$$sd_{new} = X_{unstable} + (X_{zi} - X_{unstable}) \times \lambda, \tag{14}$$

where

- $(X_{zi} - X_{unstable})$: The difference vector; and
- λ: A random number $\lambda \in [0, 1]$.

3.3. Switch-Based Prediction Model Design

To address the issue of imbalanced data between the unstable and stable classes, a switch-based neural network prediction algorithm is proposed, as detailed in Algorithm 2.

The environmental factors, including rainfall, soil moisture, and slope gradient, are used to calculate the FS values, $x_{i,fs}$, using the SHALSTAB model. Given the environmental samples $D_{landslide} = \{x_i\}$, where $x_i = < x_{i,r}, x_{i,sm}, x_{i,slope} >$, $x_{i,r}$ is the rainfall, $x_{i,sm}$ is the soil moisture, and $x_{i,slope}$ is the slope gradient, the FS values, $x_{i,fs}$, are calculated given the set of all training samples, T_{sample}, where $T_{sample} = \{X_i | X_i = < x_{i,fs}, x_{i,r}, x_{i,sm}, x_{i,slope} >\}$. Then, the corresponding class labels y_i, $T_{class} = \{X_i, y_i\}$, can be constructed and classified by Equations (7) and (8). To construct the BPNN models for different data patterns, the calculated FS need to be classified in two subsets, as follows:

$$T_{class1} = \{x_{i,fs} | x_{i,fs} \in X_i, x_{i,fs} \in \mathit{Stable\ Class}\}, \tag{15}$$

$$T_{class2} = \{x_{i,fs} | x_{i,fs} \in X_i, x_{i,fs} \in \mathit{Unstable\ Class}\}, \tag{16}$$

where T_{class1} is the set of FS values for $x_{i,fs}$, in $X_i \in \mathit{StableClass}$, and T_{class2} is the set of FS values for $x_{i,fs}$, in $X_i \in \mathit{UnstableClass}$; so that $T_{class1} \bigcap T_{class2} = \phi$ and $T_{class1} \bigcup T_{class2} = x_{i,fs}$.

Algorithm 2: Switch-based Neural Networks Prediction Algorithm.

Input:
$D_{landslide}$: $\{x_i | x_i =< x_{i,r}, x_{i,sm}, x_{i,slope} >\}$;
P_{test}: $\{x_t | x_t =< x_{t,fs}, x_{t,r}, x_{t,sm}, x_{t,slope}, y_t >\}$ for prediction;
Output:
P_{output}: $\{FS_{predict}\}$;
Variable:
T_{sample}: $\{X_i | X_i =< x_{i,fs}, x_{i,r}, x_{i,sm}, x_{i,slope} >\}$;
T_{class}: $\{< y_i, x_{i,fs}, x_{i,r}, x_{i,sm}, x_{i,slope} >\}$;
T_{class1}: $\{x_{i,fs} | y_i = 0\}$;
T_{class2}: $\{x_{i,fs} | y_i = 1\}$;
T_{ADASYN}: $\{< x_{new,fs}, x_{new,r}, x_{new,sm}, x_{new,slope}, y_i = 1 >\}$;
P_{class}: $\{y_{predict}\}$;

`// SHALSTAB model`
$T_{sample} = calculateSHALSTAB(D_{landslide})$; `// Equation (1)`
`// Classification`
$[T_{class}, T_{class1}, T_{class2}] = calculateClass(T_{sample})$; `// Equations (7) and (8)`
`// Data pre-processing`
$T_{ADASYN} = calculateADASYN(T_{class})$; `// Equation (14)`
`// Construct each prediction model`
$Feed - Forward\ BPNN_{Stable} \leftarrow PredictionModel(T_{class1})$;
$Feed - Forward\ BPNN_{Unstable} \leftarrow PredictionModel(T_{class2})$;
$Event - class\ predictor \leftarrow PredictionModel(T_{class} + T_{ADASYN})$;
`// Model switch`
$P_{class} = Event - class\ predictor(P_{test})$;
if $P_{class} == Stable$ **then**
  `// Prediction class is stable`
  $P_{output} = Feed - Forward\ BPNN_{Stable}(P_{test})$
else
  `// Prediction class is unstable`
  $P_{output} = Feed - Forward\ BPNN_{Unstable}(P_{test})$
return P_{output};

Here, the ADASYN algorithm is used to produce new synthetic samples for the unstable class, in order to balance the sizes of the two classes. The processed dataset, T_{ADASYN}, is used to predict the future class using a BPNN model. The event-class predictor can switch between the different models, according to the predicted class. As shown in Figure 4, the steps of the event-class predictor are as follows.

First, T_{class} is selected to construct the synthetic dataset $T_{ADASYN} = \{< sd_{new}, y_{new} >\}$ for balancing class distribution by the ADASYN algorithm, where sd_{new} are the new synthetic samples, $sd_{new} =< x_{new,fs}, x_{new,r}, x_{new,sm}, x_{new,slope} >$, and $y_{new} = 1$ represents that the synthetic class label is unstable class. The synthetic sd_{new} include $x_{new,fs}$, the new synthetic FS value; $x_{new,r}$, the new synthetic rainfall; $x_{new,sm}$, the new synthetic soil moisture; and $x_{new,slope}$, the new synthetic slope gradient. Both of the classes T_{class} and T_{ADASYN} are integrated into the training set of the event-class predictor. Second, the event-class predictor is constructed using the BPNN model. Finally, the event-class predictor is used to predict the future class label, $P_{class} \in \{0, 1\}$, for the testing phase, where $P_{class} = 0$ represents that the prediction class label is stable and $P_{class} = 1$ represents that the prediction class label is unstable.

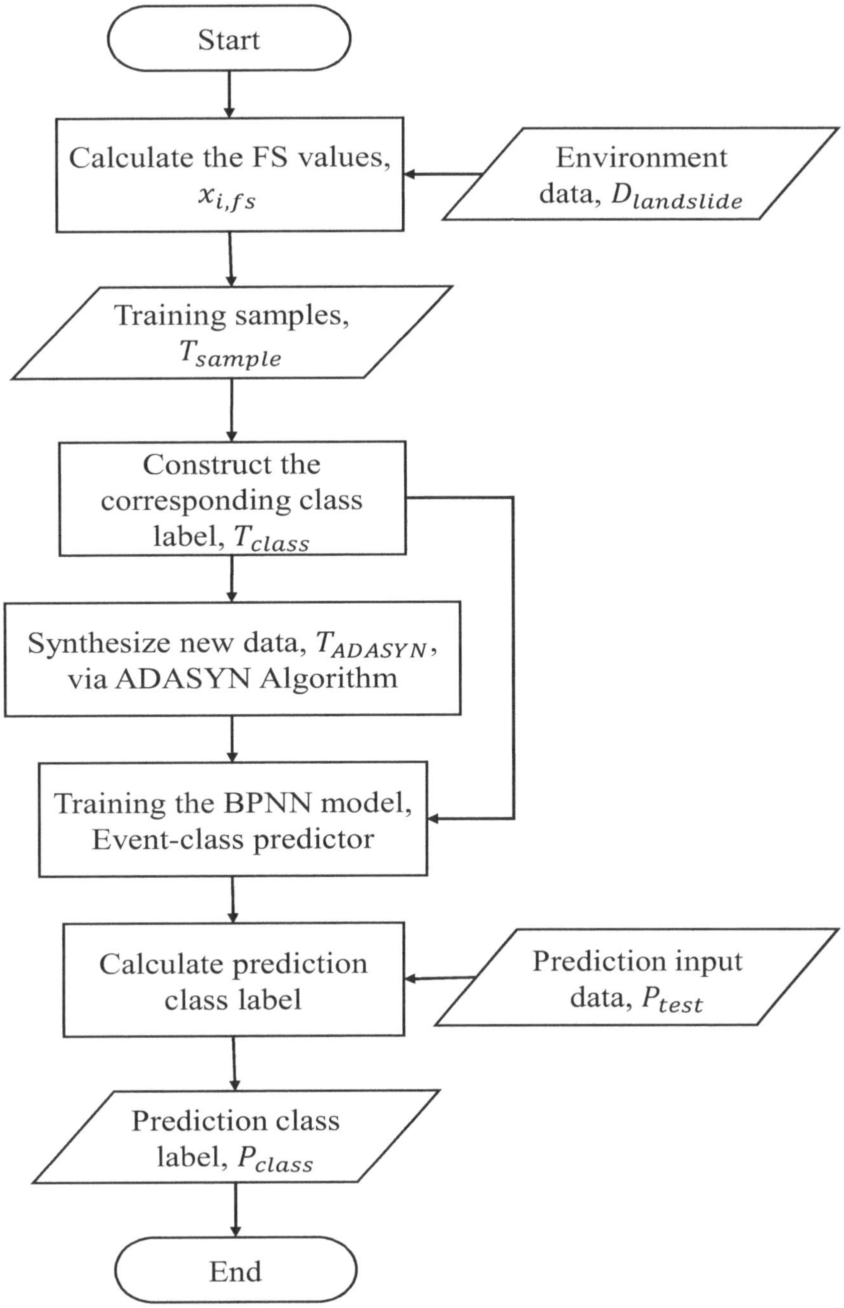

Figure 4. Flow of constructing the event-class predictor.

The aim of our proposed method is to construct different pattern predictors, as shown in Figure 5. The steps of different pattern predictors are as follows. First, T_{class1} is used as training data to train a BPNN model. After training, the BPNN model is the stable pattern of $x_{i,fs}$ (i.e., $Feed - Forward\ BPNN_{Stable}$). On the other side, T_{class2} is applied as training data to train another BPNN model. After training, the BPNN model is the unstable pattern of $x_{i,fs}$ (i.e., $Feed - Forward\ BPNN_{Unstable}$). Thus, two BPNNs are built to deal with different patterns. As shown in Figure 6, this procedure can switch different pattern predictors, according to the predicted class, P_{class}, that is obtained by the event-class predictor. When $P_{class} = 0$, the testing data, $P_{test} = \{x_t\}$, is applied to predict the future FS using $Feed - Forward\ BPNN_{Stable}$, where $x_t =< x_{t,fs}, x_{t,r}, x_{t,sm}, x_{t,slope}, y_t >$, $x_{t,fs}$ is the testing data of the FS value, $x_{t,r}$ is the testing data of the rainfall, $x_{t,sm}$ is the testing data of the soil moisture, $x_{t,slope}$ is the testing data of the slope gradient, and y_t is the testing data of the corresponding class label. On the other side, when $P_{class} = 1$, P_{test} is employed to predict the future FS using $Feed - Forward\ BPNN_{Unstable}$. Finally, the predicted FS, $P_{output} = \{FS_{predict}\}$, can be obtained by using the proposed Switch-based Prediction Model.

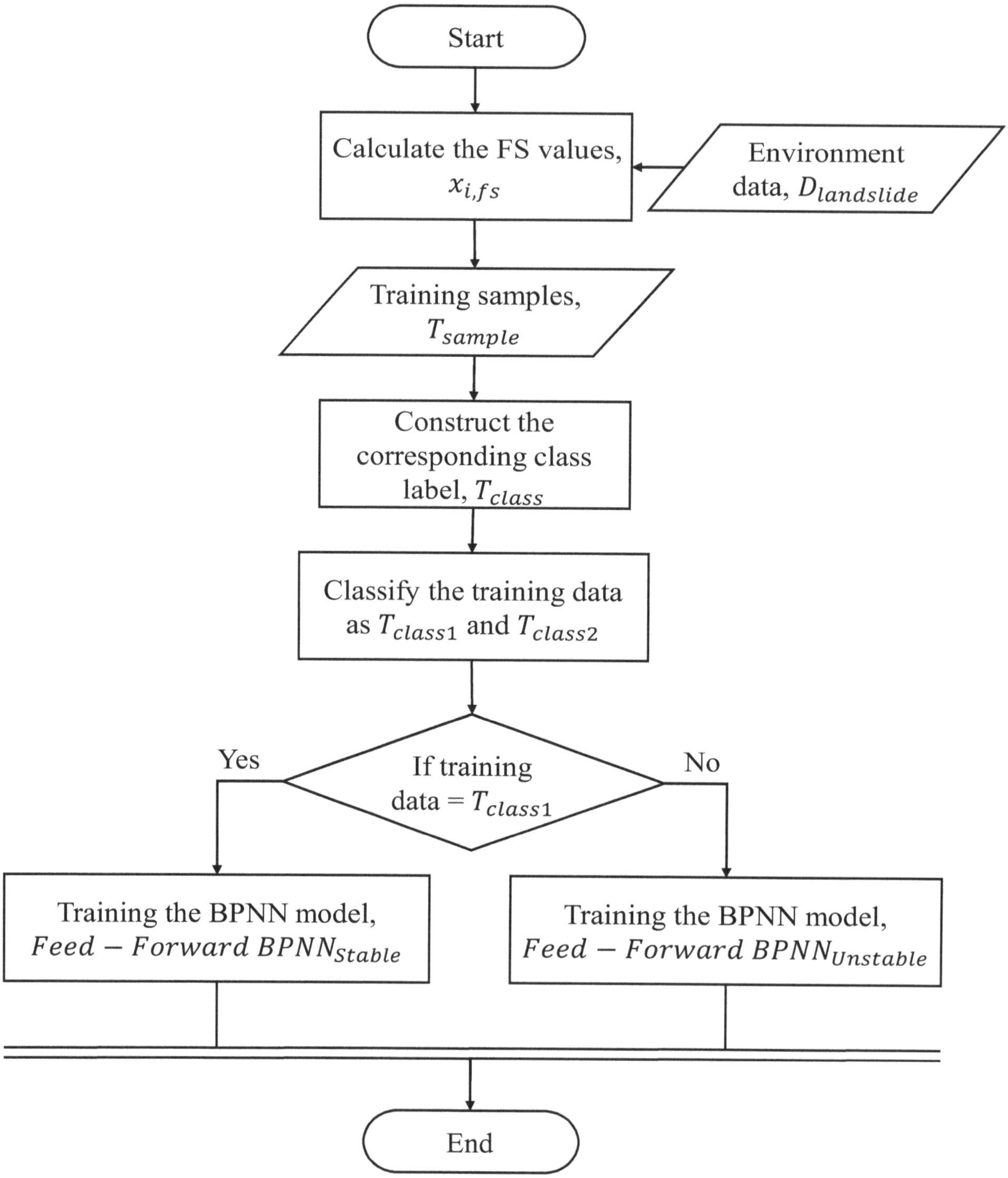

Figure 5. Flow of constructing different pattern predictors.

The main contribution of this work is that the proposed method can make highly accurate predictions, even in the case of highly imbalanced data. Two techniques were employed, ADASYN and an event-class predictor.

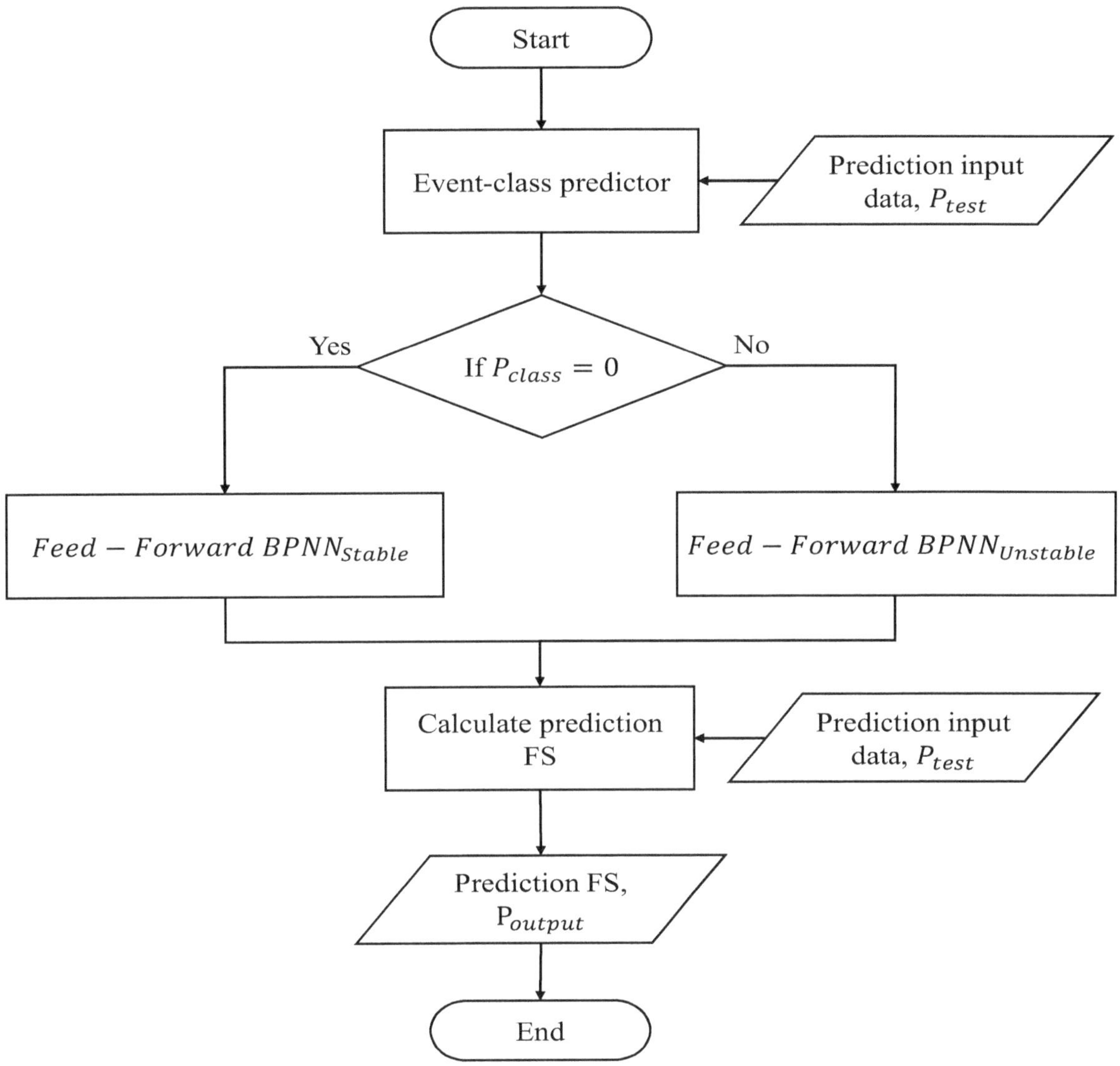

Figure 6. Flow of the switching strategy.

4. Accurate Early Warning System Design

To ensure the switch-based neural networks prediction model can be precise in a changing environment, an accurate early warning system is designed, as shown in Figure 7. It is divided into two parts: A learning-based re-training flow, as described in Section 4.1, and a Prediction Horizon tuning flow, as shown in Section 4.2.

4.1. Learning-Based Re-Training Flow

Figure 8 shows the flow of learning-based re-training. The determination of ret-raining is based on the error estimation. The error estimation process calculates the average error of all $(FS_{actual}, FS_{predict})$ pairs in an error-estimation window (EEW). For each period of the EEW, this procedure compares the average error, $AVGE$, of two error estimation windows, EEW_{now} and EEW_{prev}, and the accumulated error, $ACCE$, to check the two conditions for re-training. Equations (17) and (18) give the two conditions under which the prediction model needs to be re-trained, where C_{IE} is the coefficient of interval error used to specify the short-term tolerable error range, which is equal to the size of the prediction horizon in our work. If the difference of $AVGE$ between two continuous EEW is too large, the prediction model will be re-trained, due to the high variability of the input pattern that the original prediction model could not predict. Here, C_{AE} is the coefficient of accumulated error used to specify the long-term tolerable error range (here, long-term means since epoch). If the $ACCE$, compared to average error of the prediction model, $AVGE_{Model}$, is too large, the prediction model will also be re-trained, as the prediction results are becoming inaccurate. This further implies that the environment is changing with time, and that adaptation is needed. $AVGE_{now}$ and $ACCE_{now}$ are calculated using Equations (19) and (20), respectively.

$$AVGE_{now} > C_{IE} \times AVGE_{prev}, \tag{17}$$

$$ACCE_{now} > C_{AE} \times AVGE_{Model}, \tag{18}$$

$$AVGE_{now} = \frac{\sum |FS_{actual} - FS_{predict}|}{SIZE_{EEW}}, \tag{19}$$

$$\begin{aligned} ACCE_{now} &= ACCE_{prev} + \Delta AVGE \\ &= ACCE_{prev} + (AVGE_{now} - AVGE_{prev}) \end{aligned}. \tag{20}$$

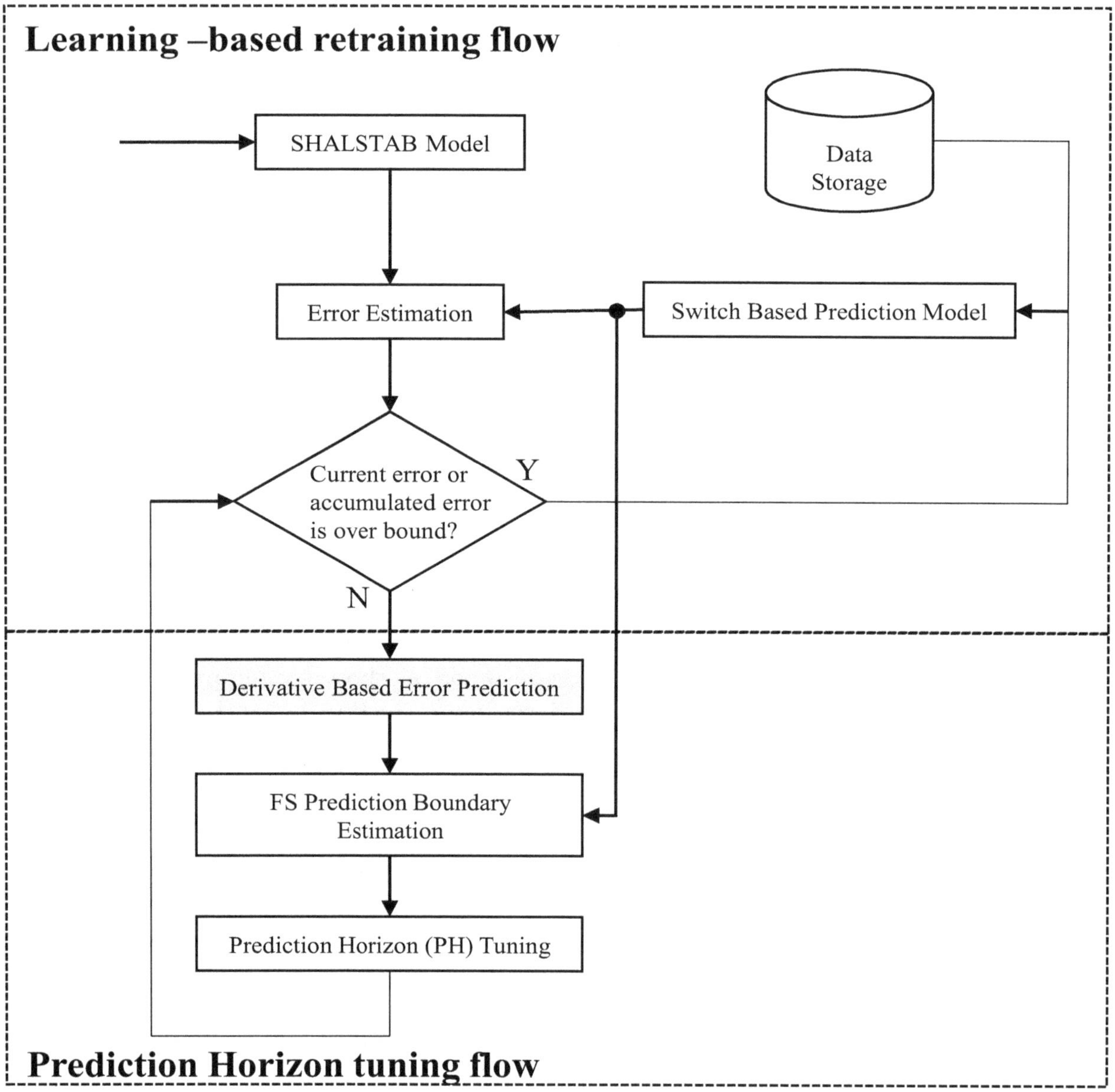

Figure 7. Overall Flowchart of Prediction Model Analysis.

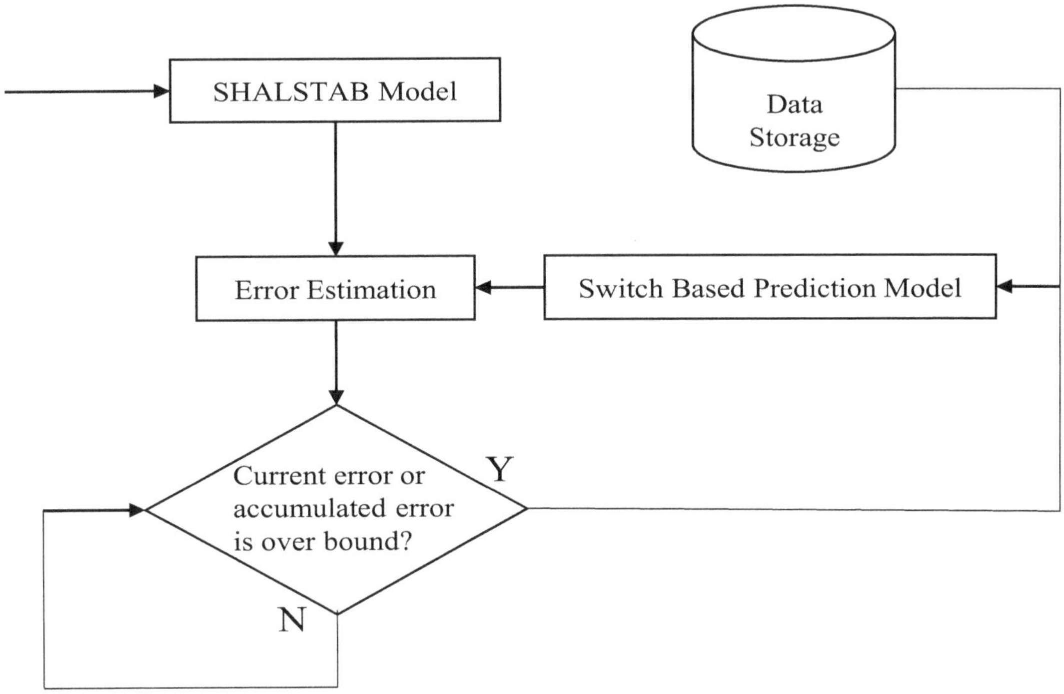

Figure 8. Learning-based re-training flow.

4.2. Prediction Horizon Tuning

In the re-training flow, the error estimation results are utilized further to tune the prediction horizon. The advantages of a variable-length prediction horizon (PH) are as follows:

- The occurrence of landslides can be predicted earlier; and
- The number of false predictions can be reduced.

The prediction horizon tuning flow is shown in Figure 9. Prediction errors are non-linear, and the prediction model is applied (as described in Section 3.1) in order to learn the inherent pattern for predicting future errors, $ERR_{predcit}$. In this way, the size of the PH is set to predict the occurrence of landslides earlier. Then, the future target ranges, P_{future}, can be determined by Equation (21). If the range of $P_{future} < 1$, this system can send alerts in advance.

$$FS_{predict} - |ERR_{predict}| \leq P_{future} \leq FS_{predict} + |ERR_{predict}|. \tag{21}$$

To decide whether the size of the PH is tuned, error boundary estimation is needed. As the program already has the predicted error, $ERR_{predict}$, and the predicted FS, $FS_{predict}$, then the predicted lower bound, $Bound_{low}$, can be estimated by Equation (22). According to the estimated results, the tuning is performed based on the following rules:

- If there is no $Bound_{low}$ lower than the lower bound of Stable Class (i.e., 1.3), for every time point in the prediction horizon, the size of the prediction horizon is increased by 1; and
- If there exists one $Bound_{low}$ lower than the lower bound of Stable Class (i.e., 1.3), for every time point in the prediction horizon, the size of prediction horizon is reset to the default value.

$$Bound_{low} = FS_{predcit} - |ERR_{predict}|. \tag{22}$$

Based on the above rules, we make our method a little more flexible for the landslide prediction scenario with variable-length prediction horizon.

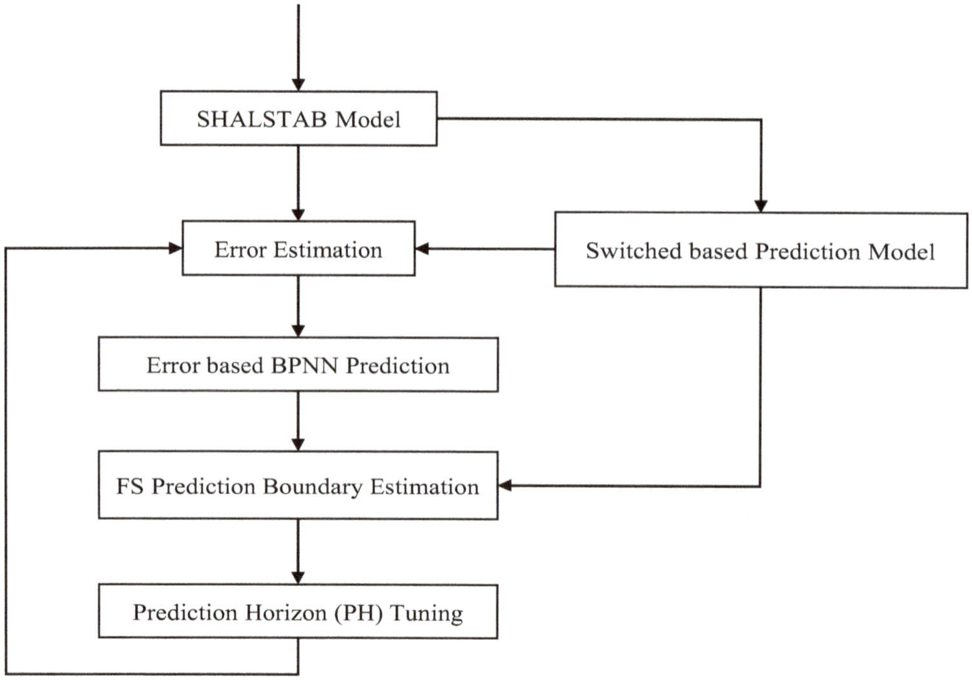

Figure 9. Prediction horizon tuning.

5. Experiments

In this section, evaluations of the proposed method for landslide prediction are presented. First, the experimental datasets used for the experiments are introduced. Then, the experimental results are illustrated. All experiments were carried out using the MATLAB® programming language on a PC with an Intel® Core™ i7-3770 CPU @ 3.40GHz and 16 GB RAM, running the Windows® 10 (64-bit) OS.

5.1. Experimental Datasets

Historical environmental monitoring datasets from the Shen-Mu station [16] were selected as a case study. Landslides are mainly influenced by rainfall and soil moisture. Using the FS, the occurrence of a landslide is estimated. The monitoring dataset of the Shen-Mu station is shown in Figures 10 and 11. Figures 10 and 11 show the relationships between rainfall and FS, and between soil moisture and FS, respectively. When rainfall and/or soil moisture increase, the FS decreases and the slope becomes unstable.

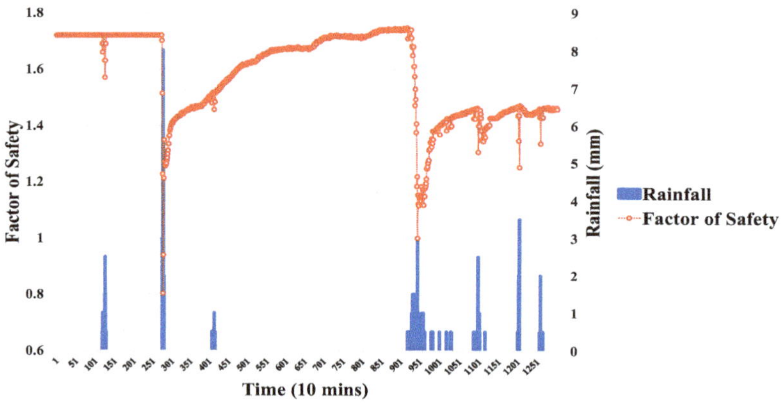

Figure 10. Monitoring curves of Factor of Safety and rainfall at Shen-Mu station.

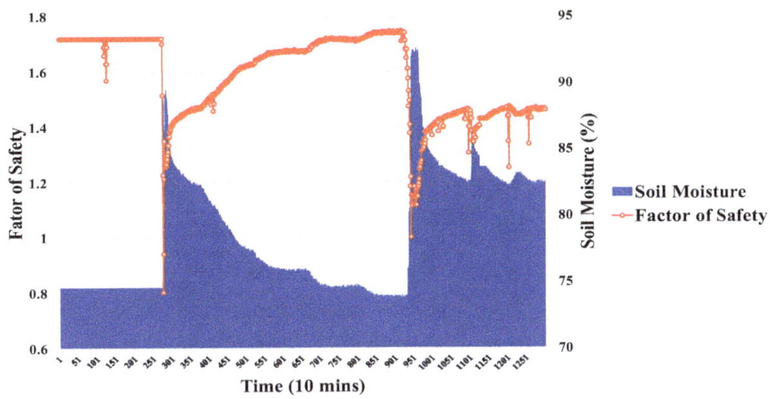

Figure 11. Monitoring curves of Factor of Safety and soil moisture at Shen-Mu station.

In the Shen-Mu datasets [16], data was recorded once per 10 min, and was collected in 2016. The program randomly selected 10 sets of samples, where each set had 1300 samples. Each dataset was further divided into two parts, the training set (75%) and the test set (25%). Note that the original data were used as input; that is, they were not normalized, because we needed to calculate the FS according to the SHALTAB model, as given in Equation (1).

Firstly, an evaluation of imbalanced-class prediction is described. Then, it is shown how the prediction accuracy is increased due to the proposed MoSLaPS.

5.2. Evaluation of Event-Class Prediction

ADASYN [1] was applied for data pre-processing in this program. It synthetically created new samples from the unstable class to balance the distribution of the data, if required. In the ADASYN algorithm, the desired level of balance, β, could be adjusted to control the number of new synthetic samples, which were needed when $0 \leq \beta \leq 1$.

Our event-class predictor used ADASYN [1] for imbalanced data processing. After data pre-processing, the processed dataset was used to train the event-class predictor (i.e., a BPNN model). To evaluate the event-class predictor, several performance indicators were applied and defined, as follows:

- True Positive Rate (TPR) is defined as in Equation (23);
- False Positive Rate (FPR) is defined as in Equation (24); and
- Accuracy (ACC) is defined as in Equation (25).

If the prediction class was unstable and the actual class was also unstable, then the result was said to be a True Positive (TP). If the prediction class was stable and the actual class Was also stable, then the result was said to be a True Negative (TN). If the prediction class was unstable and the actual class was stable, then the result was said to be a False Positive (FP). If the prediction class was stable and the actual class was unstable, then the result was said to be a False Negative (FN). Table 3 shows the classification of the above four different categories.

$$TPR = \frac{TP}{TP + FN} \times 100\%, \tag{23}$$

$$FPR = \frac{FP}{FP + TN} \times 100\%, \tag{24}$$

$$ACC = \frac{TP + TN}{TP + FN + FP + TN} \times 100\%. \tag{25}$$

Table 3. Confusion matrix.

		Actual Class	
		Unstable	**Stable**
Prediction Class	**Unstable**	True Positive	False Positive
	Stable	False Negative	True Negative

In our experiment, Figure 12 shows the evaluation results of the ADASYN algorithm for different β levels. When β was greater than 0.45, the TPR was more than 0.98. Therefore, the balance coefficient $\beta = 0.45$ was selected. Table 4 shows the number of new synthetic samples. Majority represents the size of the stable class and Minority represents the size of the unstable class.

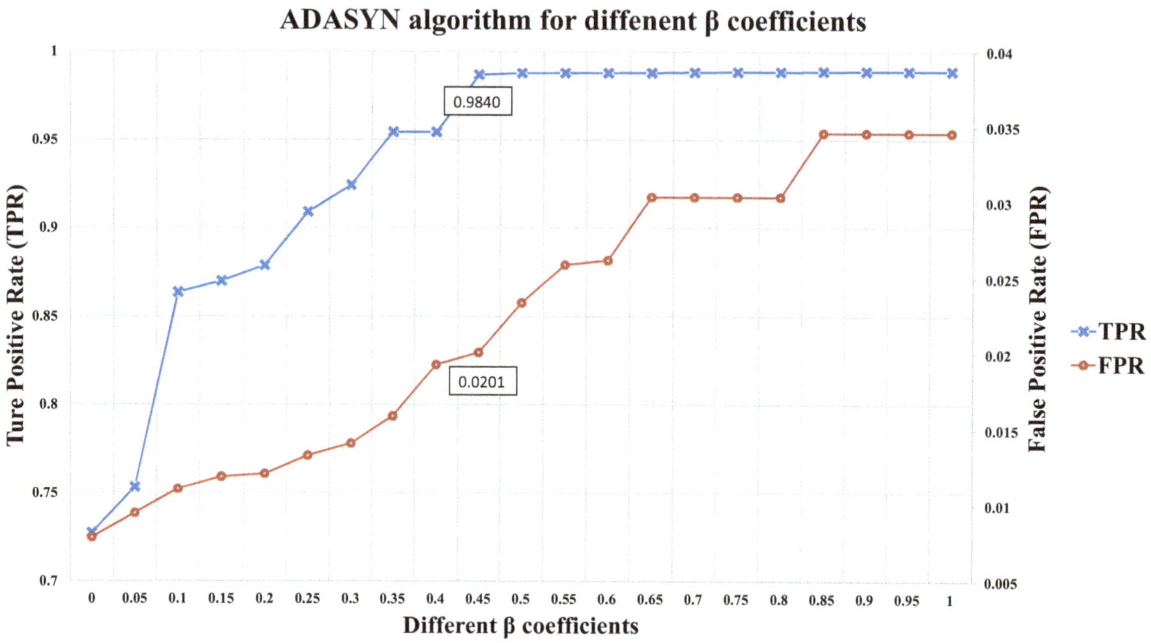

Figure 12. The evaluation result of the ADASYN algorithm for different values of the coefficient β.

Table 4. Number of synthetic new samples generated by the ADASYN algorithm ($\beta = 0.45$).

All Dataset	Majority	Minority	ADASYN
1300	1280	20	449
1300	1199	101	389
1300	1222	78	424
1300	1181	119	374
1300	1199	101	420
1300	1291	9	464
1300	1227	73	414
1300	1224	76	408
1300	1056	244	299
1300	1140	160	341

The evaluation results of the event-class predictor are shown in Table 5. There were 10 sets of testing samples. The average ACC was 97.94%, the average TPR was 98.40%, and the average FPR was 2.01%. Due to the high accuracy of event-class predictor, it can be used to choose a decision to switch between the models of different data patterns.

Table 5. The evaluation results of the event-class predictor.

No.	ACC	TPR	FPR	TP	FP	TN	FN
1	96.43%	100.00%	3.60%	2	9	241	0
2	97.62%	94.44%	2.14%	17	5	229	1
3	98.41%	95.45%	1.30%	21	3	227	1
4	98.41%	100.00%	1.64%	8	4	240	0
5	96.03%	100.00%	3.98%	1	10	241	0
6	100.00%	100.00%	0.00%	16	0	236	0
7	96.03%	100.00%	4.03%	4	10	238	0
8	99.21%	100.00%	0.83%	12	2	238	0
9	98.41%	100.00%	1.68%	14	4	234	0
10	98.81%	94.12%	0.85%	16	2	233	1
Average	97.94%	98.40%	2.01%				

Table 6 shows comparisons of the event-class predictor with other common classifiers, such as BPNN, Support Vector Machine (SVM), and Adaboost. The ACC of all classifiers were greater than 90%. A good classifier needs to have high TPR and low FPR. Although the FPR of our classifier, compared with BPNN and Adaboost, was a little higher, the TPR of our proposed classifier was much higher than that of the others. This means that our classifier exhibited a higher ratio of correct classification.

Table 6. Comparison of event-class predictor with other common classifiers.

Method	ACC	TPR	FPR
Event-Class	97.94%	98.40%	2.01%
BPNN	97.16%	57%	0.42%
SVM	90.42%	78.06%	9.02%
Adaboost	98.27%	75.1%	0.96%

5.3. *Evaluation of Model Switched Landslide Prediction System*

In the following experiments, the same datasets as used in Section 5.1 were employed to evaluate our landslide prediction model. To evaluate the proposed MoSLaPS model, several performance indicators were used and defined as follows:

- Mean Absolute Percent Error (MAPE) is defined as in Equation (26), where n is the number of predicted data, A_t is the actual value, and P_t is the predicted value.

$$MAPE = \frac{1}{n}\sum_{t=1}^{n}\frac{|A_t - P_t|}{P_t}. \tag{26}$$

- Root Mean Squared Error (RMSE) is defined as in Equation (27), where n is the number of predicted data samples, y_i is the actual FS value, and $\widehat{y_i}$ is the predicted FS value.

$$RMSE = \sqrt{\frac{1}{n}\sum_{i=1}^{n}(y_i - \hat{y}_i)^2}. \tag{27}$$

- Normalized Root Mean Squared Error (NRMSE) is defined as in Equation (28), where $\bar{y}$ is the mean of the actual values.

$$NRMSE = \frac{RMSE}{\bar{y}}. \tag{28}$$

Figure 13 shows the actual FS and predicted FS for every 10 min. A BPNN prediction model is not able to learn the pattern of unstable class, as the size of the unstable class is much smaller than that of the stable class. If the original past data were used, the BPNN was not able to predict landslide occurrences. However, after processing the training data using the ADASYN algorithm to re-balance the distribution of classes, a single BPNN prediction model was still not able to learn the pattern of unstable class perfectly. This is because the data of the stable and unstable classes affected each other. As a solution, two BPNN prediction models Were proposed, one to learn the pattern of the stable class and another to learn the pattern of the unstable class. To switch between two BPNN models, an event-class predictor was constructed that can deal with imbalanced data distribution to predict the future class as a decision. Therefore, our proposed MoSLaPS method could learn the patterns of both the stable and unstable classes.

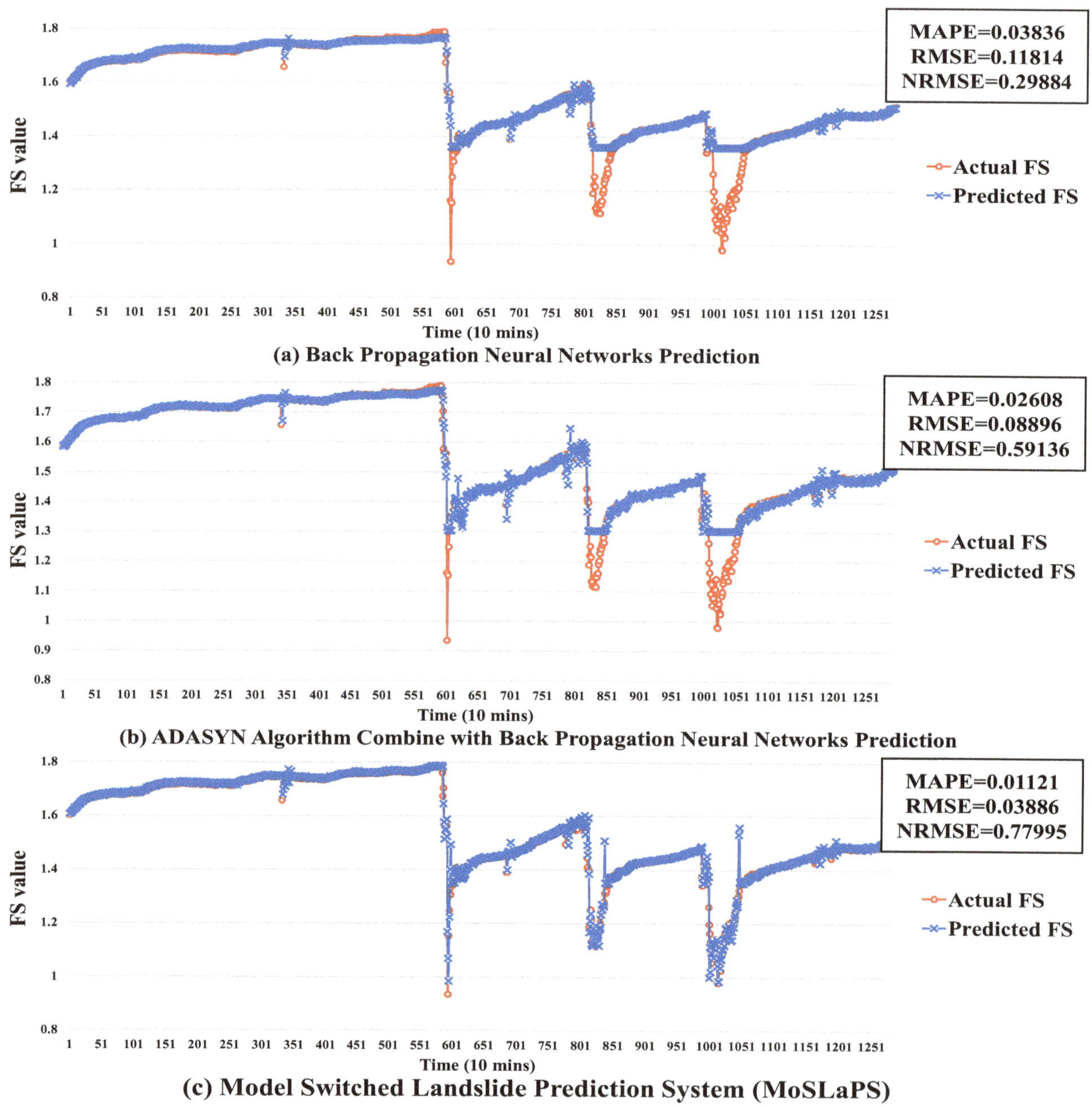

Figure 13. Comparison of the proposed method with other BPNN methods.

We compared our proposed MoSLaPS method with the above-mentioned methods, including a single BPNN and ADASYN+BPNN. A single BPNN was described, in detail, in Section 3.1; and ADASYN+BPNN use the ADASYN algorithm to re-balance the training data. After re-balancing, the processed training data were applied to train the BPNN.

The error metrics were evaluated for MoSLaPS, ADASYN+BPNN, and BPNN. Compared with BPNN and ADASYN + BPNN, our method resulted in much smaller MAPE and RMSE. This means that the prediction accuracy of our method was higher. The NRMSE of our method was also larger than that of the others, which means that our prediction results were closer to the real situation.

Although MoSLaPS was more accurate than the other methods, it requires a little more computation time and resources. As shown in Table 7, the simulation results of the different methods shows that the speed of MoSLaPS was a little slower than that of BPNN and ADASYN+BPNN, and its CPU and memory usages were higher than the others. This is because MoSLaPS took extra computational time and resources to deal with the imbalanced data classification and switching between the different predictors. In our experiment, the time interval was 10 mins; so there was ample time to deal with the process.

Table 7. Comparisons of time consumption and resource usage.

Method	Time (s)	CPU Usage	Memory
MoSLaPS	1.205	23.90%	972 kb
ADASYN+BPNN	0.739	20.30%	192 kb
BPNN	0.639	19.90%	148 kb

5.4. *Evaluation of Landslide Pre-Alarm*

The same datasets as in Section 5.1 were applied to evaluate our landslide pre-alarm method. Table 8 shows the prediction time (PT) for ten different experiments. For each experiment, the minimum PT, maximum PT, and average PT were recorded. From Table 8, the program was able to observe that the prediction time for dataset ♯5 was the longest (52.2 min); while that for dataset ♯10 was the shortest (38.4 min). As a shorter prediction time indicates that the change of FS is intense and quick, dataset ♯10 represented a higher probability of landslide occurrence. Taking the average of all timings, it can be seen that the proposed MoSLaPS method could warn of the occurrence of a landslide an average of 44.2 min in advance.

Table 8. Prediction time in advance for different datasets.

No.	Min. PT (min)	Max. PT (min)	Avg. PT (min)
1	10	80	40
2	10	70	33.3
3	10	50	35.7
4	20	80	47.6
5	10	80	52.2
6	10	80	51.8
7	10	80	42.6
8	10	70	45.7
9	10	80	43.3
10	10	80	38.4
Avg.	10.7	76.4	44.2

Further, we take a best-case example, to demonstrate how the proposed method can warn of landslide occurrence far in advance. At time point 255, the 8th prediction result is $FS < 1$; that is, a landslide will occur after eight time units, as shown in Figure 14. In our experiments, the interval between two time points is 10 min. Hence, the landslide occurrence could be predicted and warned about 80 min in advance.

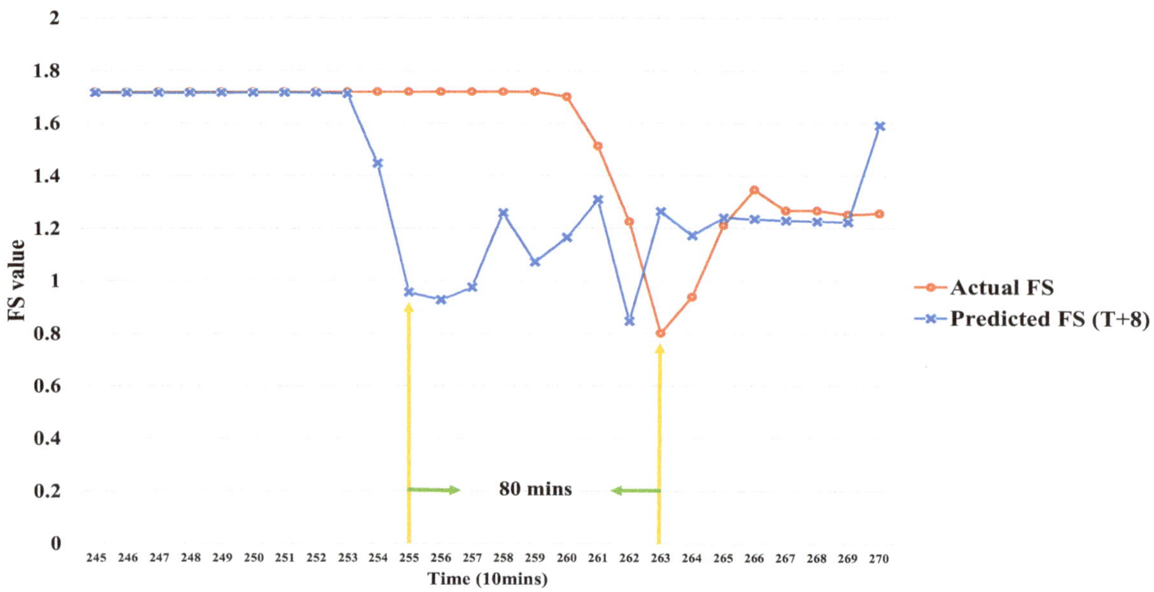

Figure 14. Landslide early warning time point.

6. Conclusions

To address the problems of imbalanced data, low true positive rate for learning, determining the prediction horizon, and the time for model re-training, MoSLaPS has been proposed as a novel method for landslide prediction. MoSLaPS employs the ADASYN method to balance the stable class (no landslide) with the unstable class (landslide), where the classification is based on the factor of safety calculated using the SHALSTAB model. To solve the problem of low true positive rate, a BPNN-based event-class predictor was proposed and two BPNN predictors were designed to learn the stable class pattern and the unstable class pattern. A novel prediction horizon tuning method was proposed, along with a learning-based model re-training technique. All of these optimizations contribute towards the goal of the proposed MoSLaPS; that is, accurate early warning of landslides.

Compared with BPNN and Adaboost, though our event-class predictor has a higher FPR, it also has a much higher TPR of 98.40%. This means our classifier has a higher ratio of correct classification. According to the predicted class, our system can switch between different predictors to adapt to the environmental state. In addition, BPNN is employed to construct the error model to predict the future errors of our proposed MoSLaPS and compensate for these errors in the prediction phase. As a result, MoSLaPS has much lower MAPE and RMSE than BPNN and ADASYN+BPNN, which means that MoSLaPS is more accurate. In addition, the NRMSE of our method is larger than the NRMSE of the other methods, which means that our method is closer to the actual conditions. Statistically, our landslide prediction system could send warnings an average of 44.2 min prior to the actual occurrence of a landslide.

In the future, we will further consider other advanced imbalanced learning algorithms to improve the performance of BPNN-based event-class predictors. For different applications, a larger number of categories (cases) can be considered for model switching. The maximum range of errors allowed in the re-training phase can also be limited, so that prediction models are more stable. As a result, the model switching strategy will be more accurate. Moreover, well-known time-series deep learning technologies, such as Recurrent Neural Networks (RNNs) or Long-Short Term Memory (LSTM) blocks, will be used to combine all three BPNN models into one.

Author Contributions: Conceptualization, P.-A.H.; methodology, D.U.; software, S.-F.C.; validation, S.-F.C. and D.U.; resources, P.-A.H.; data curation, S.-F.C.; writing—original draft preparation, S.-F.C.; writing—review and editing, D.U.; supervision, P.-A.H.; project administration, P.-A.H.; funding acquisition, P.-A.H.

Acknowledgments: This research was supported partially by the project grant MOST-107-2221-E-194-001-MY3 from the Ministry of Science and Technology, Taiwan.

Abbreviations

The following abbreviations are used in this manuscript:

ANN	Artificial Neural Network
WSN	Wireless Sensor Network
FS	Factor of Safety
MoSLaPS	Model Switched Landslide Prediction System
LMA	Levenberg–Marquardt algorithm
ADASYN	Adaptive Synthetic Sampling
SMOTE	Synthetic Minority Oversampling Technique
EEW	Error Estimation Window
PH	Prediction Horizon
TPR	True Positive Rate
FPR	False Positive Rate
ACC	Accuracy
TP	True Positive
TN	True Negative
FP	False Positive
FN	False Negative
MAPE	Mean Absolute Percentage Error
RMSE	Root Mean Squared Error
NRMSE	Normalized Root Mean Squared Error
PT	Prediction Time

References

1. He, H.; Bai, Y.; Garcia, E.A.; Li, S. ADASYN: Adaptive synthetic sampling approach for imbalanced learning. In Proceedings of the IEEE International Joint Conference on Neural Networks (IEEE World Congress on Computational Intelligence), Hong Kong, China, 1–8 June 2008; pp. 1322–1328.
2. Jiang, T.; Wand, D. A landslide stability calculation method based on Bayesian network. In Proceedings of the Instrumentation and Measurement, Sensor Network and Automation, Toronto, ON, Canada, 23–24 December 2013; pp. 905–908.
3. Devi, S.R.; Venkatesh, C.; Agarwa, P.; Arulmozhivarman, P. Daily rainfall forecasting using artificial neural networks for early warning of landslides. In Proceedings of the International Conference on Advances in Computing, Communications and Informatics, New Delhi, India, 24–27 September 2014; pp. 2218–2224.
4. Kavzoglu, T.; Sahin, E.K.; Colkesen, I. Selecting optimal conditioning factors in shallow translational landslide susceptibility mapping using genetic algorithm. *Eng. Geol.* **2015**, *192*, 101–112. [CrossRef]
5. Huang, J.C.; Kao, S.J.; Hsu, M.L.; Liu, Y.A. Influence of specific contributing area algorithms on slope failure prediction in landslide modeling. *Nat. Hazards Earth Syst. Sci.* **2007**, *7*, 781–792. [CrossRef]
6. Popescu, M. A suggested method for reporting landslide remedial measures. *Bull. Eng. Geol. Environ.* **2001**, *60*, 69–74. [CrossRef]
7. Lian, C.; Zeng, Z.; Yao, W.; Tang, H. Multiple neural networks switched prediction for landslide displacement. *Eng. Geol.* **2015**, *186*, 91–99. [CrossRef]
8. Lian, C.; Chen, C.L.P.; Zeng, Z.; Yao, W.; Tang, H. Prediction Intervals for Landslide Displacement Based on Switched Neural Networks. *IEEE Trans. Reliab.* **2016**, *65*, 1483–1495. [CrossRef]
9. Huang, J.C.; Kao, S.J. Optimal estimator for assessing landslide model performance. *Hydrol. Earth Syst. Sci.* **2006**, *10*, 957–965. [CrossRef]

10. Hagan, M.T.; Demuth, H.B.; Beale, M.H.; De Jesús, O. *Neural Network Design*; 2014. Available online: http://hagan.okstate.edu/NNDesign.pdf (accessed on 3 May 2019).
11. Levenberg Marquardt Algorithm. 2017. Available online: https://en.wikipedia.org/w/index.php?title=Levenberg%E2%80%93Marquardt_algorithm&oldid=771936514 (accessed on 3 May 2019).
12. Jacobian Matrix and Determinant. 2017. Available online: https://en.wikipedia.org/w/index.php?title=Jacobian_matrix_and_determinant&oldid=776906112 (accessed on 3 May 2019).
13. Palit, A.K.; Popovic, D. *Computational Intelligence in Time Series Forecasting: Theory and Engineering Applications*; Springer Science & Business Media: London, UK, 2006.
14. Ertekin, S.; Huang, J.; Bottou, L.; Giles, L. Learning on the border: active learning in imbalanced data classification. In Proceedings of the 16th ACM Conference on Conference on Information and Knowledge Management, Lisbon, Portugal, 6–10 November 2007; pp. 127–136.
15. Chawla, N.V.; Bowyer, K.W.; Hall, L.O.; Kegelmeyer, W.P. SMOTE: Synthetic minority over-sampling technique. *J. Artif. Intell. Res.* **2002**, *16*, 321–357. [CrossRef]
16. Soil and Water Conservation Bureau Platform for Monitoring Data. 2016. Available online: http://monitor.swcb.gov.tw (accessed on 3 May 2019).

Permissions

All chapters in this book were first published in MDPI; hereby published with permission under the Creative Commons Attribution License or equivalent. Every chapter published in this book has been scrutinized by our experts. Their significance has been extensively debated. The topics covered herein carry significant findings which will fuel the growth of the discipline. They may even be implemented as practical applications or may be referred to as a beginning point for another development.

The contributors of this book come from diverse backgrounds, making this book a truly international effort. This book will bring forth new frontiers with its revolutionizing research information and detailed analysis of the nascent developments around the world.

We would like to thank all the contributing authors for lending their expertise to make the book truly unique. They have played a crucial role in the development of this book. Without their invaluable contributions this book wouldn't have been possible. They have made vital efforts to compile up to date information on the varied aspects of this subject to make this book a valuable addition to the collection of many professionals and students.

This book was conceptualized with the vision of imparting up-to-date information and advanced data in this field. To ensure the same, a matchless editorial board was set up. Every individual on the board went through rigorous rounds of assessment to prove their worth. After which they invested a large part of their time researching and compiling the most relevant data for our readers.

The editorial board has been involved in producing this book since its inception. They have spent rigorous hours researching and exploring the diverse topics which have resulted in the successful publishing of this book. They have passed on their knowledge of decades through this book. To expedite this challenging task, the publisher supported the team at every step. A small team of assistant editors was also appointed to further simplify the editing procedure and attain best results for the readers.

Apart from the editorial board, the designing team has also invested a significant amount of their time in understanding the subject and creating the most relevant covers. They scrutinized every image to scout for the most suitable representation of the subject and create an appropriate cover for the book.

The publishing team has been an ardent support to the editorial, designing and production team. Their endless efforts to recruit the best for this project, has resulted in the accomplishment of this book. They are a veteran in the field of academics and their pool of knowledge is as vast as their experience in printing. Their expertise and guidance has proved useful at every step. Their uncompromising quality standards have made this book an exceptional effort. Their encouragement from time to time has been an inspiration for everyone.

The publisher and the editorial board hope that this book will prove to be a valuable piece of knowledge for researchers, students, practitioners and scholars across the globe.

List of Contributors

Marj Tonini
Institute of Earth Surface Dynamics, Faculty of Geosciences and Environment, University of Lausanne, CH-1015 Lausanne, Switzerland

Mirko D'Andrea, Guido Biondi, Silvia Degli Esposti, Andrea Trucchia and Paolo Fiorucci
CIMA Research Foundation, 17100 Savona, Italy

Josephine Morgenroth, Usman T. Khan and Matthew A. Perras
Department of Civil Engineering, Lassonde School of Engineering, York University, Toronto, ON M3J 1P3, Canada

Omer Saud Azeez and Helmi Z. M. Shafri
Department of Civil Engineering, Faculty of Engineering, University Putra Malaysia (UPM), Serdang 43400, Malaysia

Biswajeet Pradhan, Nagesh Shukla and Hossein Mojaddadi Rizeei
Centre for Advanced Modelling and Geospatial Information Systems (CAMGIS), Faculty of Engineering and Information Technology, University of Technology Sydney, Sydney 2007, NSW, Australia

Chang-Wook Lee
Division of Science Education, Kangwon National University, Chuncheon Campus, 1 Gangwondaehakgil, Chuncheon-si, Gangwon-do 24341, Korea

Yun-Jae Choung
GIS Research Center, Geo C&I Co., Ltd., Daegu 41165, Korea
Global Land Satellite Information Center, Kyungpook National University, Daegu 41566, Korea

Jin-Man Kim
Geotechnical Engineering Research Institute, Korea Institute of Civil Engineering and Building Technology, Goyang-si, Gyeonggi-do 10223, Korea

Ya'nan Zhang and Keqi Fan
College of Information Science and Engineering, Shandong Agricultural University, 61 Daizong Road, Taian 271000, China

Chengming Zhang and Xiaoxia Yang
College of Information Science and Engineering, Shandong Agricultural University, 61 Daizong Road, Taian 271000, China
Shandong Technology and Engineering Center for Digital Agriculture, 61 Daizong Road, Taian 271000, China

Shuai Gao
Chinese Academy of Sciences, Institute of Remote Sensing and Digital Earth, 9 Dengzhuangnan Road, Beijing 100094, China

Feng Li
Shandong Climate Center, Mountain Road, Jinan 250001, China

Maorui Yue
Taian Agriculture Bureau, Naihe Road, Taian 271000, China

Yingjuan Han and Hui Zhao
Key Laboratory for Meteorological Disaster Monitoring and EarlyWarning and Risk Management of Characteristic Agriculture in Arid Regions, CMA, 71 Xinchangxi Road, Yinchuan 750002, China

Jihua Meng
Key Laboratory of Digital Earth Sciences, Institute of Remote Sensing and Digital Earth, Chinese Academy of Sciences, Beijing 100101, China

Huihui Mao, Fujiang Ji and Huiting Fang
Key Laboratory of Digital Earth Sciences, Institute of Remote Sensing and Digital Earth, Chinese Academy of Sciences, Beijing 100101, China
University of Chinese Academy of Sciences, Beijing 100049, China

Qiankun Zhang
Key Laboratory of Digital Earth Sciences, Institute of Remote Sensing and Digital Earth, Chinese Academy of Sciences, Beijing 100101, China
College of Geomatics and Geoinformation, Guilin University of Technology, Guilin 541004, China

Dae Kyo Seo
Department of Advanced Technology Fusion, Konkuk University, Seoul 05029, Korea

Yong Hyun Kim
Department of Civil and Environmental Engineering, Seoul National University, Seoul 08826, Korea

Yang Dam Eo
Department of Technology Fusion Engineering, Konkuk University, Seoul 05029, Korea

Wan Yong Park
Agency for Defense Development, Daejeon 34060, Korea

Hyun-Joo Oh
Geo-Environmental Hazard Research Center, Korea Institute of Geoscience and Mineral Resources (KIGAM), 124, Gwahak-ro Yuseong-gu, Deajeon 34132, Korea

Mutiara Syifa
Division of Science Education, Kangwon National University, Chuncheon Campus, 1 Gangwondaehakgil, Chuncheon-si, Gangwon-do 24341, Korea

Saro Lee
Geoscience Platform Research Division, Korea Institute of Geoscience and Mineral Resources (KIGAM), 124, Gwahak-ro Yuseong-gu, Daejeon 34132, Korea
Department of Geophysical Exploration, Korea University of Science and Technology, 217 Gajeong-ro Yuseong-gu, Daejeon 34113, Korea

Guoyin Xu
Department of Hydraulic Engineering, Tsinghua University, Beijing 100084, China

Zhongjing Wang
Department of Hydraulic Engineering, Tsinghua University, Beijing 100084, China
State Key Laboratory of Hydro-Science and Engineering, Tsinghua University, Beijing 100084, China
State Key Lab of Plateau Ecology and Agriculture, Qinghai University, Xining 810016, China

Ting Xia
China Renewable Energy Engineering Institute, Beijing 100120, China

Fei Deng and Shengliang Pu
School of Geodesy and Geomatics, Wuhan University, Wuhan 430079, China

Soyoung Park
BK21 Plus Project of the Graduate School of Earth Environmental Hazard System, Pukyong National University, Busan 48513, Korea

Jinsoo Kim
Department of Spatial Information Engineering, Pukyong National University, Busan 48513, Korea

Darmawan Utomo, Shi-Feng Chen and Pao-Ann Hsiung
Computer Science and Information Engineering, National Chung Cheng University, No. 168, Sec. 1, University Rd., Minhsiung, Chiayi 62102, Taiwan

Index

S

T

V

www.ingramcontent.com/pod-product-compliance
Lightning Source LLC
LaVergne TN
LVHW082142200926
848484LV00008B/2095
9798888360736